UPCO's Review of PHYSICS

Herbert H. Gottlieb
Former Chairman
Physical Science Department
Martin Van Buren High School
Queens Village, New York 11427

CONTRIBUTING EDITORS

Joseph Drenchko
Physics Teacher
North Syracuse Central Schools

John D. FitzGibbons
Physics Teacher
Cazenovia Central School

cover design by
Gyula Madarasz

United Publishing Company, Inc.
76 Exchange Street
Albany, N.Y. 12205

ISBN 0-937-323-02-0

1 2 3 4 5 6 7 8 9 0

CONTENTS

MEASUREMENT AND MATHEMATICS REVIEW

The mathematics taught in elementary algebra and plane geometry is usually adequate for high school physics. Skills that may need review are presented here.

I. Measurement

Since no measurement can ever be exact, it is important to indicate the amount of uncertainty that was involved in gathering the data. To do this, every measurement should include:

A. A number read from the measuring device.
B. The units of measurement.
C. An indication of the certainty of the measurement.

II. Significant Digits

The certainty of measurement is expressed using a convention called **significant digits** (sometimes called significant figures). A large number of significant digits indicates that a measurement was made with great precision and certainty. When there is less precision and certainty, a smaller number of significant digits are reported for the measurement. For example, suppose that a room is 5 meters long and a rough measurement is made, with an error of +/−0.5 meter. Its length would be reported with only one significant digit as *5 m* On the other hand, if the measurement were made with better instruments and more care that would guarantee an error of no more than +/−.001 meter, it would be reported as *5.000 m* long.

A. Rules for Counting Significant Digits

1. All digits which are not zeros *are* counted.
13,426 km has five significant digits.

2. Initial zeros at the beginning of a number *are not* counted.
   0.00025 mm has only two significant digits.
3. If a number has no decimal, final zeros *are not* counted.
   22,500 cm^2 has only three significant digits.
4. If a number has a decimal, final zeros *are* counted.
   2.2500 nm has five significant digits.
   22,500. N has five significant digits.
5. Zeros in the middle of a number *are* counted.
   60,002 g has five significant digits.
6. When using scientific notation, *all digits* before the X *are* counted.
   6.09020×10^4 kg has six significant digits.

B. Operations and Significant Digits

1. Addition and Subtraction

When numbers are added or subtracted, the answer should not contain more digits than those in the original data. Therefore, 4.0 cm + 7.122 cm = 11 cm. since 4.0 cm has only two significant digits, the answer must be rounded off to only two significant digits also. In subtracting 14.3 mm from 73.04 mm, the difference is *58.7* mm. Only three significant digits are to be included in the answer.

2. Multiplication and Division

When numbers are multiplied, the product must not contain any more significant digits than the value with the least number of significant digits.

$$3.0 \text{ m} \times 2.205 \text{ m} = 6.6 \text{ m}^2$$

Only two digits appear in the answer because 3.0 m has only two significant digits. The same rule applies when numbers are divided:

$$4.414 \text{ m}^2 \div 2.0 \text{ m} = 2.2 \text{ m}$$

3. Problems involving a series of operations

When performing a series of operations one extra digit should be maintained in each individual operation. The extra digit is dropped from the final answer after all operations have been completed. In rounding off to three digits, a value of 3.168 km becomes 3.17 km; 4.496 m^2 becomes 4.50 m^2.

III. Scientific Notation

When writing numbers that are very large or very small, it is convenient to express the value in scientific notation using powers of ten. To write a number in scientific notation, express it as a number *between* one and ten (at least one but not more than nine) multiplied by ten to the appropriate power. For example:

$$3620.\text{km} = 3.620 \times 10^3 \text{ km}$$
$$0.014\text{m} = 1.4 \times 10^{-2} \text{ m}$$

Once a number is written in scientific notation, *all digits are significant*. This is one of the major advantages of scientific notation.

A. Operations with Scientific Notation

1. **Adding or subtracting**, using scientific notation require that numbers must be written with the **same power of ten**. For example, in order to add 3.68×10^5 to 4.53×10^6, we must change the 3.68×10^5 to $.368 \times 10^6$ and *then* add them:

$$\begin{array}{r} .368 \times 10^6 \text{ m} \\ + 4.53 \times 10^6 \text{ m} \\ \hline 4.898 \times 10^6 \text{ m, or } \underline{4.90 \times 10^6 \text{ m}} \end{array}$$

2. **Multiplying** values expressed in scientific notation involves multiplying the significant digits and **adding** the exponents.

   Examples:

   a. $(6.1 \times 10^6)(2.8 \times 10^3 \text{ m}) = 17.08 \times 10^9 \text{ m}^2$
   or $1.7 \times 10^{10} \text{ m}^2$
   b. $(5.0 \times 10^5 \text{ cm})(2.1 \times 10^{-8} \text{ cm}) = 10.5 \times 10^{-3} \text{ cm}^2$
   or $\underline{1.1 \times 10^{-2} \text{ cm}^2}$

3. **Dividing** in scientific notation requires dividing the significant digits and **subtracting** the exponents.

   Examples:

   a. $6.1 \times 10^2 \text{ m}^2 \div 1.8 \times 10^{-4} \text{ m} = 3.388 \times 10^6 \text{ m}^3$
   or $3.4 \times 10^6 \text{ m}^3$
   b. $1.5 \times 10^5 \text{ m} \div 3.0 \times 10^2 \text{ m} = .50 \times 10^3$
   or 5.0×10^2

   When using a calculator with an exponent key, simply follow the directions of the manufacturer and round off the answer to the appropriate number of significant digits.

IV. Conversion of Units

Measurements of a quantity can be converted from one unit to another by using the appropriate **conversion factor**. If the conversion factor is written as fractions equal to unity, we may multiply or divide by this fraction without changing the value of the quantity. For example, $3 \text{ m} \times \frac{100 \text{ cm}}{1 \text{ m}} = 300 \text{ cm}$

Sample Conversions:

A. Convert 30.4 meters to centimeters.

conversion factor: $1 \text{ m} = 100 \text{ cm} = 1 \times 10^2 \text{ cm}$

solution: $30.4 \text{ m} \times \frac{1 \times 10^2 \text{ cm}}{1 \text{ m}} = \underline{3.04 \times 10^3 \text{ cm}}$

B. Convert 3.5×10^{-6} meters to nanometers.

conversion factor: $1 \text{ m} = 1 \times 10^9 \text{ nm}$
or $1 \text{ nm} = 1 \times 10^{-9} \text{ m}$

solution: $3.5 \times 10^{-6} \text{ m} \times \frac{1 \text{ nm}}{1 \times 10^{-9} \text{ m}} = \underline{3.5 \times 10^3 \text{ nm}}$.

V. Graphs

In dealing with variables in scientific experiments, a variable that is being deliberately changed, is called an *independent* variable. The one that changes as a result of the experiment is called a *dependent variable*. The values of the dependent variable are usually plotted on the y, or vertical, axis. The values of the independent variable are plotted on the x, or horizontal axis.

A. Direct Relationships exist when the ratio of the two variables is constant. This kind of relationship can be shown as a straight line passing through the origin of a graph. The constant ratio of the change in y (Δy) to the change in x (Δx) is known as the *slope* ($\Delta y/\Delta x$).

1. **Direct Proportionality.** Two variables are said to be directly proportional when y and x both change as their ratio remains constant. The equation most often used is $y = mx$, where m is the slope.
2. **Direct Square** relationships refer to cases in which y change as the square of x. When plotted on a graph the result will be a parabola. The general equation is $y = kx^2$, where k is a constant.

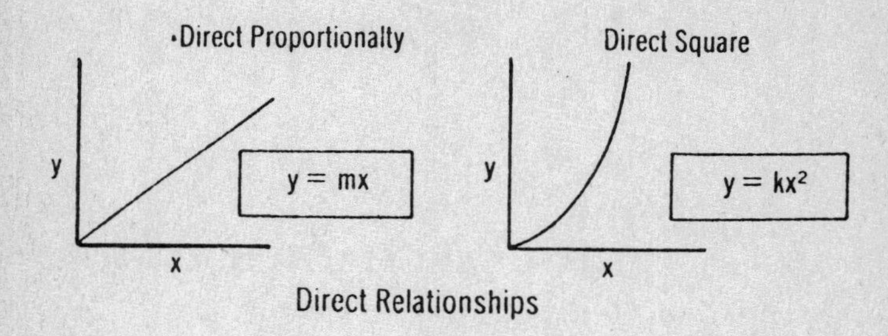

Direct Relationships

B. Inverse Relationships refer to those in which one variable increases as the other decreases.

1. **Inversely Proportional.** The two variables are said to be inversely proportional when $xy = k$, where k is a constant.
2. **Inverse Square.** An inverse square relationship exists when one variable decreases as the square of the other increases. For example: $x^2y = k$, where k is a constant.

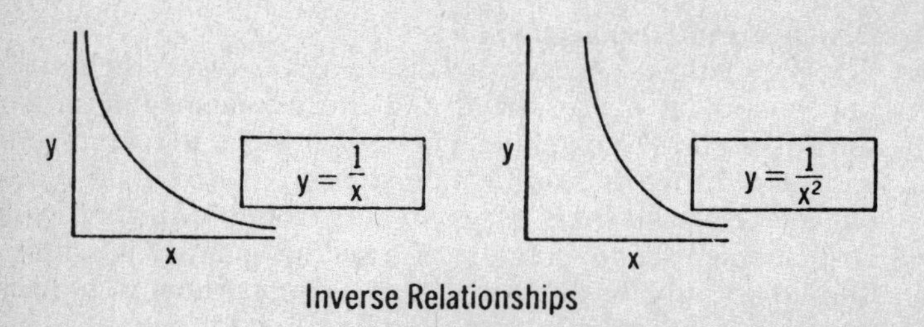

Inverse Relationships

UNIT I. MECHANICS

I. Kinematics

Kinematics is the description of motion without considering the factors that caused the motion.

A. Linear Motion

Motion in a straight line is called **linear motion**. How rapidly an object is moving in a specified direction is called its **velocity**. When a moving object changes its speed or direction, its velocity changes. The time rate of velocity change is called **acceleration**.

1. Distance and Displacement

It is sometimes necessary to distinguish between the **distance** that an object has traveled and the **displacement** from its starting point. For example, if a person walks 10 meters north and then 2 meters south, the **distance** covered is 12 meters but the **displacement** is only 8 meters from the starting point. Displacement is an example of a **vector quantity**. Quantities that do not involve direction, such as the number of atoms in a molecule, are called **scalar quantities**. Vector quantities indicate direction as well as magnitude. Therefore, when adding or subtracting them, special methods must be used.

In describing motion, physicists find it convenient to use the **meter** for length and the **second** for time, which are SI units. This system is used consistently throughout this course.

2. Velocity and speed

The terms velocity and speed both apply to how fast an object is moving but they have different meanings in the language of kinematics. Velocity is a vector quantity representing the rate at which the object is changing its displacement. Speed is a scalar quantity used to signify the magnitude of the velocity.

An object starting at rest and moving with uniform speed in a straight line has the same displacement each second and we may write:

$$v = \frac{s}{t}$$

where "v" is velocity, "s" is displacement and "t" is time. To express this relationship with a graph:

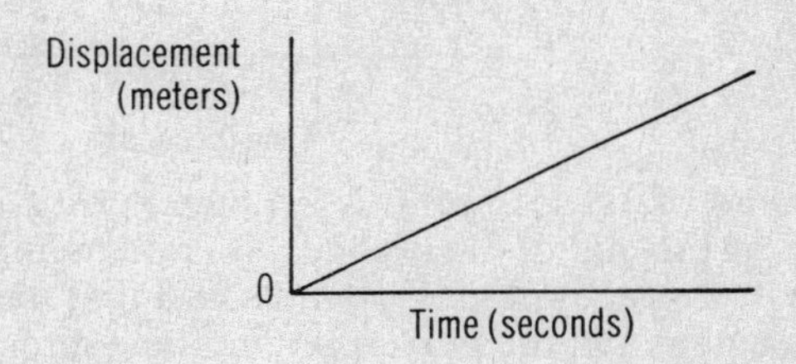

The graph is a straight line passing through the origin. Therefore s/t or v is a constant, the slope of the graph.

If the speed is changing, and we wish to describe the motion mathematically, we must define a new term. We can define the **average velocity** during any interval of time as the displacement (s) divided by the time interval (t). **Average** quantities are noted by placing a bar over the symbol. Average velocity is written $\bar{v}$ and is read "vee bar." The equation is:

$$\bar{v} = \frac{s_f - s_i}{t_f - t_i}$$

Where: "s_f" is the final displacement;
"s_i" is the initial displacement;
"t_f" is the final time and;
"t_i" is the initial time.

This equation may also be written:

$$\bar{v} = \frac{\Delta s}{\Delta t}$$

A graph representing the motion of an object whose velocity is changing might look like this:

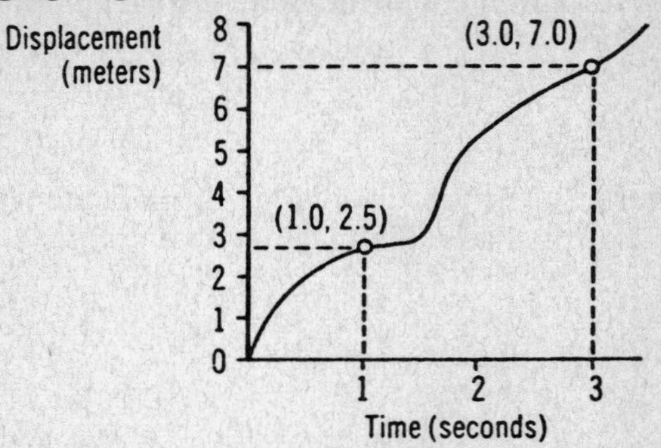

At the end of 1.0 second, it is 2.4 meters from its starting point and at the end of 3.0 seconds, it is 7.0 meters away. To calculate its average velocity we must first find the change in its displacement and the corresponding time interval.

$$\Delta s = 7.0\text{m} - 2.5\text{m} = 4.5\text{m}$$
$$\Delta t = 3.0\text{s} - 1.0\text{s} = 2.0\text{s}$$

using the equation

$$\bar{v} = \frac{\Delta s}{\Delta t}, \bar{v} = \frac{4.5\text{m}}{2.0\text{s}} = \underline{2.3\text{ m/s}}$$

Instantaneous velocity

Even though the velocity of an object may be changing, it has a definite value at any moment. This value is called the instantaneous velocity. It is the quantity shown by the speedometer of a car at any given time.

When the speed is changing, the instantaneous speed at any given time can be found by drawing a tangent to the curve at that point. The slope of this tangent is the instantaneous velocity

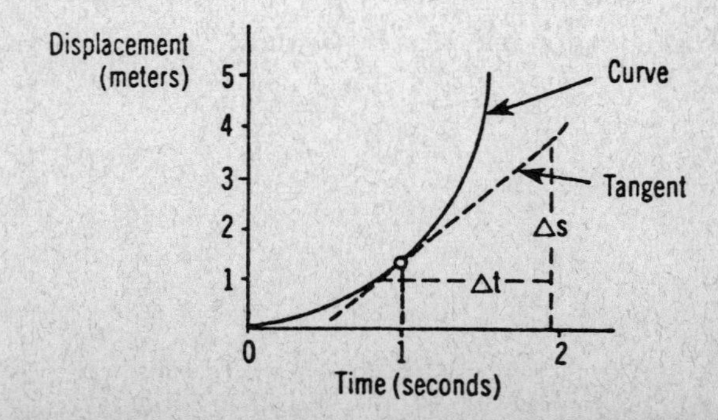

3. Acceleration

Acceleration is a vector quantity that represents the rate at which the velocity of an object is changing.

When the velocity changes at a constant rate, the acceleration is described as uniform and the object is said to be **uniformly accelerated**.

Acceleration, a, is equal to the change in the velocity divided by the time it has taken for the change.

$$a = \frac{y_2 - y_1}{x_2 - x_1} = \frac{\Delta v}{\Delta t}$$

If an object increases its speed from 2m/s to 8m/s in 3 seconds, without changing its direction, its velocity has changed by 6m/s /3s, or 2m/s/s (read "two meters per second, per second"). The acceleration (a) is 2m/s/s or $2m/sec^2$.

Sample Problem

An object moving in a straight line at the rate of 6m/s begins to accelerate uniformly. At the end of 5 seconds it's velocity is 36m/s. find its acceleration during that time interval.

$$a = \frac{\Delta v}{\Delta t} = \frac{36\ m/s - 6\ m/s}{5\ s}$$

$$a = 6\ m/s/s \text{ or } 6\ m/s^2$$

A graph of this condition would look like this:

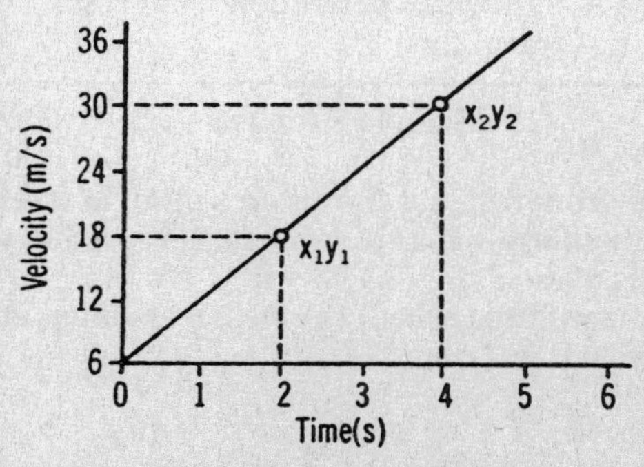

$$\text{The slope} = \frac{y_2 - y_1}{x_2 - x_1} = \frac{30 - 18}{4 - 2} = \frac{12\ m/s}{2\ s} = 6\ m/s^2$$

4. Displacement of a Uniformly Accelerating Object

The displacement, or distance traveled by any object moving in a straight line is equal to the product of its average speed and the travel time.

$$s = \bar{v}t$$

If the object is accelerating uniformly, find $\bar{v}$ as follows:

Add initial speed (v_i) and final speed (v_f) during the travel time and divide by two:

$$\bar{v} = \frac{v_i + v_f}{2}$$

Since many of the problems encountered that involve uniform acceleration require the description of straight-line motion that starts or ends at rest, it's useful to derive a special group of equations for this condition.

Consider an object that starts at rest. Since the initial speed (v_i) is zero, the change of velocity Δv is also its final velocity (v_f).

Then, $a = v_f/t$ and $\underline{v_f = at}$

However, we know that $\bar{v} = s/t$ so we can substitute s/t for v_f and write: $s/t = v_f/2$

and, solving for s, $\underline{s = v_f t/2}$

Substituting v_f/a for t in this equation: $s = v_f^2$, or $v_f^2 = 2\,as$

Substituting (at) for v_f,

$$(at)^2 + 2\,as$$
$$\text{or } \underline{s = \tfrac{1}{2}\,at^2}$$

Where s = displacement
a = uniform acceleration from rest
t = travel time

Sample Problem

A train starts from rest and travels on a straight track increasing its speed uniformly for six seconds. Its speed at the end of six seconds is 60m/s.

Find (a) the average speed, (b) the acceleration and (c) the distance traveled

Solution: (a) $\bar{v} = \frac{v_f}{2} = \frac{60\text{ m/s}}{2} = 30\text{ m/s}$

(b) $a = \frac{v_f}{t} = \frac{60\text{ m/s}}{6\text{ s}} = 10\text{ m/s}^2$

(c) $s = \tfrac{1}{2}\,at^2 = \tfrac{1}{2}\,(10\text{ m/s}^2)\,(6\text{ s})^2 = 180\text{ m}$

The distance traveled can also be determined by finding the area under a velocity-time curve.

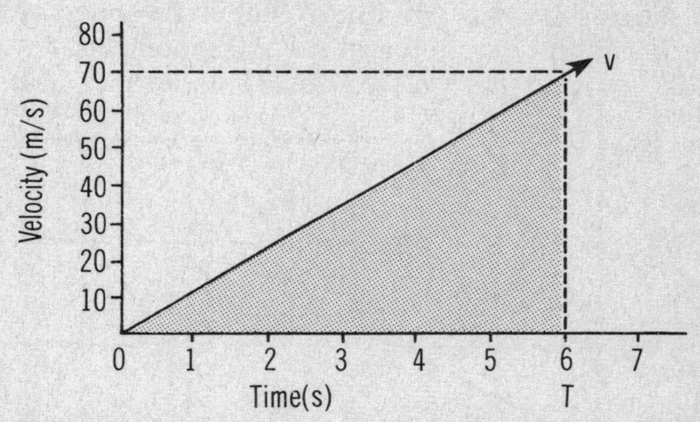

The shaded triangle is the area under the velocity-time curve. Since the area of a triangle is one half the base times the altitude,

$$\text{Area} = (\tfrac{1}{2}\,\text{OT})\,(\text{VT})$$
$$= (\tfrac{1}{2}\,6\text{ s})\,(60\text{ m/s})$$
$$s = 180\text{ m}$$

Another useful equation to find the displacement of an object after accelerating uniformly from rest is obtained by combining the equations $s = \frac{1}{2}at^2$ and $a = \Delta v^2/2s$ to give:

$$s = \frac{\Delta v^2}{2\,a}$$

If the object is not accelerating from rest but has an initial velocity, v_i, the displacement will be greater than it would be if the object started at rest. When the initial velocity must be taken into consideration the equations for uniformly accelerated motion are:

$$s = v_i t + \tfrac{1}{2}at^2$$

$$v_f^2 = v_i^2 + 2\,as$$

$$s = \frac{v_f + v_i}{2}(t) \text{ or } \overline{v} = \frac{s}{t} = \frac{v_f + v_i}{2}$$

QUESTIONS

1. Acceleration is a vector quantity that represents the time-rate of change in (1) momentum (2) velocity (3) distance (4) energy

2. A moving body must undergo a change of (1) velocity (2) acceleration (3) position (4) direction
3. The distance-time graph below represents the position of an object moving in a straight line. What is the speed of the object during the time interval $t = 2.0$ seconds to $t = 4.0$ seconds? (1) 0.0 m/s (2) 5.0 m/s (3) 7.5 m/s (4) 10. m/s

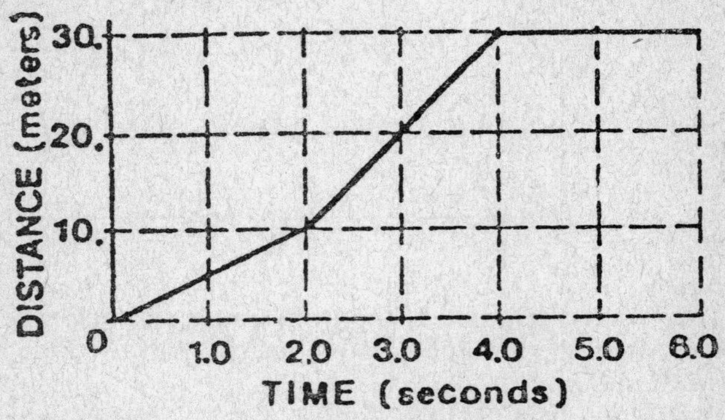

4. If an object is traveling east with a decreasing speed, the direction of the object's acceleration is (1) north (2) south (3) east (4) west
5. Which statement about the movement of an object with zero acceleration is true? (1) The object must be at rest. (2) The object must be slowing down. (3) The object may be speeding up. (4) The object may be in motion.
6. The graph below shows the relationship between speed and time for two objects, A and B. Compared with the acceleration of object B, the acceleration of object A is (1) one-third as great (2) twice as great (3) three times as great (4) the same

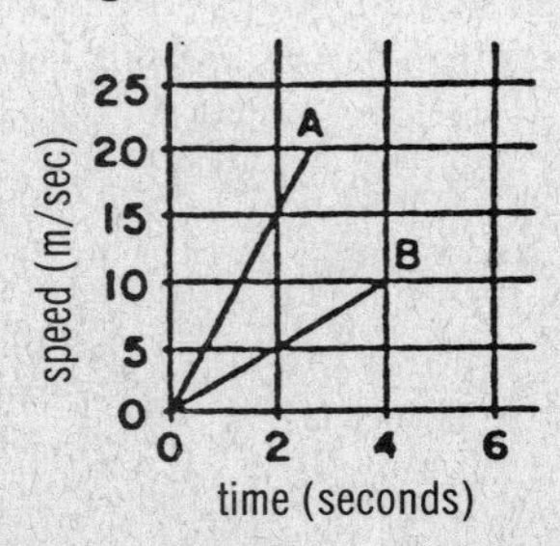

7. An object travels for 8.00 seconds with an average speed of 160. meters per second. The distance traveled by the object is (1) 20.0 m (2) 200. m (3) 1,280 m (4) 2,560 m

8. Which quantity has both magnitude and direction?
(1) distance (2) speed (3)mass (4) velocity

9. The graph below represents the relationship between velocity and time for an object moving in a straight line. What is the acceleration of the object? (1) 0 m/sec² (2) 5 m/sec² (3) 3 m/sec² (4) 15 m/sec²

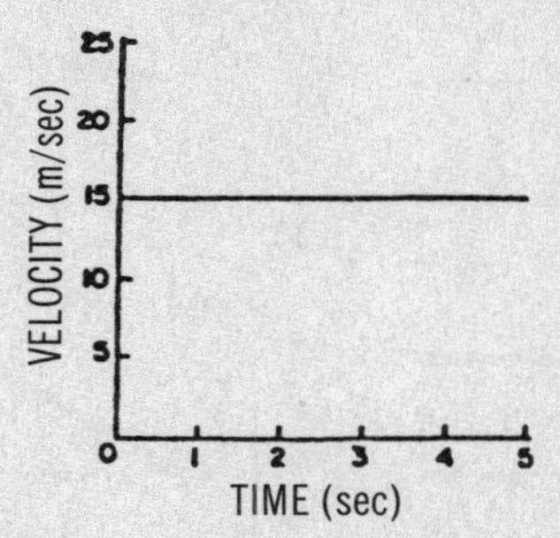

10. If a man walks 17 meters east then 17 meters south, the magnitude of the man's displacement is (1) 17 m (2) 24 m (3) 30. m (4) 34 m

11. The average speed of a runner in a 400.-meter race is 8.0 meters per second. How long did it take the runner to complete the race? (1) 80. sec (2) 50. sec (3) 40. sec (4) 32. sec

12. Which graph best represents the relationship between velocity and time for an object which accelerates uniformly for 2 seconds, then moves at a constant velocity for 1 second, and finally decelerates for 3 seconds?

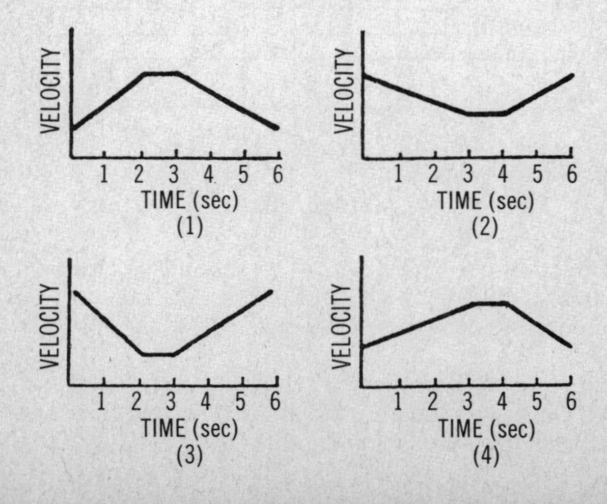

13. Which graph represents an object moving at a constant speed for the entire time interval?

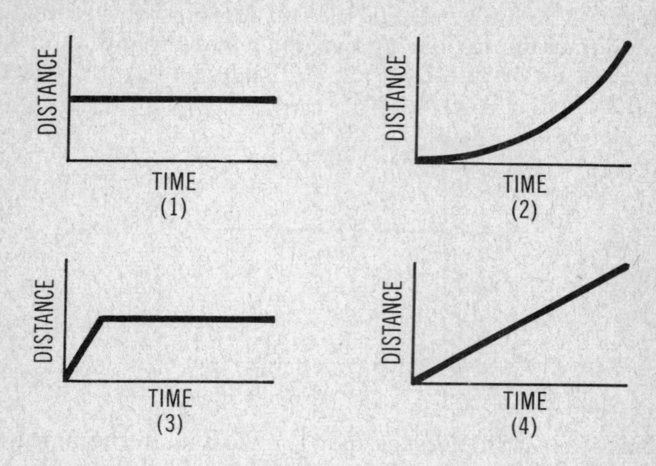

Base your answers to questions 14 through 18 on the accompanying graph which represents the motions of four cars on a straight road.

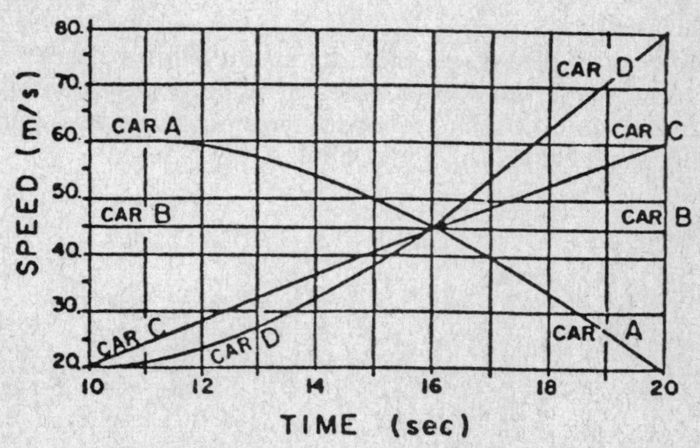

14. The speed of car C at time t = 20 seconds is closest to (1) 60 m/s (2) 45 m/s (3) 3.0 m/s (4) 600 m/s

15. Which car has zero acceleration? (1) A (2) B (3) C (4) D

16. Which car is decelerating? (1) A (2) B (3) C (4) D

17. Which car moves the greatest distance in the time interval t = 10 seconds to t = 16 seconds? (1) A (2) B (3) C (4) D

18. Which graph best represents the relationship between distance and time for car C?

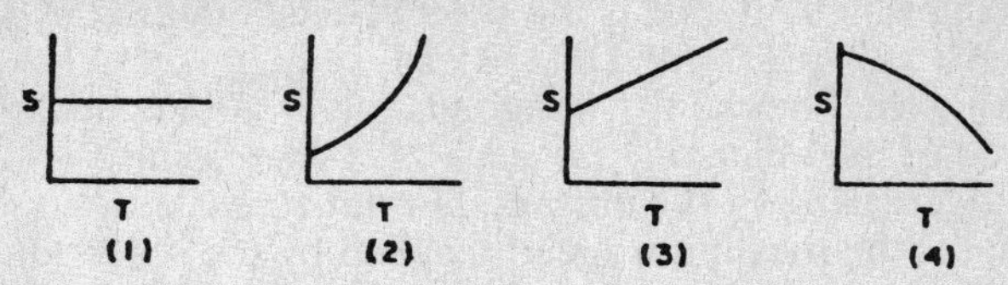

19. The graph below represents the motion of an object traveling in a straight line as a function of time. What is the average speed of the object during the first four seconds?
(1) 1 m/s (2) 2 m/s (3) 0.5 m/s (4) 0 m/s

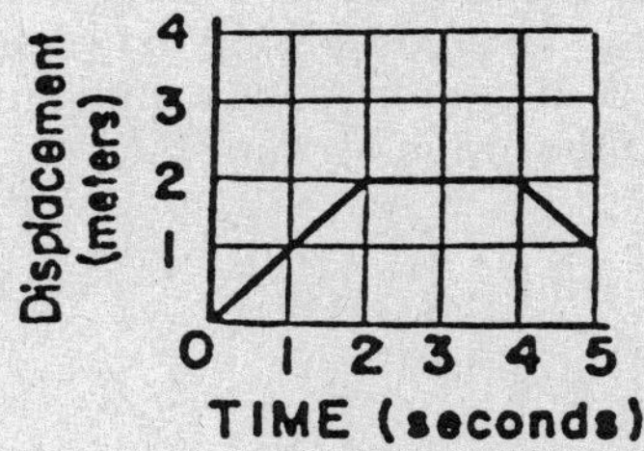

Base your answers to questions 20 through 25 on the graph below which represents the relationship between speed and time for an object in motion along a straight line.

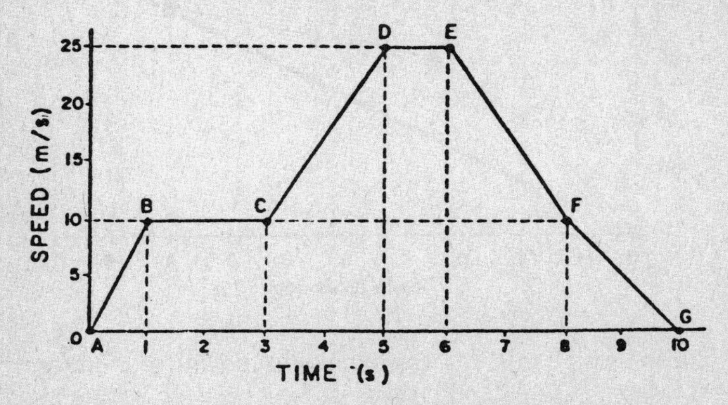

20. What is the acceleration of the object during the time interval t = 3 seconds to t = 5 seconds? (1) 5.0 m/s (2) 7.5 m/s (3) 12.5 m/s (4) 17.5 m/s

21. What is the average speed of the object during the time interval t = 6 seconds to t = 8 seconds? (1) 7.5 m/s (2) 10 m/s (3) 15 m/s (4) 17.5 m/s

22. What is the total distance traveled by the object during the first 3 seconds? (1) 15 m (2) 20 m (3) 25 m (4) 30 m

23. During which interval is the object's acceleration the greatest? (1) AB (2) CD (3) DE (4) EF

24. During the interval t = 8 seconds to t = 10 seconds, the speed of the object is (1) zero (2) increasing (3) decreasing (4) constant, but not zero

25. What is the maximum speed reached by the object during the 10 seconds of travel? (1) 10 m/s (2) 25 m/s (3) 150 m/s (4) 250 m/s

Base your answers to questions 26 through 32 on the graph below which represents the relationship between displacement and time for an object in motion along a straight line. Line xy is a tangent drawn to the graph at point E.

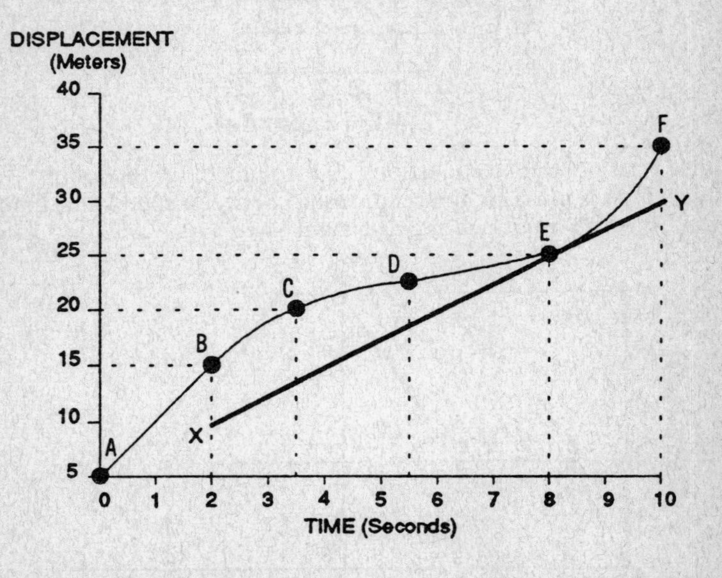

26. During which interval was the object moving at a uniform velocity? (1) AB (2) BC (3) CD (4) DF

27. During which interval was the speed of the object increasing? (1) AB (2) BC (3) CD (4) DF

28. During which interval was the acceleration the greatest? (1) AB (2) BC (3) CD (4) DF

29. During which interval was the change of displacement the greatest? (1) AB (2) BC (3) CD (4) DF

30. During which interval was the average velocity the greatest? (1) AB (2) BC (3) CD (4) DF

31. How fast was the object moving at point E? (1) 2.5 m/s (2) 3.1 m/s (3) 3.5 m/s (4) 8.0 m/s

32. How fast was the object moving at time t = 1.0 second? (1) 0 m/s (2) 5.0 m/s (3) 7.5 m/s (4) 15 m/s

Questions 33 to 37 are free response questions based on the motion of an object that travelled in a straight line for a period of 10.0 seconds. At the beginning of this period, its velocity was 2.0 m/s. At the end of the period, its velocity was 20.0 m/s.

33. Draw and label the axes of a velocity-time graph that covers the entire period of motion.

34. Draw a line on your graph that shows the object moving at a constant speed during the first 5-second interval (t = 0 to t = 5.0 seconds).

35. Continue the graph line for the second interval (t = 5.0 to t = 10.0 seconds). During this interval, the object accelerated uniformly until it reached a final velocity of 20.0 m/s.

36. Calculate the acceleration of the object during the second interval.

37. Calculate the total displacement of the object during its entire 10.0 seconds of motion.

5. Free Falling Objects

Objects are said to be freely falling when they are falling under the influence of gravity only. This condition can exist in a vacuum where there is no air resistance to retard the motion of objects. In a vacuum, objects fall at the same acceleration regardless of their shape or weight. In air, however, large light weight objects will accelerate less than small heavy objects.

A freely falling object, near the surface of the earth accelerates at $9.8 m/s^2$. This value is called the "acceleration due to gravity" and is identified as "g". Equations for velocity (v) and displacement (s) of freely falling objects simply replace "a" with the constant "g"

$$v_f = gt$$
$$\text{and } s = \tfrac{1}{2} gt^2$$

Sample Problem

A coin is dropped in a vacuum tube. Find (a) its velocity and (b) distance traveled at the end of 0.30 sec.

Solution:

(a) $v = gt$

$= (9.8\text{ m})\ (0.30\text{ sec})$

$= \underline{2.9\text{ m/s}}$

(b) $s = \frac{1}{2} gt^2$

$= (\frac{1}{2})\ (9.8\text{ m/s})\ (0.30\text{ s})^2$

$= \underline{0.45\text{ m}}$

For a summary of motion, the following graphs are useful:

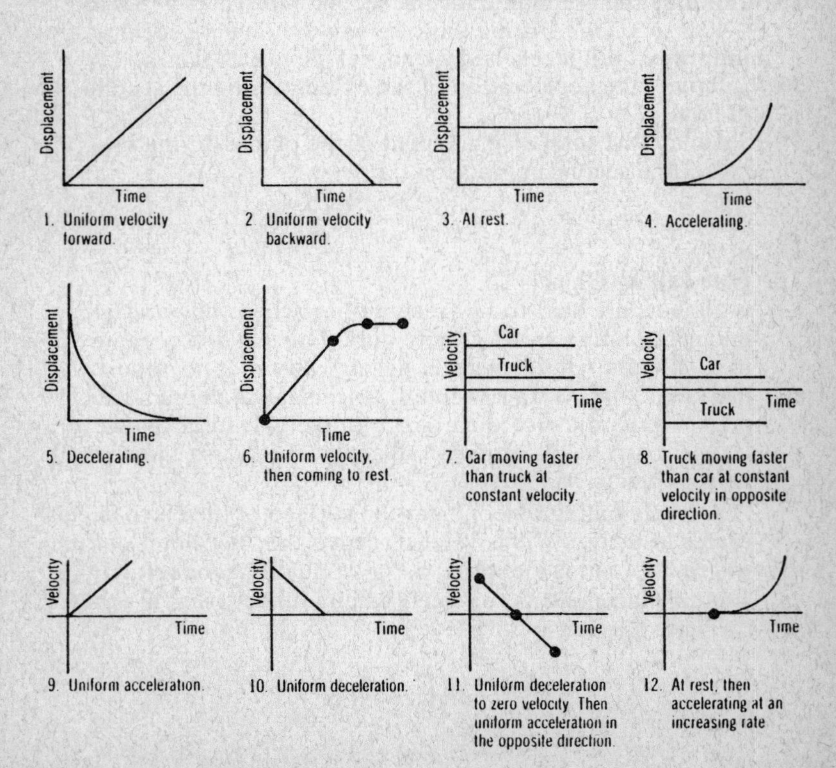

QUESTIONS

1. Starting from rest, an object rolls freely down an incline that is 10 meters long in 2 seconds. The acceleration of the object is approximately (1) 5 m/s (2) 5 m/s^2 (3) 10 m/s (4) 10 m/s^2

2. As a body falls freely near the surface of the Earth, its acceleration (1) decreases (2) increases (3) remains the same

3. An astronaut drops a stone near the surface of the Moon. Which graph best represents the motion of the stone as it falls toward the Moon's surface?

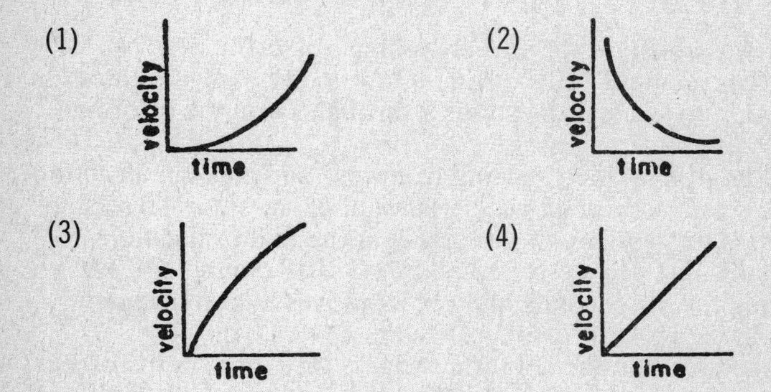

4. A rock is thrown horizontally from the top of a cliff at 12 meters per second. Approximately how long does it take the rock to fall 45 meters vertically? [Assume negligible air resistance.] (1) 1.0 s (2) 5.0 s (3) 3.0 s (4) 8.0s

5. An object starting from rest falls freely near the surface of the Earth. Which graph best represents the motion of the object?

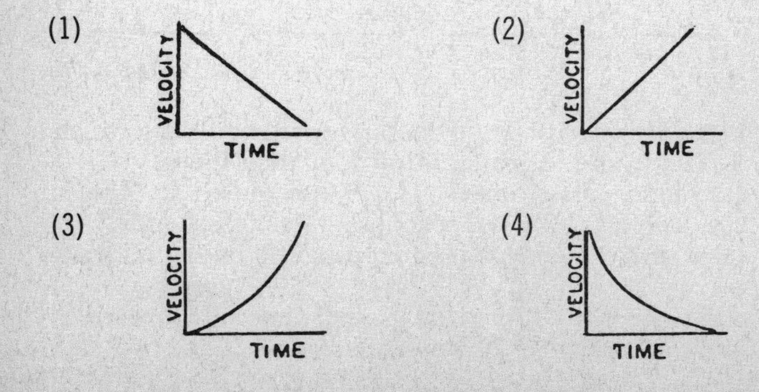

6. An object, initiall at rest, falls freely near the Earth's surface. How long does it take the object to attain a speed of 98 meters per second? (1) 0.1 s (2) 10 s (3) 98 s (4) 960 s

7. Which graph best represents the motion of an object sliding down a frictionless inclined plane?

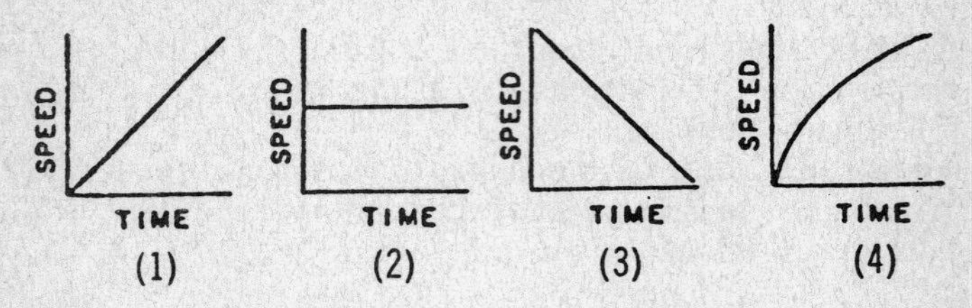

8. Which is constant for a freely falling object?
(1) displacement (2) speed (3) velocity (4) acceleration

Base your answers to questions 9 through 14 on the information below.

A 10. kilogram object, starting from rest, slides down a frictionless incline with a constant acceleration of 2.0 m/s^2 for 4.0 seconds.

9. What is the velocity of the object at the end of the 4.0 seconds? (1) 16 m/s (2) 2.0 m/s (3) 8.0 m/s (4) 4.0 m/s

10. During the 4.0 seconds, the object moves a total distance of (1) 32 m (2) 16 m (3) 8.0 m (4) 4.0 m

11. To produce this acceleration, what is the force on the object? (1) 10. N (2) 2.0×10^1 N (3) 5.0 N (4) 20.0×10^2 N

12. What is the approximate weight of the object? (1) 1 newton (2) 10 newtons (3) 100 newtons (4) 1,000 newtons

13. Which graph best represents the relationship between acceleration (a) and time (t) for the object?

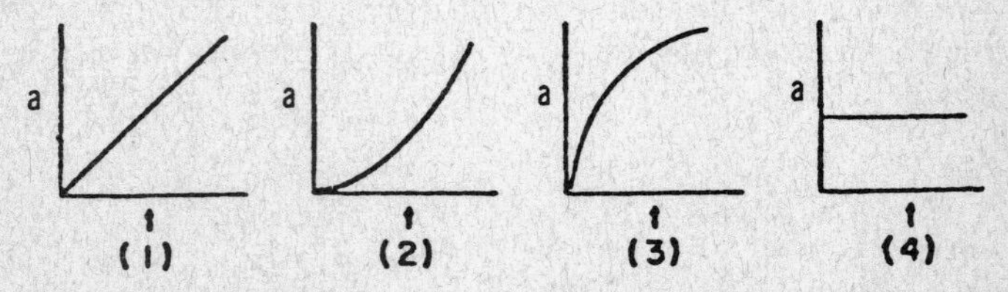

14. A student drops an object from the top of a building which is 19.6 meters from the ground. How long does it take the object to fall to the ground? (1) 19.6 seconds (2) 2.00 seconds (3) 3.00 seconds (4) 4.00 seconds

Base your answers to questions 15 through 19 on the information below.

A toy projectile is fired from the ground vertically upward with an initial velocity of +29 meters per second. The projectile arrives at its maximum altitude in 3.0 seconds. [Neglect air resistance.]

15. The greatest height the projectile reaches is approximately (1) 23 m (2) 44 m (3) 87 m (4) 260 m

16. What is the velocity of the projectile when it hits the ground? (1) 0. m/s (2) −9.8 m/s (3) −29 m/s (4) +29 m/s

17. What is the displacement of the projectile from the time it left the ground until it returned to the ground? (1) 0. m (2) 9.8 m (3) 44 m (4) 88 m

18. Which graph best represents the relationship between velocity (v) and time (t) for the projectile?

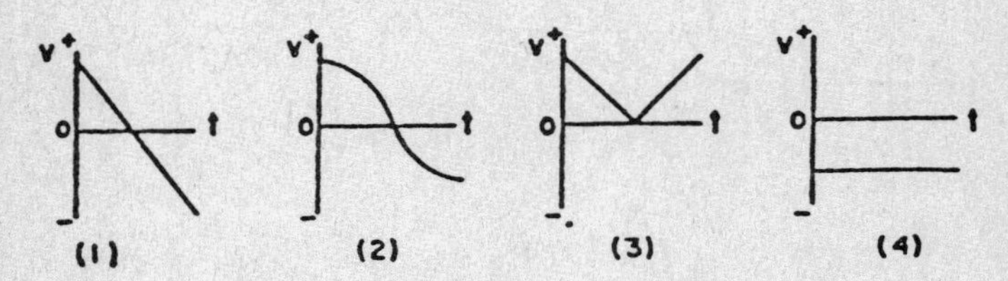

19. As the projectile rises and then falls back to the ground, its acceleration (1) decreases, then increases (2) increases, then decreases (3) increases, only (4) ramains the same

20. An object is allowed to fall freely near the surface of a planet. The object falls 54 meters in the first 3.0 seconds after it is released. The acceleration due to gravity on that planet is (1) 6.0 m/s^2 (2) 12 m/s^2 (3) 27 m/s^2 (4) 108 m/s^2

II. Force

A force is a vector quantity that may be loosely defined as a push or a pull. Four kinds of force are **gravitational, electromagnetic, nuclear** and **weak** interactions. The region in space in which a force acts is known as the "field" of the force. Units of force include pounds (lb) in the English system and newtons (N) in the SI system.

A. Vector Addition of Concurrent Forces

Any forces acting at the same point are said to be concurrent. We can determine the combined effect of such forces by an operation called composition of forces. The single vector obtained is called the *resultant* of the vectors. Since, by definition, vectors have *direction* as well as magnitude, some special rules must be followed to find the resultant of two or more concurrent forces.

In many instances, graphic methods are applied. Vector quantities can be represented graphically as arrows. The length of the arrow is made proportional to the magnitude and its orientation shows direction.

1. Finding the Resultant

The resultant of two or more concurrent forces can be determined by using the head-to-tail method. The direction of the resultant can be found by measuring the angle that the resultant makes with one of the original forces.

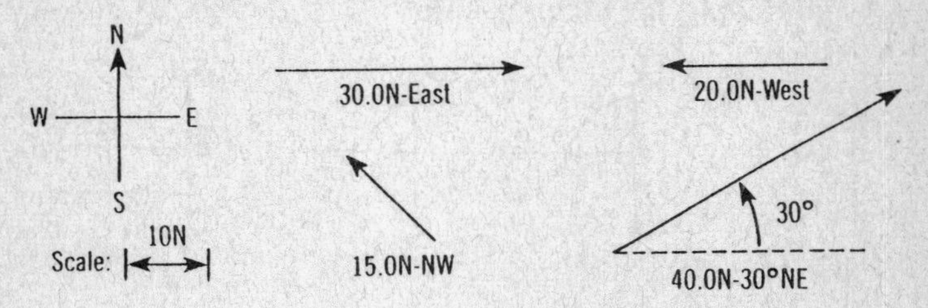

a. **Forces acting in the same direction** act at an angle of 0° to each other. The resultant has a magnitude equal to the arithmetic sum of the magnitudes of the original vectors and acts in their common direction. For example 6N east + 2N east = 8N east.

b. **Forces acting in opposite directions** form an angle of 180° to each other. Their resultant has a magnitude equal to the arithmetic difference between the magnitudes of the original vectors and acts in the direction of the larger force. For example 6N east + 2N west = 4N east.

c. **Forces Acting on Angles other than 0° and 180°.** First, it should be noted that when forces act at any angle, their resultant cannot be larger than their sum (when they are at 0°) or smaller than their difference (when they are at 180°).

2. Finding the Equilibrant

The **equilibrant** force is equal to the resultant in magnitude and opposite in direction. It is a single vector that can balance two or more concurrent vectors.

Sample Problem

Find the resultant and the equilibrant of 3 newtons pulling an object to the west while 4 newtons is pulling it north.

Solution:

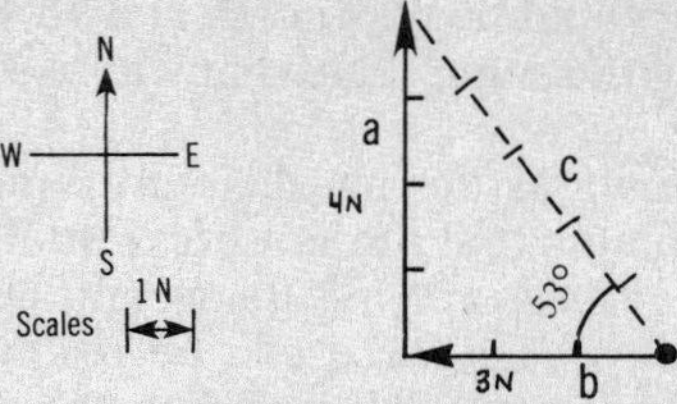

The resultant (c) according to the scale is equal to 5N and is found by protractor to be acting 53°N of W. In the special case of forces acting at right angles, such as this one, the Pythogorean theorem may be used. ($c^2 = a^2 + b^2$)

Solving for c:

$$c = \sqrt{a^2 + b^2}$$

$$c = \sqrt{(3\ N)^2 + (4\ N)^2}$$

$$c = \sqrt{25\ N^2}$$

$$c = \underline{5\ N}$$

The equilibrant is 53°S of E.

B. Resolution of Forces

By the process of *resolution*, any given force may be broken into two or more *component* forces which could act together and have the same effect as the original force. Any single force may be resolved into an unlimited number of components. We shall confine our discussion to resolution of a force into *two* components that are at right angles to each other.

Suppose we wish to find the north-south and east-west components, of a 60.N force acting in a 37° north of east direction. The north-south component is shown on the y axis of a graph and the east-west component on the x axis. The magnitudes of the components are found by drawing perpendiculars from the end of the 60.N vector and measuring the distances from the graph origin to the respective points of intersection.

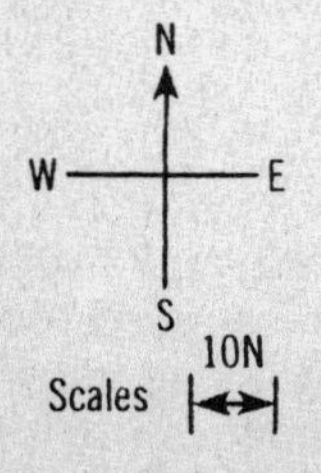

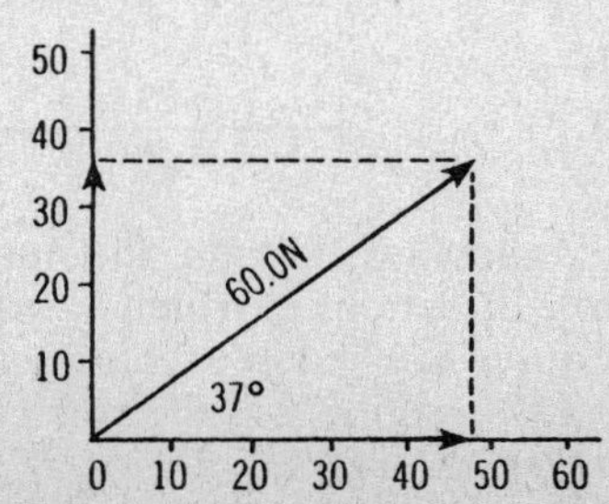

The components would be 36N north and 48N east; found along the graph axes using the same scale that was used to draw the 60.N force vector.

This method, known as the parallelogram method can also be used to find components that are not horizontal or vertical.

Suppose a car is parked on a hill. Its weight can be considered a force acting vertically. This force can be resolved into a component pushing the car down hill parallel to the road and a second component acting perpendicular to the road.

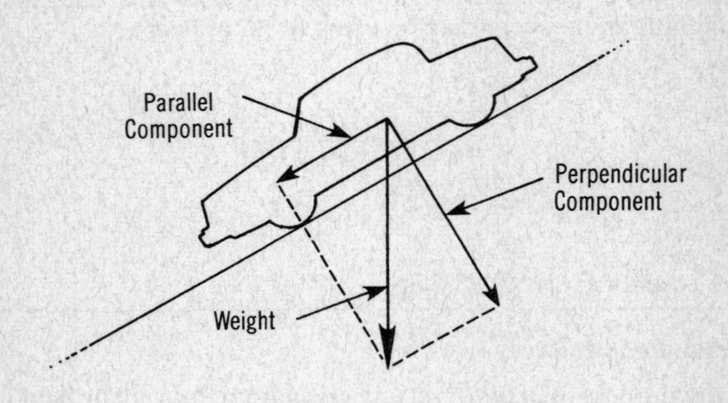

QUESTIONS

1. A force of 3 newtons and a force of 5 newtons act concurrently to produce a resultant of 8 newtons. The angle between the forces must be (1) 0° (2) 60° (3) 90° (4) 180°
2. The diagram below represents two forces acting concurrently on an object. The magnitude of the resultant force is closest to (1) 20. N (2) 40. N (3) 45. N (4) 60: N

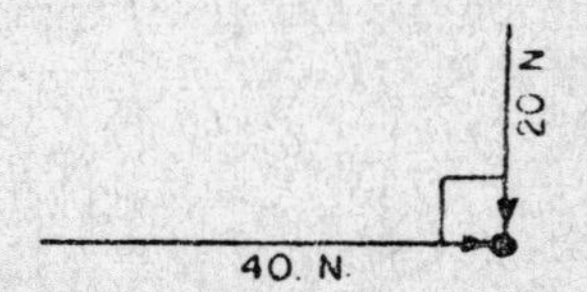

3. In the following diagram, the numbers 1, 2, 3, and 4 represent possible directions in which a force could be applied to a cart. If the force applied in each direction has the same magnitude, in which direction will the vertical component of the force be the *least*? (1) 1 (2) 2 (3) 3 (4) 4

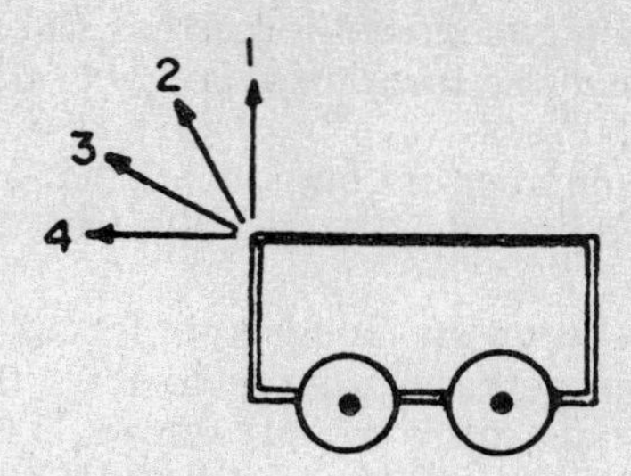

4. What is the average speed of an object that travels 6.00 meters north in 2.00 seconds and then travels 3.00 meters east in 1.00 seconds? (1) 9.00 m/s (2) 0.333 m/s (3) 3.00 m/s (4) 2.24 m/s

5. A bullet is fired from a rifle with a muzzle velocity of 100. meters per second at an angle of 30.° up from the horizontal. What is the magnitude of the vertical component of the muzzle velocity? (1) 0.0 m/s (2) 50. m/s (3) 87 m/s (4) 100 m/s

6. Two forces act on an object concurrently. The resultant will be greatest when the angle between the forces is (1) 0° (2) 60° (3) 90° (4) 180°

7. If a woman runs 100 meters north and then 70 meters south, her total displacement will be (1) 30 m north (2) 30 m south (3) 170 m north (4) 170 m south

8. The diagram below represents two concurrent forces acting on a point. Which vector best represents their resultant?

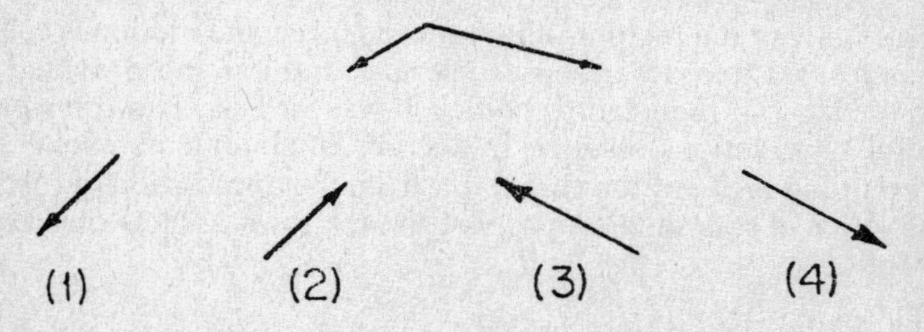

9. The resultant of two forces acting on the same point at the same time will be greatest when the angle between the forces is (1) 0° (2) 45° (3) 90° (4) 180°

10. Two concurrent forces act at right angles to each other. If one of the forces is 40 newtons and the resultant of the two forces is 50 newtons, the magnitude of the other force must be (1) 10 newtons (2) 20 newtons (3) 30 newtons (4) 40 newtons

11. If two 10.-newton concurrent forces have a resultant of zero, the angle between the forces must be (1) 0° (2) .45° (3) 90° (4) 180°

12. The maximum number of components that a single force may be resolved into is (1) one (2) two (3) three (4) unlimited

13. As the angle between two concurrent forces of 5.0 newtons and 7.0 newtons increases from 0° to 180°, the magnitude of their resultant changes from (1) 0 N to 35 N (2) 2.0 N to 12 N (3) 12 N to 2.0 N (4) 12 N to 0 N

14. Two 10.0-newton forces act currently on a point at an angle of 180° to each other. The magnitude of the resultant of the two forces is (1) 0.00 N (2) 10.0 N (3) 18.0 N (4) 20.0 N

15. What is the magnitude of the vertical component of the velocity vector shown below? (1) 10. m/s (2) 69 m/s (3) 30. m/s (4) 40. m/s

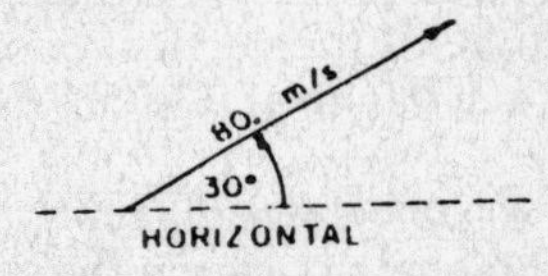

16. Forces of 6.0N north and 8.0N west act concurrently. What is the magnitude and direction of the *equilibrant* force? (1) 10N Northwest (2) 10N Southeast (3) 14N Northeast (4) 14N Southwest

III. Dynamics

The study of the relationships between force and motion is called dynamics. Galileo described accelerated motion and provided us with the laws of freely falling bodies. It was Sir Isaac Newton whose careful observations resulted in a series of statements explaining different kinds of motion that result from the application of forces. This series of statements has come to be known as Newton's Laws of Motion.

A. Mass, Force and Acceleration

1. The First Law of Motion

Newton's first law of motion is also called the Law of Inertia. It is often stated; An object at rest, will remain at rest; an object in motion will retain this motion in a straight line unless acted upon by an unbalanced force. Inertia, then, is that property of matter which resists changes in velocity; in other words, inertia is resistance to acceleration.

2. The Second Law of Motion

When the vector sum of all the forces acting on an object is not zero, it is customary to refer to their resultant as an unbalanced force. The second law states that an unbalanced force acting on an object causes it to change its velocity (to accelerate) in the direction of the force. The amount of change is directly proportional to the majnitude of the force. Newton observed that when a constant force is applied to objects of different mass, those with the largest mass accelerated least and those with the smallest mass accelerated most. Acceleration is inversely proportional to the mass of the object.

A summary statement of these observations would be, Acceleration is directly proportional to the magnitude of an unbalanced force and inversely proportional to the mass of the object. The acceleration is always in the direction of the force."

Written as an equation:

$$a = \frac{F}{m}$$

Solving for m, $m = F/a$, provides a definition for inertial mass. It is the ratio of the force applied to the resulting acceleration of the object. Inertial mass is a scalar quantity.

a. **The Kilogram** in SI units, the Kilogram (kg) is the fundamental unit of mass. The unit of acceleration is meters per second per second expressed as m/s^2.

b. **The Newton** in SI units, the newton (N) is the derived unit of force. One newton is the force needed to cause an object whose mass is 1 Kg to accelerate $1 m/sec^2$

Written as an equation: $1\ N = 1\ kg \bullet m/s^2$.

3. The Third Law of Motion

Newton's third law of motion is often stated. "For every action, there is an equal and opposite reaction. Another way of describing this is to state that whenever two objects (A and B) interact, the force on B exerted by A is equal and opposite to the force on A exerted by B. It should be noted that the two forces in Newton's third law cannot both act on the same object. The propulsion of a rocket through space is based on this law. The gases produced by the burning fuel are ejected at high speed form the rear of the rocket. If this is considered the result of the action, then the forward motion of the rocket results from the reaction to it. Either force may be called the action and the other the reaction to it. Action and reaction should not be considered cause and effect.

B. Free Body Diagrams

A free body diagram is a drawing which represents an object isolated from its surroundings. Vectors are drawn to show the direction of each force that may be acting on the object at a given time.

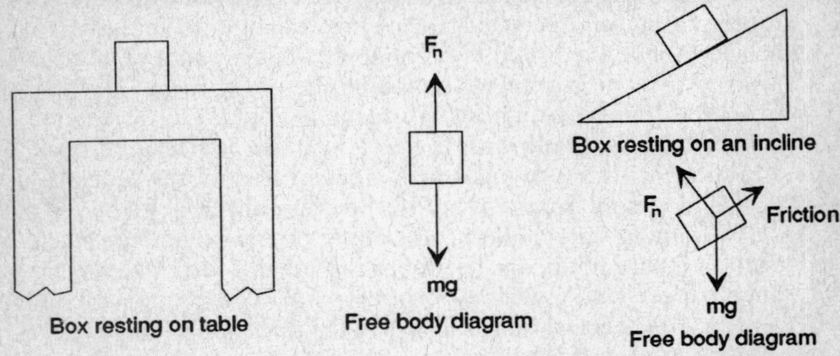

Box resting on table Free body diagram Free body diagram

When a box is resting motionless on a table, two forces are acting on it. One is the downward force due to gravity (mg). The other is the upward force exerted by the table (F_N). Since the upward force is perpendicular to the surface of the table, it is called a *normal* force.

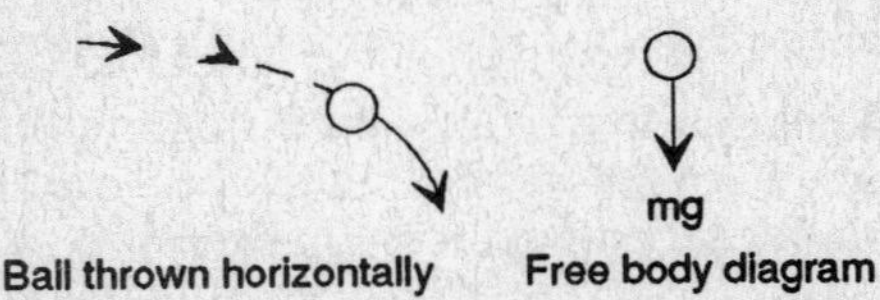

Ball thrown horizontally Free body diagram

A horizontal force must be applied to start the ball moving. If we neglect the air resistance, the only force that is acting on the ball after release is the force of gravity (mg). According to Newton's first law of motion, no force is needed to keep the ball moving at a constant speed in the horizontal direction.

QUESTIONS

1. As the vector sum of all the forces acting on an object increases, the acceleration of the object (1) decreases (2) increases (3) remains the same

2. If the mass of a moving object could be doubled, the inertia of the object would be (1) halved (2) doubled (3) unchanged (4) quadrupled

3. An object accelerates at 2.5 meters per second^2 when an unbalanced force of 10. newtons acts on it. What is the mass of the object? (1) 1.0 kg (2) 2.0 kg (3) 3.0 kg (4) 4.0 kg

4. A rocket in space can travel without engine power at constant speed in the same direction. This condition is best explained by the concept of (1) gravitation (2) action-reaction (3) acceleration (4) inertia

5. A table exerts a 2.0-newton force on a book lying on the table. The force exerted by the book on the table is (1) 20. N (2) 2.0 N (3) 0.20 N (4) 0 N

6. Two frictionless blocks, having masses of 8.0 kilograms and 2.0 kilograms, rest on a horizontal surface. If a force applied to the 8.0-kilogram block gives it an acceleration of 5.0 m/sec^2, then the same force will give the 2.0-kilogram block an acceleration of (1) 1.2 m/s^2 (2) 2.5 m/s^2 (3) 10. m/s^2 (4) 20. m/s^2

7. A 5.0-kilogram cart moving with a velocity of 4.0 meters per second is brought to a stop in 2.0 seconds. The magnitude of the average force used to stop the cart is (1) 20. newtons (2) 2.0 newtons (3) 10. newtons (4) 4.0 newtons

8. Which term represents a fundamental unit? (1) watt (2) newton (3) joule (4) meter

9. An unbalanced force of 10.0 newtons causes an object to accelerate at 2.0 m/sec^2. What is the mass of the object? (1) 0.2 kg (2) 5.0 kg (3) 8.0 kg (4) 20 kg

The diagram below represents a constant force F acting on a box located on a frictionless horizontal surface.

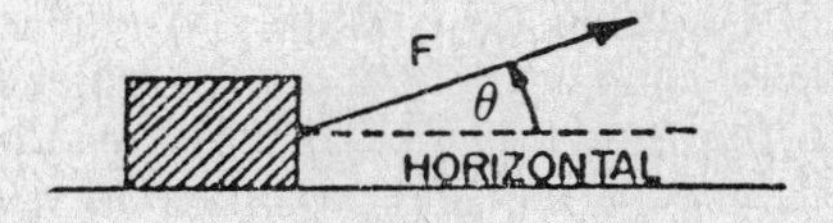

10. As the angle θ between the force and the horizontal increases, the acceleration of the box will (1) decrease (2) increase (3) remain the same

11. An object with a mass of 2 kilograms is accelerated at 5 m/s^2 The net force act on the mass is (1) 5 N (2) 2 N (3) 10 N (4) 20 N

12. An unbalanced force of 160 newtons acts on a 320-kilogram mass for 5 seconds. The acceleration of the mass is (1) 0.5 m/s^2 (2) 2 m/s^2 (3) 40 m/s^2 (4) 200 m/s^2

13. The fundamental units for a force of one newton are (1) meters/second^2 (2) kilograms (3) meters/second^2/kilogram (4) kilogram-meters/second^2

14. Which of the following free body diagrams shows the forces acting on a box as it slides down a frictionless slope?

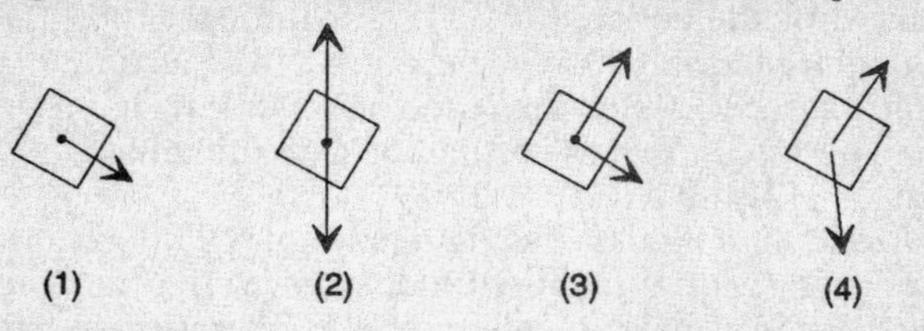

15. A rocket is accelerating to the right in horizontal flight near the ground and encounters air resistance. Which force represents the air resistance?

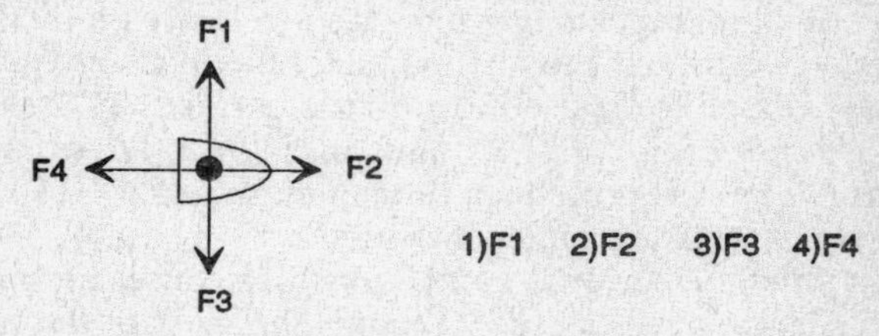

Newtons Law of Universal Gravitation

Newton was able to clearly define force and mass through his first law of motion. Afterward, he became concerned with the motions of the moon and the planets. His studies resulted in his Law of Universal Gravitation; "Any two objects in the universe attract each other with a force which is directly proportional to the product of their masses and inversely proportional to the square of the distance between them."

In the form of a proportion.

$$F_g \propto \frac{m_1 m_2}{r^2}$$

Where Fg is the Force of gravitational attraction
$\propto$ is the proportionality symbol
m_1 is the mass of one object
m_2 is the mass of the other object
r is the distance between centers of mass of the two objects

Adding a proportionality constant "G" which later was found to be 6.67×10^{-11} N m^2/kg^2 enables us to write the *equation*:

$$F_g = \frac{G\, m_1\, m_2}{r^2}$$

The law can only be applied to objects that are very small in comparison to the distance between them or spheres with uniform mass distribution. It also represents our first encounter with two concepts that are very important in nature, namely, "fields" and the "inverse square" law.

Sample Problem

Two spheres, each having a mass of 10.kg are separated by 2.0 meters. Calculate the gravitational force of attraction between them.

Given: $m = 10.\text{kg}, d = 2.0\text{m}$

Find: F_g

Equation: $F_g = \dfrac{Gm_1 m_2}{r^2}$

Solution

$$F_g = \frac{\left(6.67 \times 10^{-11} \frac{\text{Nm}^2}{\text{kg}^2}\right)(10.\text{kg})(10.\text{kg})}{(2.0\text{m})^2}$$

Answer $= 1.7 \times 10^{-9}$ N

The inverse square law—holds for gravitation and other forces where fields are involved. Prime examples are the attraction and repulsion of electric charges and magnetic poles. This law also applies to variations in intensity of light and sound with distances from the source.

Practical application of the inverse square law allows us to predict that if the distance between two objects is doubled, the gravitational force will be changed by $\frac{1}{2}^2$ or ¼. Tripling the distance, the force changes by $1/3^2$ or 1/9.

Fields—are regions in space through which forces act. It may be stated that surrounding every object there is a gravitational field through which a gravitational force acts on other objects. The field can be considered a "sphere of influence."

For example, both the earth and the moon are surrounded by gravitational fields. These two fields interact to keep the earth and the moon together.

The magnitude of the strength of a gravitational field at any point is the force per unit mass at that point in the gravitational field. It can be calculated using the equation:

$$g = F/m$$

where F is the gravitational force and m is mass.

The direction of the gravitational field at any point is the direction of the gravitational force on any mass placed at that point.

Sample Problem

A satellite having a mass of 25 kilograms experiences a gravitational force of 200. newtons. Calculate the strength of the gravitational field at this location.

Given: $F = 200.\ N,\ m = 25\ kg$

Find: g

Equation: $g = F/m$

Solution $g = 200.\ N/25\ kg$

Answer = 8.0 N/kg

Weight

The weight of an object is determined by its mass and the gravitational acceleration. Since weight is a force, it is a vector quantity. Applying Newton's second law, $F = ma$ and substituting w (weight) for F and g (acceleration due to gravity) for a, an equation for weight can be written $w = mg$.

Therefore, at a given location the weight of an object is directly proportional to its mass. However, it is important to be constantly alert to the fact that mass is an unchanging property of a stationary object: Weight is a force that acts on the object, the weight changes as its location changes. The weight of a 1 kg object on the surface of the earth is found from the equation

$$w = mg$$

Since $g = 9.8\ m/s^2$

$$w = (1\ kg)\ (9.8\ m/s^2) = 9.8\ kg\ m/s^2$$

Using the conversion factor $1N = \frac{1\ kg\ m}{s^2}$

$$9.8\ kg\ m/s^2 \left(\frac{(1N)}{1\ kg\ m/s^2}\right) = 9.8N$$

QUESTIONS

1. What is the weight of a 5.00-kilogram mass at the Earth's surface? (1) 5.00 N (2) 14.7 N (3) 49.0 N (4) 147 N

2. As a satellite is accelerated away from the Earth by a rocket, the satellite's mass (1) decreases (2) increases (3) remains the same

3. Two objects of fixed mass are moved apart so that they are separated by three times their original distance. Compared to the original gravitational force between them, the new gravitational force is (1) one-third as great (2) one-ninth as great (3) three times greater (4) nine times greater

4. A 50-kilogram student, standing on the Earth, attracts the Earth with a force closest to (1) 0 newtons (2) 5 newtons (3) 50 newtons (4) 500 newtons

5. Two point masses that are equal are separated by a distance of 1 meter. If one mass is doubled, the gravitational force between the two masses would be (1) one-half as great (2) two times greater (3) one-fourth is great (4) four times greater

6. What is the weight of a 10. kilogram object at the surface of the Earth? (1) 10. kg (2) 49N (3) 98 N (4) 49 kg

7. Which two quantities are measured in the same units? (1) velocity and acceleration (2) weight and force (3) mass and weight (4) force and momentum

8. An 800-newton person is standing in an elevator. If the upward force of the elevator on the person is 600 newtons, the person is (1) at rest (2) accelerating upward (3) accelerating downward (4) moving downward at constant speed

9. Two objects of equal mass are a fixed distance apart. If the mass of each object could be tripled, the gravitational force between the objects would (1) decrease by one-third (2) decrease by one-ninth (3) triple (4) increase 9 times

10. A space probe has a mass of 5.0 kg on the surface of the earth. What is its mass when it is at a place where the gravitational field strength is 2.0 N/kg? 1) 2.5 kg 2) 5.0 kg 3) 7.0 kg 4) 10. kg

11. An astronaut has a mass of 60. kilograms and weighs 12 newtons out in space. What is the gravitational field strength at her location? 1) 0.20 N/kg 2) 5.0 N/kg 3) 48 N/kg 4) 72 N/kg

Friction

Friction is a force that opposes motion between particles of solids, liquids, or gases that come into contact. In solids it results whenever irregular surfaces touch. The magnitude of this frictional force depends on the nature of the surfaces in contact and on the amount of force which is pressing their surfaces together.

Friction may be reduced by waxing skiis or coating a frying pan with teflon. It may be increased by putting sand on icy highways and rubbing rosin on the fingers when throwing a baseball.

A. Static Friction

Two surfaces at rest will resist sliding over each other until the force called *static friction* is overcome. This makes it difficult to start an object sliding over another surface. However, once the sliding has been started, the static friction force disappears.

Kinetic Friction

Kinetic or sliding friction replaces static friction as soon as motion has started. The coefficient of kinetic friction is less than that of starting friction. Also, it is practically independent of the areas in contact or the relative velocity of the objects. Thus, it is difficult to start sliding an object over another but once the motion has started it is much easier to continue.

Sample Problem

A 5.0 kg steel block is resting on a horizontal steel table. The coefficient of static friction of steel on steel is 0.74 and the coefficient of kinetic friction is 0.57.

a) What is the minimum horizontal force that is needed to start the block sliding.

b) What is the minimum horizontal force that is needed to keep the block moving at a constant speed?

Given: $m = 5.0 \text{ kg}, u_s = 0.74, u_k = 0.57$

Find: a) f_s b) f_k

Equations: a) $f_s = u_sF_N = u_smg$
b) $f_k = u_kF_N = u_kmg$

Solution a) $f_s = u_sF_N = u_smg$
$= (0.74)(5.0 \text{ kg})(9.8 \text{ m/s}^2)$
$= 36 \text{ kg m/s}^2$

Answer = 36 newtons

b) $f_k = u_k F_N = u_k mg$
$= (0.57)(5.0\ kg)(9.8\ m/s^2)$
$= 28\ kg\ m/s^2$

Answer = 28 newtons

QUESTIONS

1. A horizontal force is applied to a wooden box resting on a table top. As soon as the box begins to slide, the force of friction will (1) decrease (2) increase (3) remain the same
2. As the normal force between two surfaces in contact is increased, the static friction will (1) decrease (2) increase (3) remain the same

Momentum

Momentum is a vector quantity and is defined as the product of mass and velocity. Its directions is the same as the velocity.

$$p = mv$$

Where: p is the momentum,
m is the mass, and
v is the velocity

1. **Impulse** is another vector quantity with a magnitude equal to the product of the force and the time the force acts. The direction of the impulse is the same as that of the force.

$$J = F\Delta t$$

Where: J is the impulse,
F is the unbalanced force acting on an object, and
Δt is the change in time while the force acts

2. **Change of momentum.** When an object speeds up, slows down, or changes direction, its momentum changes. Such changes require an applied force. If it is a *uniform* force, resulting in constant acceleration, then

$$a = \frac{\Delta v}{\Delta t}$$

Since F = ma

Substituting $\Delta v/t$ for (a), $F = m\Delta v/\Delta t$, a very useful definition for force:

Force is the time rate of change of momentum.

The relationship can be written:

$$F\Delta t \text{ (impulse)} = m\Delta v \text{ (change of momentum)}$$

If the changes begin or end with the object at rest, then $Ft = mv$.

This equation means that when a force (F) is applied to an object at rest and continues for a time (t) the impulse (Ft) given to the object is equal to the momentum (mv) increase.

Sample Problem

A 1.0kg ball traveling at 4.0m/s strikes a wall and bounces straight back at 2.0m/s. What is its change in momentum? What impulse is applied to the wall?

a) **Given** $m = 1.0$ kg
$v_1 = 4.0$ m/sec
$v_2 = -2.0$ m/sec

Find Δmv
(actually $m\Delta v$, since mass doesn't change)

Equation $\Delta mv = m(v_2 - v_1)$

Solution $= 1.0\ \text{kg}\ [(2.0\ \text{m/s}) - (-4.0\ \text{m/s})]$

$= 6.0$ kg m/s

b) To find the impulse, use the conversion factor $1N = 1\ \text{kg m/s}^2$ converting to newtons

$$(6.0\ \text{kg m/s})\frac{1\ \text{N}}{1\ \text{kg m/s}^2}$$

Answer = 6.0 Ns

According to Newton's third law of action and reaction, this impulse is applied to both the ball and the wall simultaneously.

Law of Conservation of Momentum

When two objects interact the *sum* of their momenta after the interaction, is the same as it was before the interaction. Physicists are inclined to say that under these conditions, momentum is conserved. The principle is therefore referred to as the Law of Conservation of Momentum

$$m_1v_1 + m_2v_2 = m_1'v_1' + m_1'v_2'$$

where m_1v_1 and m_2v_2 are the momenta of the objects before the collision and $m_1'v_1'$ and $m_2'v_2'$ are the momenta after the collision.

Note:—These rules apply only for "perfectly elastic collisions" in which no other forces are in effect.

Sample Problem

A 3.0kg object traveling 6.0 meters per second east has a perfectly elastic collision with a 4.0kg object traveling 8.0 meters per second west. After the collision the 3.0kg object is traveling 10.0 meters per second west. What is the velocity of the 4.0Kg object after the collision?

(Let + = East and − = West)

Given
$m_1 = 3.0$ kg $m'_1 = 3.0$ kg
$v_1 = +6.0$ m/s $v'_1 = -10.0$ m/s
$m_2 = 4.0$ kg $m_2 = 4.0$ kg
$v_2 = -8.0$ m/s

Find v'_2

Equation $m, v, + m_2v_2 = m'_1v'_1 + m'_2v'_2$

$$(3.0 \text{ kg})(6.0 \text{ m/s}) + (4.0 \text{ kg})(-8.0 \text{ m/s})$$
$$= (3.0 \text{ kg})(-10.0 \text{ m/s}) + (4.0 \text{ kg})(v'_2)$$
$$18.0 \text{ kg m/s} + (-32 \text{ kg m/s}) = (30 \text{ kg m/s}) + (4.0 \text{ kg})(v'_2)$$
$$4.0 \text{ kg } v'_2 = (-14 \text{ kg m/s}) + (30 \text{ kg m/s})$$
$$4.0\, v'_2 = 16 \text{ m/s}$$
$$v'_2 = +4 \text{ m/s}$$

Answer = 4 m/s East

A good example of this principle is the missile ejection and the recoil that occurs when a rifle is fired.

Sample Problem

A 4.0 kg rifle fires a 5.0g (5.0×10^{-3} kg) bullet at a velocity of 500m/s. What velocity is acquired by the rifle?

Since the total momentum of the bullet and rifle before firing is zero, it must also be zero after it is fired. This is expressed by the equation

$$(5.0 \times 10^{-3} \text{ kg})(500 \text{ m/s}) + (4.0 \text{ kg})\, v = 0$$
$$4.0 \text{ kg } v = -2.5 \text{ kg m/s}$$
$$v = -.63 \text{ m/s}$$

The negative sign of the answer expresses the fact that the rifle travels in the opposite direction from the bullet.

QUESTIONS

1. An impulse of 30.0 newton-seconds is applied to a 5.00-kilogram mass. If the mass had a speed of 100. meters per second before the impulse, its speed after the impulse could be (1) 250. m/s (2) 106 m/s (3) 6.00 m/s (4) 0 m/s

2. If a 3.0-kilogram object moves 10. meters in 2.0 seconds, its average momentum is (1) 60. kg-m/s (2) 30. kg-m/s (3) 15 kg-m/s (4) 10. kg-m/s

3. Two carts of masses of 5.0 kilograms and 1.0 kilogram are pushed apart by a compressed spring. If the 5.0-kilogram cart moves westward at 2.0 meters per second, the magnitude of the velocity of the 1.0-kilogram cart will be (1) 2.0 kg-m/s (2) 2.0 m/s (3) 10. kg-m/s (4) 10. m/s

4. A baseball bat moving at high velocity strikes a feather. If air resistance is neglected, compared to the force exerted by the bat on the feather, the force exerted by the feather on the bat will be (1) smaller (2) larger (3) the same

5. A 2-kilogram car and 3-kilogram car are originally at rest on a horizontal frictionless surface as shown in the diagram below. A compressed spring is released, causing the cars to separate. The 3-kilogram car reaches a maximum speed of 2 meters per second. What is the maximum speed of the 2-kilogram car? (1) 1 m/s (2) 2 m/s (3) 3 m/s (4) 6 m/s

6. The product of an object's mass and velocity is equal to (1) force (2) weight (3) kinetic energy (4) momentum

7. Two disk magnets are arranged at rest on a fricitonless horizontal surface as shown in the diagram below. When the string holding them together is cut, they move apart under a magnetic force of repulsion. When the 1.0-kilogram disk reaches a speed of 3.0 meters per second, what is the speed of the 0.5-kilogram disk? (1) 1.0 m/s (2) 1.5 m/s (3) 3.0m/s (4) 6.0m/s

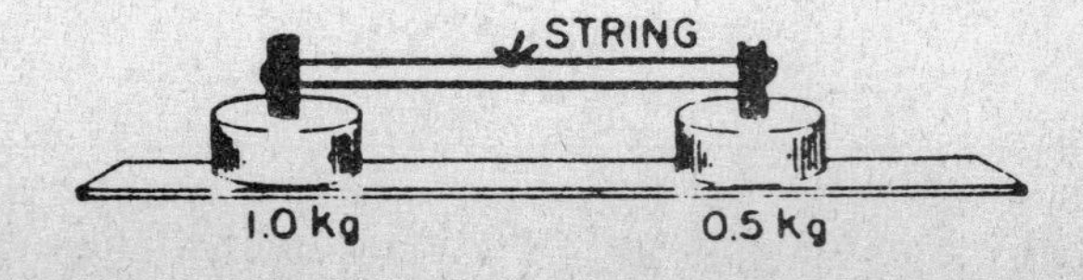

8. A net force of 12 newtons acting north on an object for 4.0 seconds will produce an impulse of (1) 48 kg-m/s north (2) 48 kg-m/s south (3) 3.0 kg-m/s north (4) 3.0 kg-m/s south

9. A force of 10. newtons acts on an object for 0.010 second. What force, acting on the object for 0.050 second, would produce the same impulse? (1) 1.0 newton (2) 2.0 newtons (3) 5.0 newtons (4) 10. newtons

10. A 2-kilogram object traveling 10 meters per second north has a perfect elastic collision with a 5-kilogram object traveling 4 meters per second south. What is the total momentum after collision? (1) 0 kg-m/s (2) 20 kg-m/s north (3) 20 kg-m/s south (4) 40 kg-m/s east

11. A 1.0-kilogram mass changes speed from2.0 meters per second to 5.0 meters per second. The change in the object's momentum is (1) 9.0 kg-m/s (2) 21 kg-m/s (3) 3.0 kg-m/s (4) 29 kg-m/s

12. A 15-newton force acts on an object in a direction due east for 3.0 seconds. What will be the change in momentum of the object? (1) 45 kg-m/s due east (2) 45 kg-m/s west (3) 5.0 kg-m/s east (4) 0.20 kg-m/s west

13. An object traveling at 4.0 meters per second has a momentum of 16 kilogram-meters per second. What is the mass of the object? (1) 64 kg (2) 20 kg (3) 12 kg (4) 4.0 kg

14. Two carts resting on a frictionless surface are forced apart by a spring. One cart has a mass of 2 kilograms and moves to the left at a speed of 3 meters per second. If the second cart has a mass of 3 kilograms, it will move to the right at a speed of (1) 1 m/s (2) 2 m/s (3) 3 m/s (4) 6 m/s

15. A force of 80. newtons pushes a 50.-kilogram object across a level floor for 8.0 meters. The work done is (1) 10. joules (2) 400 joules(3) 640 joules (4) 3,920 joules

16. A 20.-kilogram mass moving at a speed of 3.0 meters per second is stopped by a constant force of 15 newtons. How many seconds must the force act on the mass to stop it? (1) 0.20 s (2) 1.3 s (3) 5.0 s (4) 4.0 s

17. A 5.0-newton force imparts an impulse of 15 newton-seconds to an object. the force acted on the object for a period of (1) 0.33 s (2) 20. s (3) 3.0 s (4) 75 s

Torque (or moment) is a measure of the rotation effects that can occur when a force is applied to an object.

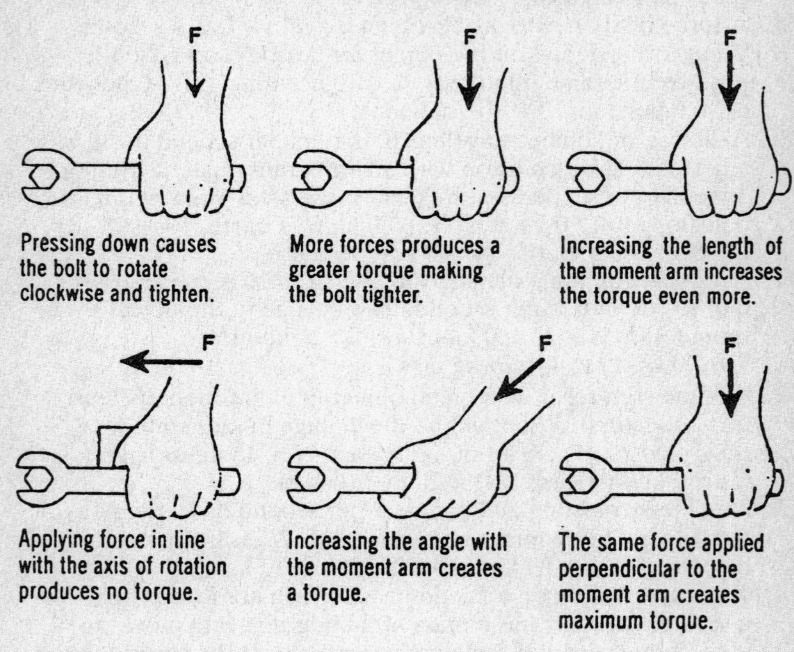

Pressing down causes the bolt to rotate clockwise and tighten.

More forces produces a greater torque making the bolt tighter.

Increasing the length of the moment arm increases the torque even more.

Applying force in line with the axis of rotation produces no torque.

Increasing the angle with the moment arm creates a torque.

The same force applied perpendicular to the moment arm creates maximum torque.

Torque (τ greek letter tau) is defined as the product of a force and the perpendicular distance along the moment arm to the axis of rotation.

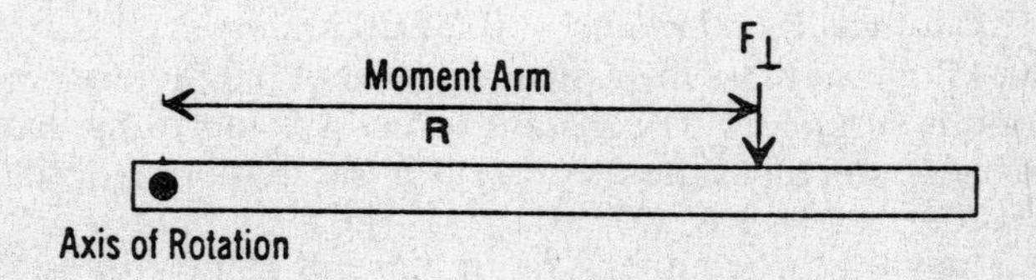

$$\tau = F_\perp R$$

Where: τ is the torque in newton-meters
$F_\perp$ is the perpendicular force in newtons, and
R is the distance along the moment arm in meters

UNIT II. ENERGY

Work and Energy

The total energy of a system is changes when work is done on or by the system. Energy is required to get the work done.

A. **Work.** In the language of Physics, **work** is done only when a force causes a displacement of an object. Work is a scalar quantity. It is defined as the product of the component of force acting in the direction of motion and the displacement of the object.

$$W = Fs$$

W = work done
F = Force
s = displacement

The joule is the SI unit of work. It is the work done when an object is displaced one meter by a force of one newton. Remember that a newton is defined as the force needed to accelerate 1kg of mass $1 m/s^2$ and the meter is the SI unit for displacement. Therefore, a joule may be expressed as a kgm/s^2 (m) or kgm^2/s^2

Sample Problem

How much work is done when a force of 50 N is used to move an object 4 meters in the same direction as the force?

Given $F = 50\ N$
$s = 4\ m$
Find W
Equation $W = Fs$
Solution
$$W = 200\ Nm \times \frac{1\ joule}{1\ Nm}$$

Answer $W = 200$ joules

B. Energy

Energy is defined as the capacity for doing work. This implies that when work is done on an object, the object has gained energy. Therefore, work transfers energy from one object or system to another. Furthermore, the same units are used to measure energy that are used to measure work. Like work, energy is a sclar quantity and the SI unit is the *joule*.

Note:—An important "fact of life" is that, at least until now, there is no way to utilize all the energy of a system.

Different forms of energy are named according to changes made in objects when work is done on the objects. Work that is done in lifting an object results in an increased *gravitational potential energy*. Work that is done when a spring is stretched adds extra *spring potential energy*. Work done to accelerate an object is observed as additional *kinetic energy*. Temperature is one measure of the *internal energy* of an object based on the average kinetic energy of its particles.

1. **Potential energy** is the name given to the energy possessed by an object because of its position or its condition. Under ideal conditions the potential energy of an object is equal to the amount of work done in getting the object where it is or into its present condition.

   a. **Gravitational potential energy** is the energy that an object has because of the work that was done against a gravitational force to put the object at its present location in the gravitational force field. If we lift an object, we do work on it *against* a gravitational force. This results in an *increase* in the gravitational potential energy of the object. If we permit the object to fall, the work is done by the gravitational force and there is a *decrease* in the gravitational potential energy of the object. The equation used for gravitational potential energy is:

   $$\Delta PE = mg\Delta h$$

   This may be stated, "the change in gravitational energy (ΔPE)is equal to the product of the *weight* of the object (mg) and the vertical change of its position (Δh)." It is important to keep in mind that we must arbitrarily select some starting position as a reference, or zero level and that our formula weakens as Δh becomes large.

Sample Problem

How much potential energy is gained by a 75kg object when it is raised 2 meters straight up?

Given $m = 75$ kg
$\Delta h = 2$ m
$g = 9.8$ m/s^2

Find ΔPE

Equation $\Delta PE = mg\Delta h$

Solution $\Delta PE = (75 \text{ kg})(9.8 \text{ m/s}^2)(2 \text{ m})$

$$= 1.5 \times 10^3 \text{ kg m}^2/\text{s}^2 \times \frac{1 \text{ joule}}{1 \text{ kg m}^2/\text{s}^2}$$

Answer $= 1.5 \times 10^3$ joules

b. **Elastic Potential Energy**

Work is required to stretch or compress a spring. This work is stored in the deformed spring as *elastic potential energy*. When the spring is allowed to return to its original size, it can perform work as it loses this energy.

We have already learned that work is equal to the product of force and the distance through which the force acts. However, when a spring is stretched, the force is not constant. It is very small at first and then increases as the spring is stretched further. Thus, the work that is done in deforming a spring depends on the *average* force ($\overline{F}$) that is applied and the total change in length (the spring *displacement* (x). This is given by the relationship:

$$PE_s = \overline{F}\,x$$

where: PE_s is the elastic potential energy in joules,
$\overline{F}$ is the average force in newtons, and
x is the spring displacement in meters.

According to Hooke's Law, the force needed to stretch an ideal spring a given amount is directly proportional to its displacement. This is shown in the diagram below. A displacement of 0.1 meter requires a force of 2 newtons. Further 0.1 meter displacements each require the addition of another 2 newton force. The ration F/x is the same for any displacement of a given spring. This value (F/x) is called the *spring constant*).

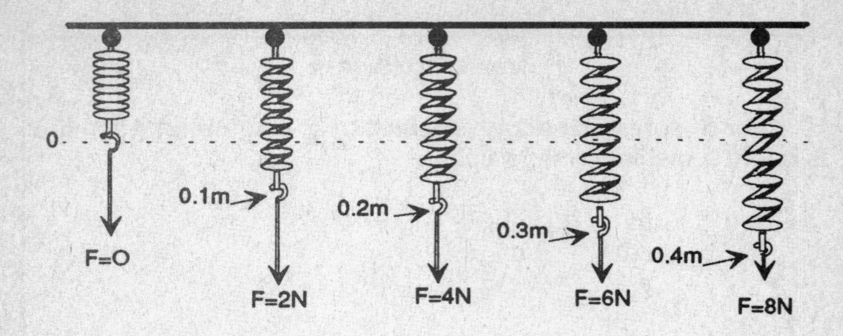

The constant of the spring shown in the diagrams above can be calculated using any of the diagrams where the spring is stretched and the relationship:

$$k = F/x$$
$$= 6\ \text{N}/0.3\text{m}$$
$$= 20\ \text{N/m}$$

Another way to determine the spring constant is to plot force vs displacement data on a graph. The slope of the graph is the spring constant.

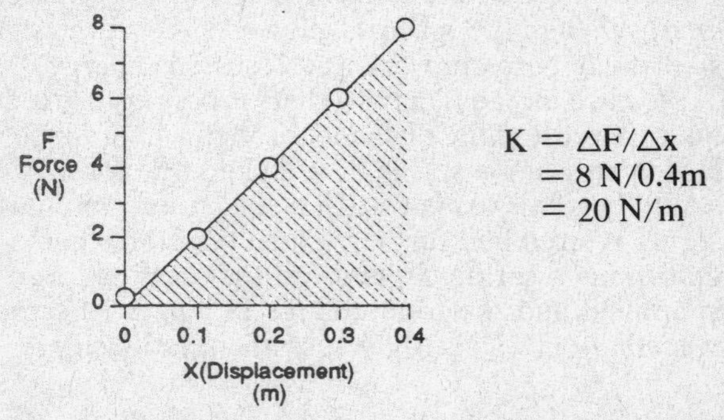

$$K = \Delta F/\Delta x$$
$$= 8\ \text{N}/0.4\text{m}$$
$$= 20\ \text{N/m}$$

A force vs displacement graph is also useful to help calculate the elastic potential energy. When the applied force is given in newtons and the spring displacement is given in meters, the potential energy is equal to the area of the triangle under the graph line. This area is shaded in the diagram above.

The area of the shaded triangle $= 1/2$ base $\times$ altitude

$$PE_s = 1/2\ (8\ \text{N})\ (0.4\ \text{m})$$
$$= 1.6\ \text{Nm} = \underline{1.6\ \text{joules}}$$

There is another way to calculate the elastic potential energy of a stretched or compressed spring without having to draw a graph. If the spring constant (k) and the displacement (x) are known, use the relationship:

$$PE_s = 1/2\ kx^2$$

The equation above is derived as follows:

The spring constant is defined ...... $k = F/x$ (1)

(multiplying both sides by x) ...... $F = kx$ (2)

Spring energy is equal to work (average force × displacement)

$$PE_s = \overline{F}\ x \quad (3)$$

Since the force increases from zero to F uniformly, the average force is half as great as the maximum force

$$\overline{F} = (0 + F)/2 = 1/2\ F \quad (4)$$

substituting 1/2 F for F (Eq. 4 in Eq. 3)

$$PE_s = 1/2\ F\ x \quad (5)$$

substituting for kx for F (Eq. 2 in Eq. 5)

$$PE_s = 1/2\ kx\ x \text{ or } PE_s = 1/2\ kx^2$$

Sample Problem

A force of 6.0 N stretches a spring making it 0.15 m longer.
a) Calculate the spring constant.
b) Calculate the increase in elastic potential energy that is caused by the stretching.

Solution

a) $k = F/x$
$= 6.0\ N/0.15\ m$
$= \underline{40\ N/m}$

b) $PE_s = 1/2\ k\ x^2$
$= 1/2\ (40\ N/m)\ (0.15\ m)^2$
$= \underline{0.45 J}$

3. **Kinetic energy** is the energy an object possesses because of its motion. **Kinetic** energy is related to **motion** as **potential** energy is related to **position**. The work done in accelerating an object from rest ($v = 0$) to a given speed is equal to its kinetic energy at that speed. Applying the laws of motion, we can develop an equation that relates kinetic energy to mass and velocity.

The work done (w) is Fs (1)
Since F = ma, (2)
Then W = ma s (3)
Earlier, we found that $v^2 = 2as$ (4)

which can be written: as = $v^2/2$

Substituting ($v^2/2$) for (as) in equation (3)

$$w = \frac{mv^2}{2} = KE$$

The well known equation is: KE = ½ mv^2

Under ideal conditions, KE is equal to the work needed to stop a movng object or to bring the object from rest to a given speed.

Power is the rate of doing work. It is a scalar quantity which is a ratio of the work performed to the time it took to perform it.

$$P = \frac{W}{t}$$

The **watt** is the SI unit of power defined as one joule per second. Other useful equivalents can be used in problem solving

$$P = \frac{W}{t}, \text{ but } W = fs \text{ and so}$$

$$P = \frac{fs}{t}, \text{ but } s/t = v, \text{ and so}$$

$$P = Fv$$

Sample Problem

How much power is used when an object is brought to a speed of 15 meters per second by an applied force of 30 newtons?

Given v = 15 m/s
F = 30 N

Find P

Equation: P = Fv

Solution P = (30 N) 15 m/s

$$= 450 \text{ N m/s} \times \frac{1 \text{ joule}}{1 \text{ Nm}} \times \frac{1 \text{ watt}}{1 \text{ j/s}}$$

Answer = 450 watts

V. Conservation of Energy

Unless work is done on or by a system, its total energy does not change. The total energy in a system is the sum of its kinetic energy, plus its potential and internal energy. There is an increased trend in calling this law, the Conservation of *Mass-Energy* because of the equivalence of mass and energy under certain conditions.

A falling object provides a good illustration. Under ideal conditions, the object's gain of kinetic energy equals its loss of potential energy. No energy is created, nor lost. It has simply changed to another form.

Friction. In any real system there is always some friction that opposes the motion of two objects in contact.To move an object against this force, work must be done. This work does not increase the kinetic energy or the potential energy of the object. However, friction has the effect of raising the temperature of the objects involved. By defining a quantity called **internal energy** we can account for the work done by friction by relating it to the total energy of the particles that make up the object. In this way it is still possible to validate the Law of Conservation of Energy.

QUESTIONS

1. A motor rated at 100. watts accelerates an object along a horizontal surface with an average speed of 4.0 meters per second. What force is supplied by the motor in the direction of motion? [Assume 100% efficiency.] (1) 0.04 N (2) 25 N (3) 100 N (4) 400 N

2. A 20.-newton block falls freely from rest from a point 3.0 meters above the surface of the Earth. With respect to the surface of the Earth, what is the gravitational potential energy of the block-Earth system after the block has fallen 1.5 meters? (1) 20. joules (2) 30. joules (3) 60. joules (4) 120 joules

3. As the time required to do a certain amount of work increases, the power (1) decreases (2) increases (3) remains the same

4. A unit for kinetic energy is the (1) watt (2) joule (3) newton (4) kilogram-meter/second

5. A 2.0-newton book falls from a table 1.0 meter high. After falling 0.5 meter, the book's kinetic energy is (1) 1.0 joule (2) 2.0 joules (3) 10 joules (4) 20 joules

6. The direction of exchange of internal energy between objects is determined by their relative (1) inertias (2) momentums (3) temperatures (4) masses

7. A constant force of 20. newtons applied to a box causes it to move at a constant speed of 4.0 meters per second. How much work is done on the box in 6.0 seconds?
(1) 480 joules (2) 240joules (3) 120 joules (4) 80. joules

8. Which mass has the greatest potential energy with respect to the floor? (1) 50-kg mass resting on the floor
(2) 2-kg mass 10 meters above the floor
(3) 10-kg mass 2 meters above the floor
(4) 6-kg mass 5 meters above the floor

9. A crane raises a 200-newton weight to a height of 50 meters in 5 seconds. The crane does work at the rate of
(1) 8×10^{-1} watt (2) 2×10^{1} watts (3) 2×10^{3} watts
(4) 5×10^{4} watts

10. When the speed of an object is halved, its kinetic energy is (1) quartered (2) halved (3) the same (4) doubled

11. An object has a mass of 8.0 kilograms. A 2.0-newton force displaces the object a distance of 3.0 meters to the east, and ten 4.0 meters to the north. What is the total work done on the object? (1) 10. joules (2) 14 joules(3) 28 joules
(4) 56 joules

12. As an object slides across a horizontal surface, the gravitational potential energy of the object will (1) decrease
(2) increase (3) remain the same

13. As the power of a machine is increased, the time required to move an object a fixed distance (1) decreases
(2) increases (3) remains the same

14. If an engine rated at 5.0×10^{4} watts exerts a constant force of 2.5×10^{3} newtons on a vehicle, the velocity of the vehicle is (1) 0.050 m/s (2) $2\sqrt{10}$ m/s (3) 20.m/s
(4) 1.25×10^{8} m/s

15. If the velocity of an automobile is doubled, its kinetic energy
(1) decreases to one-half (2) doubles
(3) decreases to one-fourth (4) quadruples

16. Which quantities are measured in the same units?
(1) mass and weight (2) heat and temperature
(3) power and work (4) work and energy

17. Ten joules of work are done in accelerating a 2.0-kilogram mass from rest across a horizontal frictionless table. The total kinetic energy gained by the mass is (1) 3.2 joules
(2) 5.0 joules (3) 10. joules (4) 20.joules

18. The diagram below represents a cart traveling from left to right along a frictionless surface with an initial speed of v.

At which point is the gravitational potential energy of the cart *least?* (1) A (2) B (3) C (4) D

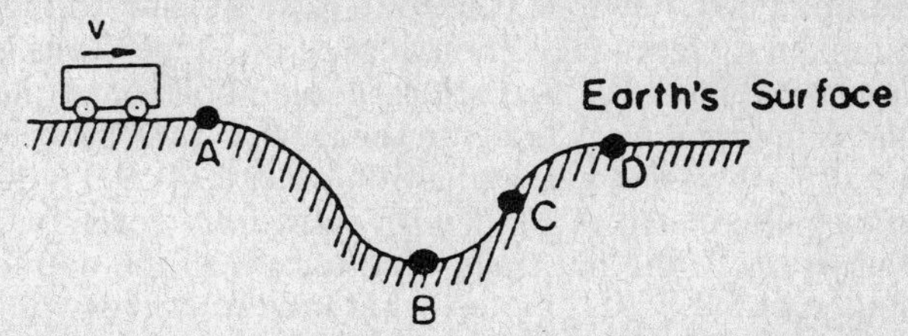

19. A 0.50-kilogram sphere at the top of an incline has a potential energy of 6.0 joules relative to the base of the incline. Rolling halfway down the incline will cause the sphere's potential energy to be (1) 0 joules (2) 12 joules (3) 3.0 joules (4) 6.0 joules

20. A 1.0-kilogram mass falls a distance of 0.50 meter, causing a 2.0-kilogram mass to slide the same distance along a table top, as represented in the diagram below. How much work is done by the falling mass? (1) 1.5 J (2) 4.9 J (3) 9.8 J (4) 14.7 J

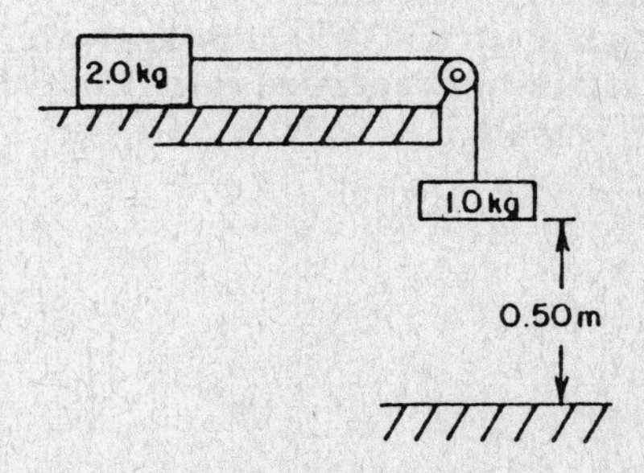

21. If 700 watts of power is needed to keep a boat moving through the water at a constant speed of 10 meters per second, what is the magnitude of the force exerted by the water on the boat? (1) 0.01 N (2) 70 N (3) 700 N (4) 7,000 N

22. A constant unbalanced force of friction acts on a 15.0-kilogram mass moving along a horizontal surface at 10.0 meters per second. If the mass is brought to rest in 1.50 seconds, what is the magnitude of the force of friction? (1) 10.0 N (2) 100. N (3) 147 N (4) 150. N

23. A crate is pulled 6.0 meters up an incline with a force of 50 newtons. If the potential energy of the box increases 250 joules, the total work done against friction in moving the box is (1) 0 J (2) 50. J (3) 250 J (4) 300 J

24. If the kinetic energy of a given mass is to be doubled, its speed must be multiplied by (1) 8 (2) 2 (3) $\sqrt{2}$ (4) 4

25. A force of 100. newtons is used to push a trunk to the top an incline 3.0 meters long. Then a force of 50. newtons is used to push the trunk for 10. meters along a horizontal platform. What is the total work done on the trunk? (1) 8.0×10^2 joules (2) 5.0×10^2 joules (3) 3.0×10^2 joules (4) 9.0×10^2 joules

26. A baseball bat strikes a ball with an average force of 2.0×10^2 newtons. If the bat stays in contact with the ball for a distance of 5.0×10^{-3} meter, what kinetic energy will the ball acquire from the bat? (1) 1.0×10^0 joules (2) 2.0×10^2 joules (3) 2.5×10^1 joules (4) 4.0×10^2 joules

27. A coiled spring is compressed 0.5 m by a force of 3.0 N. As it is compressed further, its spring constant will (1) decrease (2) increase (3) remain the same

28. Hanging a weight of 5.0 N on a coiled spring stretches it 6.0 cm. A 10 N weight will stretch this spring (1) 6.0 cm (2) 12 cm (3) 30 cm (4) 60 cm

29. Applying a force of 2.0 N causes a long spring to stretch 3.0 m. The increase of its elastic potential energy is (1) 0.66 J (2) 1.5 J (3) 3.0 J (4) 6.0 J

30. A door spring is stretched 0.30 m by an *average force* of 0.10 N. The total work expended is (1) 0.03 J (2) 0.33 J (3) 3.0 J (4) 0.009 J

UNIT III. ELECTRICITY AND MAGNETISM

I. Static Electricity

The term static electricity identifies electronic charges that are not flowing in any particular direction. The term "static" implies motionless and, of course, this is not the case. Current electricity refers to the condition in which charges are flowing. The purpose of studying static electricity is to become familiar with electric forces and the nature of electric charges.

A. Micro Structure of Matter

Atoms in turn are composed of subatomic particles. Among these are the very dense neutrons and protons which are found in the nucleus and the very light but very active electrons, found around the nucleus, forming "clouds" of various shapes

1. Each neutron has a mass approximately equal to that of a proton and carries no electrical charge.
2. Each proton has a mass value of 1 atomic mass unit and carries a positive charge.
3. Each electron has about 1/1840 the mass of an atomic mass unit and carries a negative charge.

Neutrons and protons are held together by enormous nuclear forces and are very massive compared to electrons. As a result, when objects gain or lose charges, it is because they have lost or gained electrons.

In studies dealing with static electricity (electrostatics) it is generally agreed that:

1. Each electron carries a small negative charge that cannot be subdivided. This charge is called an elementary negative charge.
2. The proton carries a small positive charge that cannot be subdivided. This charge is called an elementary positive charge.

3. The magnitudes of the electron and proton charges are equal.
4. The electron and proton charges are opposite in sign.

B. Charged Objects

1. A neutral object is one with an equal number of electrons and protons.
2. A positively charged object has a deficiency of electrons.
3. A negatively charged object has an excess of electrons.
4. A charged object loses its charge and becomes neutral when it is grounded.

In general, matter becomes charged through a transfer of electrons. An electroscope is used to detect the presence of a charge. When the leaves (usually made of gold foil) are neutral, they simply hang vertically. When they are charged, they must be carrying charges of the same sign and so they will diverge, or spread apart, because of the force of repulsion between them. The leaves of an electroscope cannot carry opposite charges.

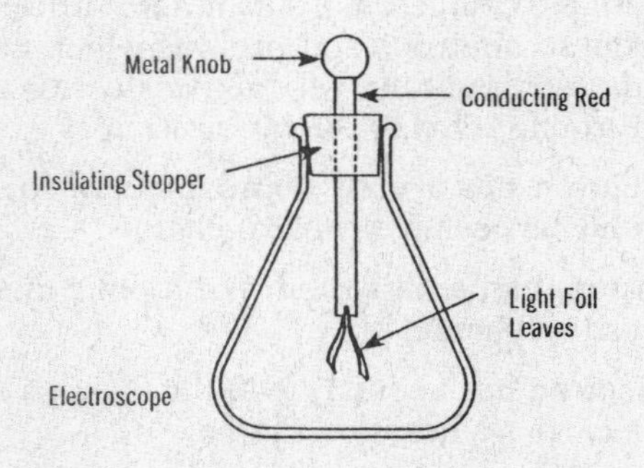

C. Transfer of Charges

Static charges may be transfered from one object to another.

1. **Conservation of charge** is a fundamental law which may be stated, "The net charge in a closed system remains constant."
2. Separation of charge by contact occurs when different neutral objects are brought together and electrons migrate from one object to the other. This process may be speeded up by rubbing the two objects in order to increase the contact. It has been found that when a hard rubber rod is rubbed with

fur, the rod gains electrons from the fur and becomes negatively charged. The fur, meanwhile, having lost electrons becomes positively charged.

3. **Conduction** is the process of charging an object by transferring electrons during contact with a charged object. Electrons actually migrate from one object to the other. The charge received is always of the same sign as that of the charging object. The electroscope can be used to demonstrate the presence of electric charges. When a positive or negative charge is placed on the knob, it spreads over the rod and onto the leaves. The leaves will repel each other and spread apart.
4. **Induction** is a charging process in which a neutral object is attracted to a nearby charged object. The charged object, in this process, causes a redistribution of the charges in the neutral object. If the charged object carries a negative charge it will repel the negative charges on the neutral object leaving a positive charge on the side nearest the charged object. If the neutral object is grounded (providing a pathway for electrons to travel toward or away from the ground), then the neutral object will acquire a charge opposite to that of the charged object.

D. **Elementary Charges** refer to the smallest charges known to exist in nature. The charge consists of whole number multiples of elementary charges. The charge on an electron (-1.6×10^{-19} coulomb) is one negative elementary charge.

E. **The Quantity of Charge** that an object has, depends on numbers of unpaired electrons or protons in the material. The SI unit of charge is the coulomb. One coulomb represents an excess or deficiency of 6.25×10^{18} electrons. Therefore, the charge of an individual electron is $1/6.25 \times 10^{18}$ coulomb $= 1.6 \times 10^{-19}$C.

F. **Coulomb's Law** is to electrostatic force what Newton's Universal Law is to gravitational force. It should be no surprise then that its form is very familiar and that it applies to **point charges** only. Point charges are those that are very small compared to the distance between them. The law may be stated, "The force between fixed point charges is directly proportional to the product of the charges and inversely proportional to the square of the distance between them."

$$F \propto \frac{q_1 q_2}{r^2} \text{ or } F = k\frac{q_1 q_2}{r^2}$$

F = force in newtons

q_1 and q_2 = the charges in coulombs

r = the distance between the objects in meters

$$k = 9 \times 10^9 \frac{\text{newton meter}^2}{\text{coulomb}^2}$$

Sample Problem

What is the electric force between two very small charged objects located 0.5 meters apart, if the charge on one object is -4×10^{-8} coulombs and the charge on the second object is $+6 \times 10^{-5}$ coulombs?

Given:
$r = 0.5$ meters
$q_1 = -4 \times 10^{-8}$ C
$q_2 = +6 \times 10^{-5}$ C
$k = 9 \times 10^9$ Nm²/C²

Find F

Equation $F = k\frac{q_1 q_2}{r^2}$

Solution
$$F = \left(9 \times 10^9 \frac{N\cancel{m^2}}{\cancel{C^2}}\right) \frac{(-4 \times 10^{-8}\,\cancel{C})(6 \times 10^{-5}\,\cancel{C})}{(.5\cancel{m})^2}$$

$$\frac{-2.16 \times 10^{-2}\text{ N}}{.25}$$

Answer = -9×10^{-2} N

G. **Electric fields** are regions in space in which electric forces act on charges. It follows that an electric field exists around every charged object. Electric field intensity is a vector quantity. The direction is defined as the direction of the force on a positive test charge placed in the field. If the object bears a negative charge, the positive test charge will be attracted to it. Therefore, the direction of the field points **toward** the object rather than away from it. The magnitude is equal to the force per unit charge exerted at that point, or

$$E = \frac{F}{q}$$

where E = electric field

F = electric force

q = charge used to test the field.

The SI unit for E is newtons per coulomb (N/C).

1. Field around a point charge is radial. The intensity of this field varies inversely with the square of the distance from the point charge. The field around a charged conducting sphere behaves as if the charge were concentrated at the center and, therefore the field is similar to that of a point charge. The field **within** a charged conducting sphere is **zero**.

Electric Field Patterns

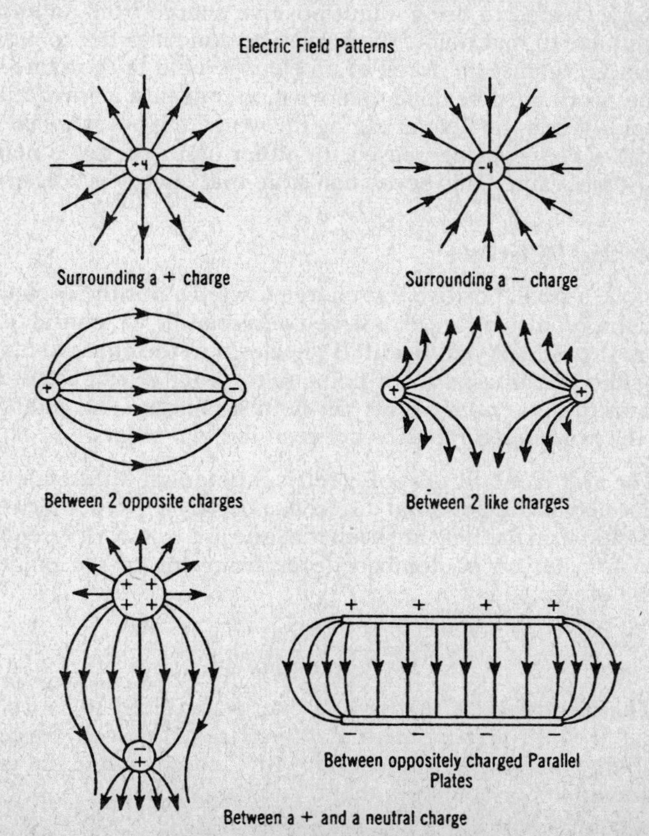

2. **The field around a uniformly charged rod** is also radially directed. Since the field lines are perpendicular to the surface of the rod they all pass through the surface of a cylinder drawn concentric with the wire. The area of such a culinder is proportional to its radius. Therefore, the intensity of the electric field around a uniformly charged rod is inversely proportional to the distance from the rod and **not** the **square** of this distance.
3. **The field between two parallel oppositely charged plates** is uniform except near the edges provided the distance between them is small, compared to their size. The force on a charged object, therefore, will be the same anywhere in this central space.
4. **The electric potential** at any point in an electric field is the work needed to bring a unit positive charge from an infinite distance to that point. Work must be done in order to move a charge against the force of an electric field in the same way that work must be done to move matter against a gravitational field. When the field is "doing the work" the potential energy of the charge is decreased. In either case, energy is neither lost nor gained and so we can state that energy is conserved.

H. Potential Difference

To move a small positive test charge toward a positively charged object, the repulsive force must be overcome. If we consider two points in the field, A and B with B the closer to the charged object, moving the test charge from point A to point B will result in a change in the potential energy of the test charge. This change is called the *potential difference* between the two points.

1. **The volt** is the SI unit of electrical potential difference. It is defined as the potential difference between two points whose positions in the field are such that one joule of work is required to transfer on coulomb of force from one of the points to the other.

$$1 \text{ volt} = \frac{1 \text{ joule}}{1 \text{ coulomb}} \text{ or } V = \frac{W}{q}$$

2. **The electron volt** (eV) is the energy required to move one electron through a potential difference of one volt. Since the charge on the electron is 1.60×10^{-19} coulomb and the volt is one joule per coulomb, then:

$$1 \text{ e V} = 1.60 \times 10^{-19} \text{ joule.}$$

The electron volt is an extremely small unit of energy and is, therefore, most useful in considering the energies of atoms and their particles.

3. Electric Field in Terms of Electric Potential

There is a relation between field (E) and potential V. This relationship may be expressed in terms of the change in potential per unit distance (ΔV/meter), This expression can be shown to be identical to newtons/coulomb.

$$\text{If } E = \frac{\text{volt}}{\text{meter}} \text{ and volt} = \frac{\text{joule}}{\text{coulomb}}$$

$$\text{then } E = \frac{\text{joule}}{\text{coulomb meter}} \text{ and, since joule} = \text{newton meter}$$

$$\text{then } E = \frac{\text{newton meter}}{\text{coulomb meter}} = \frac{\text{newton}}{\text{coulomb}}$$

In summary

$$E = \frac{V}{d},\ E = \frac{F}{q},\ \text{and}\ \frac{V}{d} = \frac{F}{q}$$

When: E is the electric field, in volts/meter
V is the electric potential, in volts
F is the force on a charged particle, in newtons and
q is the charge in coulombs

I. The Millikan Experiment and the Elementary Charge

One of the most famous experiments in physics is the oil drop experiment performed by Robert A. Millikan and reported in 1913. Millikan measured the forces on charged oil drops in a uniform electric field. He found no drop with a charge less than 1.60×10^{-19} coulomb. The charges on other drops were integral multiples of this value. This finding demonstrated that there **is** a fundamental unit of charge. This elementary charge of 1.60×10^{-19} coulomb is called the charge on a single electron.

QUESTIONS

1. One of two identical metal spheres has a charge of $+q$, and the other sphere has a charge of $-q$. The spheres are brought together and then separated. Compared to the total charge on the two spheres before contact, the total charge on the two spheres after contact is (1) less (2) greater (3) the same

Base your answers to questions 2 through 7 on the diagram below which represents two charged spheres, X and Y.

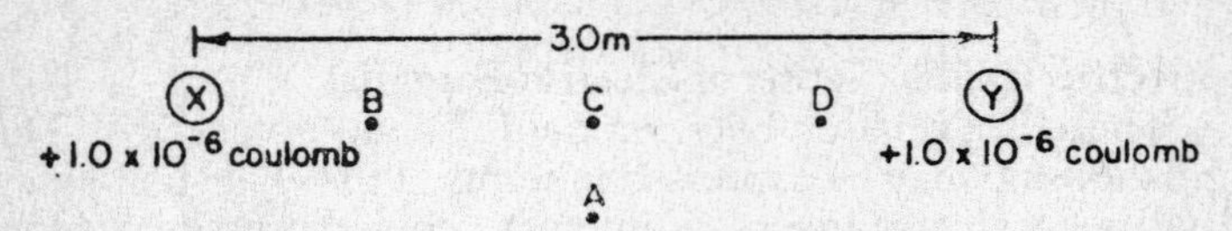

2. At which point is the magnitude of the electric field equal to zero? (1) A (2) B (3) C (4) D

3. Which arrow best represents the direction of the electric field at point A?

4. If a unit positive charge moves directly from point B to point D, the potential energy of the charge will (1) decrease, only (2) increase, only (3) decrease then increase (4) increase then decrease

5. The magnitude of the force between the two spheres is (1) 1.0×10^{-3} N (2) 1.0×10^{3} N (3) 3.0×10^{-3} N (4) 3.0×10^{3} N

6. Moving the two spheres toward each other would cause their electrical potential energy to (1) decrease (2) increase (3) remain the same

7. Compared to the force of the electric field of sphere X on sphere Y, the force of the electric field of sphere Y on sphere X is (1) less (2) greater (3) the same

8. Which particle has no charge? (1) electron (2) neutron (3) proton (4) alpha particle

9. A pith ball may become charged by losing or gaining (1) electrons, only (2) protons, only (3) protons and electrons (4) neutrons and protons

10. Two identical conducting spheres carry charges of +3 coulombs and −1 coulomb, respectively. If the spheres are brought into contact and separated, the final charge on each will be (1) +1C (2) +2C (3) −1 C (4) −2 C

11. An alpha particle with a charge of +2 elementary charges is accelerated in a vacuum through a potential difference of 10,000. volts. What is the energy acquired by the particle? (1) 3.2×10^{-15} eV (2) 2.0 eV (3) 20,000 eV (4) 40,000 eV

12. Which diagram best illustrates a neutral electroscope being charged by conduction?

13. If the magnitude of the charge on each of two positively charged objects is halved, the electrostatic force between the objects will (1) decrease to one-half (2) decrease to one-quarter (3) decrease to one-sixteenth (4) remain the same

14. The diagram below shows some of the lines of electrical force around a positive point charge. the strength of the electric field is (1) greatest at point A (2) greatest at point B (3) greatest at point C (4) equal at points A, B and C

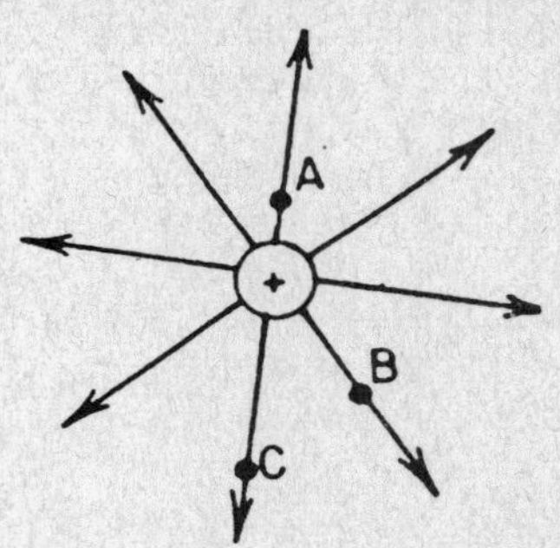

Base your answers to questions 15 through 19 on the diagram below which represents a system consisting of two charged metal spheres with equal radii.

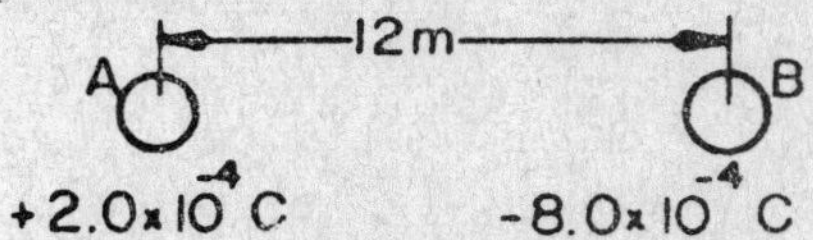

15. What is the magnitude of the electrostatic force exerted on sphere A? (1) 1.1×10^{-9} N (2) 1.3×10^{-8} N (3) 120 N (4) 10. N

16. Compared to the force exerted on sphere B at a separation of 12 meters, the force exerted on sphere B at a separation of 6.0 meters would be (1) ½ as great (2) 2 times as great (3) ¼ as great (4) 4 times as great

17. If the two spheres were touched together and then separated, the charge on sphere A would be (1) -6.0×10^{-4} C (2) 2.0×10^{-4} C (3) -3.0×10^{-4} C (4) -8.0×10^{-4} C

18. Compared to the electrical potential *energy* of the system at a separation of 12 meters, the electrical potential energy of the system at a separation of 6 meters is (1) less (2) greater (3) the same

19. If spheres A and B, as represented in the diagram, were touched together and then separated, the net charge on the two spheres would (1) decrease (2) increase (3) remain the same

20. A wool cloth becomes positively charged as it
(1) gains protons (2) gains electrons (3) loses protons
(4) loses electrons

21. A charged particle is placed in an electric field E. If the charge on the particle is doubled, the force exerted on the particle by the field E is (1) unchanged (2) doubled
(3) halved (4) quadrupled

22. As the distance between two point charges is tripled; the electrostatic force between the charges will become
(1) 1/9 as great (2) 1/3 as great (3) 3 times as great
(4) 9 times as great

23. Which is equivalent to three elementary charges?
(1) 2.4×10^{-19} C (2) 2.0×10^{-19} C (3) 4.8×10^{-19} C
(4) 5.4×10^{-19} C

24. Which diagram best represents the electric field surrounding a point positive charge?

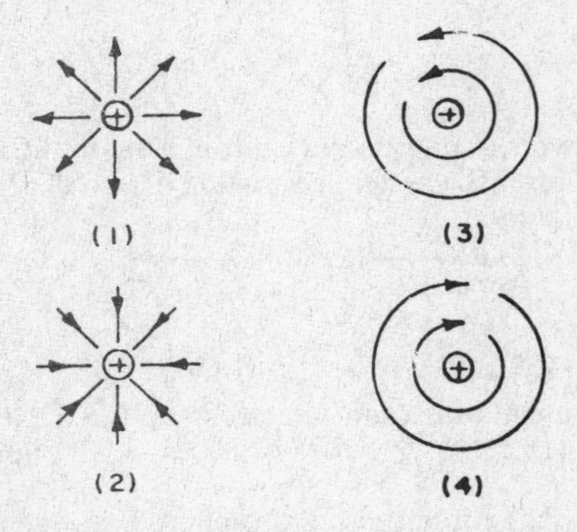

25. A negatively charged rod is held near the knob of an uncharged electroscope. Which diagram best represents the distribution of charge on the electroscope?

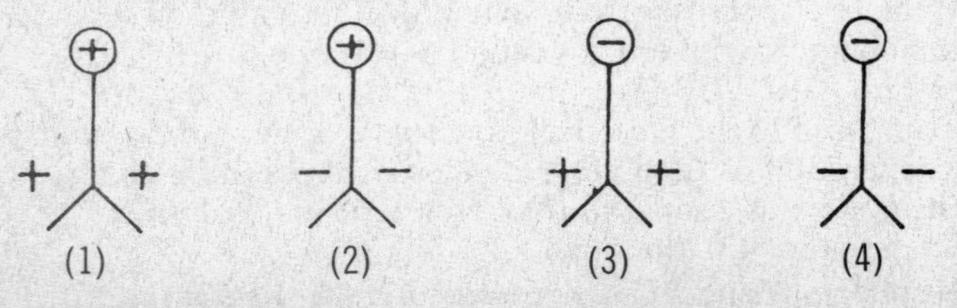

26. The coulomb is a unit of electrical (1) charge
(2) current (3) potential (4) resistance

27. After two neutral solids, A and B, were rubbed together, solid A acquired a net negative charge. Solid B, therefore, experienced a net (1) loss of protons (2) increase of protons (3) loss of electrons (4) increase of electrons

28. Which diagram best illustrates the electric field around two unlike charges?

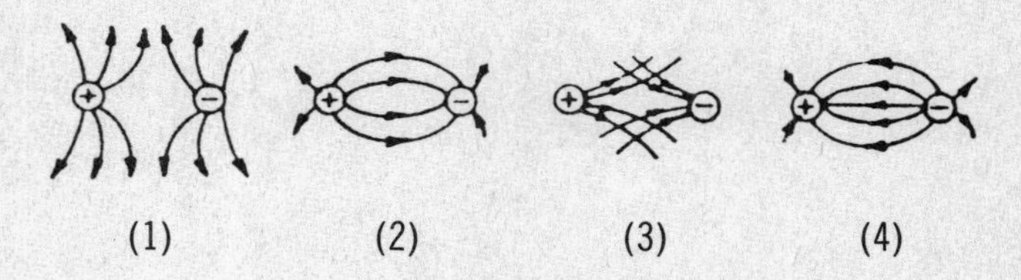

29. It it takes 12 joules of energy to move 2.0 coulombs of charge through a lamp filament, the potential difference across the filament is (1) 6.0 volts (2) 10 volts (3) 14 volts (4) 24 volts

30. If 10. joules of work must be done to move 2.0 coulombs of charge from point A to point B in an electric field, the potential difference between points A and B is (1) 5.0 V (2) 10. V (3) 12 V (4) 20. V

31. An elementary charge is accelerated by a potential difference of 9.0 volts. The total energy acquired by the charge is (1) 9.0 eV (2) 12 eV (3) 3.0 eV (4) 27 eV

32. Which electric charge is possible? (1) 8.0×10^{-20} coulomb (2) 2.4×10^{-19} coulomb (3) 3.2×10^{-19} coulomb (4) 6.32×10^{-18} coulomb

33. The electron volt is a unit of (1) charge (2) potential difference (3) current (4) energy

34. How many electrons will a neutral atom of carbon have if the carbon nucleus has 6 protons and 8 neutrons? (1) 6 (2) 2 (3) 8 (4) 14

35. Which diagram shows the leaves of the electroscope charged negatively by induction?

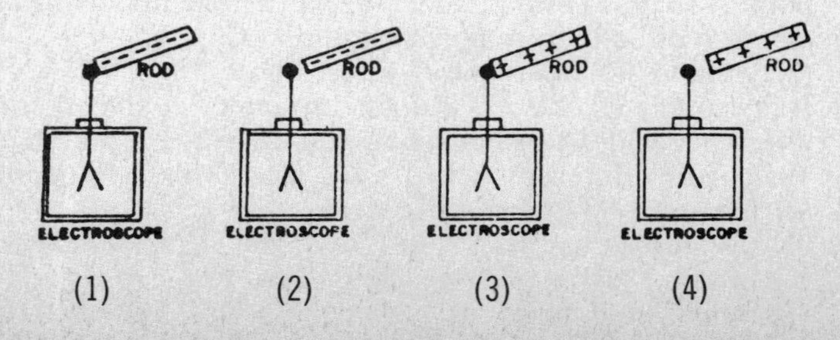

36. The diagram below represents a positive test charge located near a positively charged sphere. The greatest increase in the electric potential energy of the test charge relative to the sphere would be caused by moving the charge to point
(1) A (2) B (3) C (4) D

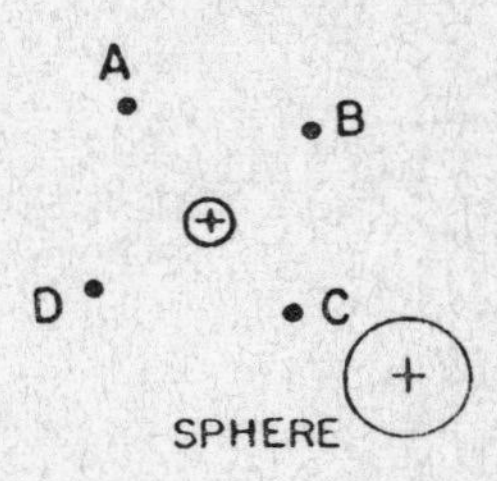

Base your answers to questions 37 through 41 on the diagram below which represents a source connected to two large, parallel metal plates. The electric field intensity between the plates is 3.75×10^4 newtons per coulomb.

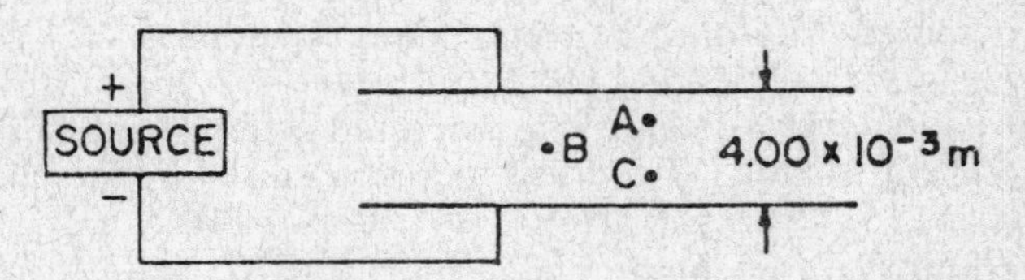

37. What is the potential difference of the source?
(1) 9.38×10^5 volts (2) 4.00×10^3 volts (3) 3.75×10^2 volts
(4) 1.50×10^2 volts

38. What would be the magnitude of the electric force on a proton at point A? (1) 1.60×10^{-19} newton
(2) 6.00×10^{-15} newton (3) 0 newtons
(4) 3.75×10^4 newtons

39. Compared to the work done in moving an electron from point A to point B to point C, the work done in moving an electron directly from point A to point C is (1) less
(2) greater (3) the same

40. If the source is replaced with one having twice the potential difference and the distance between the plates is halved, the electric field intensity between the plates will (1) decrease
(2) increase (3) remain the same

41. As a proton moves from A to B to C, the electric force on the proton (1) decreases (2) increases
(3) remains the same

42. The diagram represents two charges at a separation of d. Which would produce the greatest increase in the force between the two charges? (1) doubling charge q_1, only (2) doubling d, only (3) doubling charge q_1 and d, only (4) doubling both charges and d

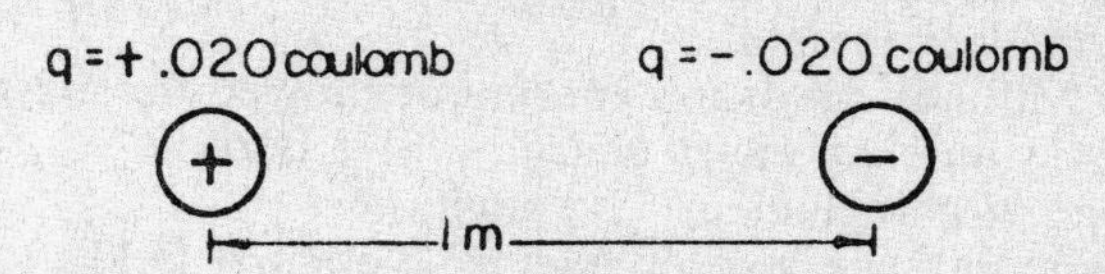

Base your answers to questions 43 through 47 on the diagram below which represents two charged metal spheres.

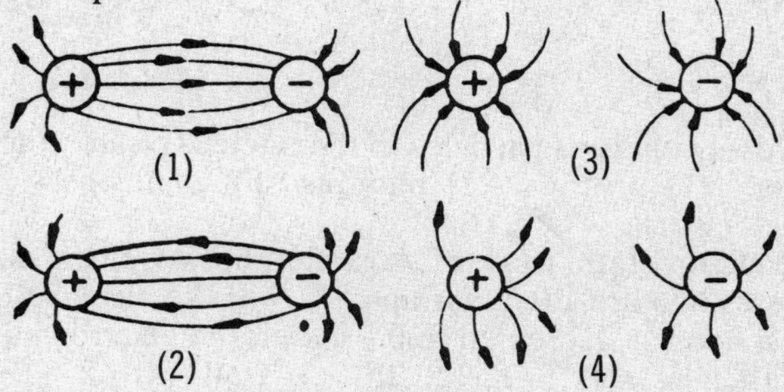

43. Which diagram below best represents the electric field around the two spheres?

(1) (2) (3) (4)

44. What is the magnitude of the force between the two spheres? (1) 3.6×10^6 N (2) 1.8×10^8 N (3) 3.6×10^9 N (4) 9.0×10^9 N

45. When a charge of 0.04 coulomb is placed at a point in the electric field, the force on the charge is 100 newtons. What is the magnitude of the electric field at that point? (1) 0.00040 N/coulomb (2) 2,500 N/coulomb (3) 100 N/coulomb (4) 4.0 N/coulomb

46. The work required to move a charge of 0.04 coulomb from one point to another point in the electric field is 200 joules. What is the potential difference between the two points? (1) 0.0002 volt (2) 8 volts (3) 200 volts (4) 5,000 volts

47. The two spheres are brought into contact with each other and then separated. After separation, what is the charge on each sphere? (1) 0.01 coulomb (2) 0.02 coulomb (3) 0 coulombs (4) 0.04 coulomb

48. Metal sphere A has a charge of −2 units and an identical sphere B has a charge of −4 units. If the two spheres are brought together and then separated, the charge on sphere A will be (1) 0 units (2) −2 units (3) − 3 units (4) +4 units

49. A charge of 100. elementary charges is equivalent to (1) 1.60×10^{-21} coulomb (2) 1.60×10^{-17} coulomb (3) 6.25×10^{16} coulombs (4) 6.25×10^{20} coulombs

50. Millikan determined the charge on an electron by observing the motion of charged oil drops in the presence of (1) sound waves (2) gold nuclei (3) standing waves (4) an electric field

51. The Millikan oil drop experiment showed the existence of (1) an elementary charge (2) a photon (3) an atomic nucleus (4) an atom

52. Which diagram best represents an electric field?

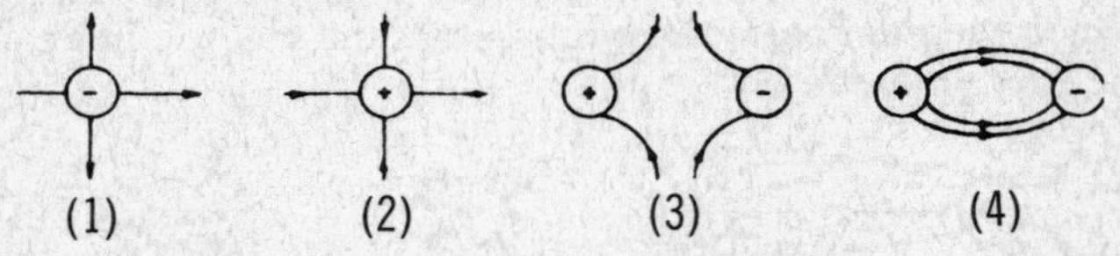

53. Positively charged particles in the nucleus of an atom are called (1) protons (2) photons (3) neutrons (4) electrons

54. The electrostatic force of attraction between two small spheres that are 1.0 meter apart is *F*. If the distance between the spheres is decreased to 0.5 meter, the electrostatic force will then be (1) F/2 (2) 2F (3) F/4 (4) 4F

55. A rod and a piece of cloth are rubbed together. If the rod acquires a charge of $+1 \times 10^{-6}$ coulomb, the cloth acquires a charge of (1) 0 coulombs (2) $+1 \times 10^{-6}$ coulomb (3) -1×10^{-6} coulomb (4) $+1 \times 10^{6}$ coulombs

56. After a neutral object loses 2 electrons, it will have a net charge of (1) −2 elementary charges (2) +2 elementary charges (3) -3.2×10^{-19} elementary charge (4) $+3.2 \times 10^{-19}$ elementary charge

57. When an object is brought near the knob of a positively charged electroscope, the leaves of the electroscope initially diverge. The charge on the object (1) must be zero (2) must be positive (3) must be negative (4) cannot be determined

Base your answers to questions 58 through 62 on the following diagram which represents an electron projected into the region between two parallel charged plates which are 10^{-3} meter apart.

The electric field intensity between the plates is 10^6 volts per meter.

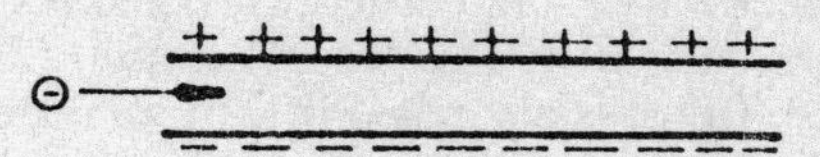

58. In which direction will the electron be deflected? (1) into the page (2) out of the page (3) toward the bottom of the page (4) toward the top of the page

59. What is the potential difference across the two plates? (1) 10^{-3} volt (2) 10^3 volts (3) 10^6 volts (4) 10^9 volts

60. What is the magnitude of the force acting on the electron when it is in the electric field? (1) 1.6×10^{-25} N (2) 1.6×10^{-13} N (3) 1.0×10^6 N (4) 1.6×10^{25} N

61. As an electron moves from the negatively charged plate to the positively charged plate, the force on the electron due to the electric field (1) decreases (2) increases (3) remains the same

62. The electron is replaced by a proton. Compared to the magnitude of the force on the electron, the magnitude of the force on the proton will be (1) less (2) greater (3) the same

63. Two charged spheres are shown in the diagram. Which polarities will produce the electric field shown? (1) A and B both negative (2) A and B both positive (3) A positive and B negative (4) A negative and B positive

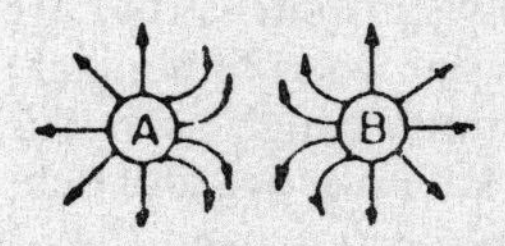

64. If the charge on one of two small charged spheres is doubled while the distance between them remains the same, the electrostatic force between the point sources will be (1) halved (2) doubled (3) tripled (4) unchanged

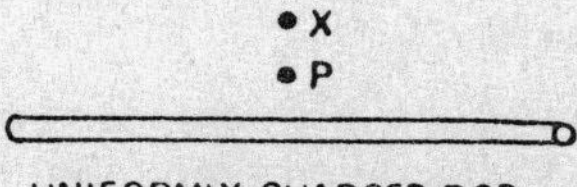

UNIFORMLY CHARGED ROD

In the diagram, point P is located 0.2 meter from the uniformly charged rod, and point X is located 0.4 meter from the rod.

65. Compared to the electric field intensity at P, the electric field intensity at X is (1) ½ as great (2) 2 times greater (3) ¼ as great (4) 4 times greater

II. Electric Current

Just as static electricity is concerned with charges at rest, *current* electricity is concerned with charges in motion. *Current* is a flow of electric charge. But it is also used in a more specific way as the *rate* of electric flow. This rate is expressed in units called amperes. A current of one ampere refers to a rate of flow of one coulomb per second.

A. Conductivity in Solids

The electron is the mobile charge in most solids. Since metals contain many electrons that are free to move, they tend to be good conductors. Although there are no perfect insulators, certain solids, such as glass or fused quartz which have very few free electrons are such poor conductors that they are classified as non-conductors or insulators. *Resistivity* is a property of materials which retard current. Good conductors, such as silver, copper and aluminum have the lowest resistivity. Certain materials, whose resistivities are somewhere between that of metals and insulators are call semiconductors.

B. Conductivity in Liquids

Conduction in most liquids depends on the motion of ions rather than free elections. Most substances that are liquids at room temperature, however, are composed of neutral particles and are poor conductors. Many substances such as acids, bases and salts when dissolved in water separate into ions and become conductors. Such substances are called *electrolytes* and their aqueous solutions are called *electrolytic solutions*. In these cases the motion of positive charges in one direction is equivalent to the motion of negative charges in the other direction.

C. Conductivity in Gases

Gases, like liquids, are normally composed of neutral molecules. They may be ionized to become conductors by exposure to high energy radiations, electric fields, or by collisions with particles. The mobile charges in ionized gases may be positive ions, negative ions or freely moving electrons. *Plasma* is a name given to ionized gases. It constitutes the fourth and most common phase of matter in the universe and is found in space, outside our atmosphere. The stars, the ions that radiate from the stars, and the Van Allen belts around the Earth are examples of plasma.

D. Conditions Necessary for a Current

In order for a current of electricity to be maintained at least two conditions are required: 1) A potential difference must be supplied by some source such as a battery or generator, and 2) A complete or closed conducting path connecting the region of high potential to that of low potential.

E. The Unit of Current

The unit of current is the ampere which is considered an elemental or fundamental unit in the SI system. If an electric charge is moving past a point at a rate of one ampere one coulomb of charge is flowing past the current at point each second. The conventional symbol is I.

$$I = \frac{q}{t}$$

Where: I is the current in amperes
q is the charge in coulombs
t is the time in seconds

F. Resistance

Virtually all conductors offer some resistance to electric current. The ratio of the potential difference between the ends of a conductor and the current can be predicted for many materials. This ratio is known as the *resistance* of the conductor and it is constant for a given conducting material at a given temperature.

1. **The unit of resistance** in the SI system is the *ohm*. One ohm is the resistance that will allow one ampere of current through a potential difference of one volt. The symbol for ohm is Ω (Greek letter omega).
2. **Resistance of conductors.** The length, cross section and resistivity of a given conductor are properties that remain constant at a given temperature. Of these resistivity is least familiar and a definition is in order. The resistivity of a substance* (ρ) in the SI system is the resistance between opposite faces of a cube with edges one meter long at a given temperature. The unit is the ohm-meter. The resistance of a conductor

*Greek letter rho = ρ

varies directly with its resistivity and its length and inversely with its cross sectional area. This is expressed by the equation:

$$R = \frac{\rho L}{A}$$

R is the resistance to ohms
ρ is the resistivity in ohm-meters
L is the length in meters
A is the cross sectional area in meters^2

These relationships, of course, require uniform cross section and composition of the conductor.

3. **Ohm's Law.** In a metallic conductor at a given temperature, the current (I) varies directly with the applied potential difference (V) and inversely with the resistance (R). This relationship is known as *Ohm's* Law and, written as an equation:

$$I = \frac{V}{R}$$

I is current in amperes
R = resistance in ohms
V is the potential difference in volts

Sample Problem

What is the current in a 60 ohm resistance when the potential difference is 120 volts?

Given $R = 60\ \Omega$
$V = 120V$

Find I

Equation $I = \frac{V}{R}$

Solution $I = \frac{120V}{60\ \Omega}$

Answer = 2A

In many cases the Ohm's law relationship is not linear. This is especially true in vacuum tubes, transistors and gas discharge tubes.

4. **Temperature.** In most metals, resistance increases with temperature. At low temperatures some materials become "super

conductors" with virtually no resistance. In non metals and in solutions, however, the reverse is more often true and resistance decreases with increasing temperature. In certain materials, notably some semiconductors, the temperature-resistance relationship is dependent upon temperature ranges.

G. Conservation of Charge and Energy in Electric Circuits

Two major laws governing electric circuits were stated by Gustave Kirchhoff. They are based on the conservation of charge and the conservation of energy.

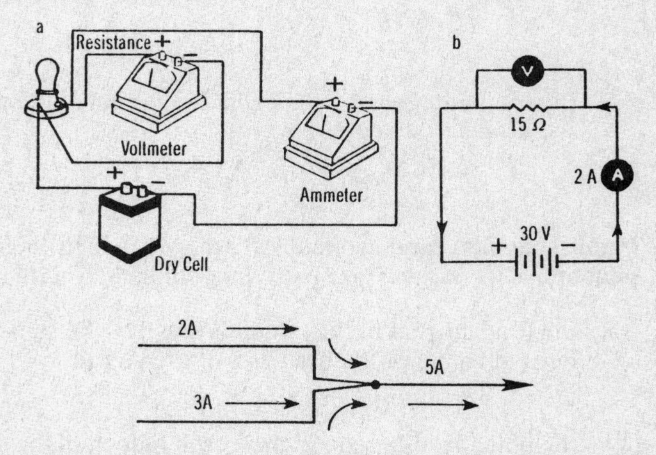

1. **Conservation of charge.** Kirchhoff's first rule is "For any point in a circuit, the total current arriving at the point must equal the total current leaving it."

$$2A + 3A = 5A$$

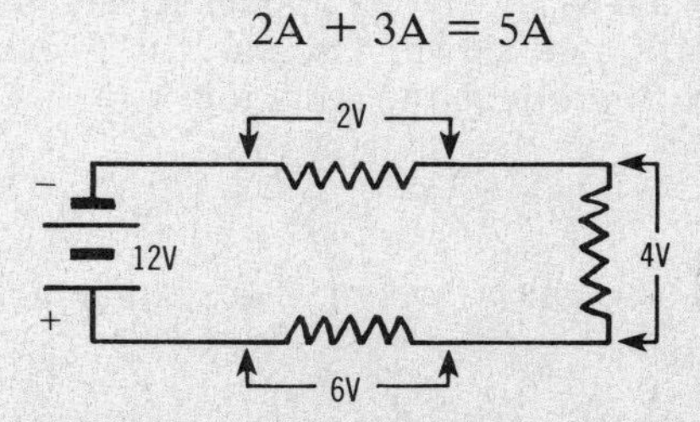

2. **Conservation of energy.** Kirchhoff's second rule states, "The algebraic sum of all the voltage drops and applied voltages around a complete circuit is equal to zero."

$$12V - 2V - 4V - 6V = 0$$

3. Series circuits. Series circuits are those in which there is only one current path. In a series circuit:

a. the current is the same in all its components.

$$I_{total} = I_1 = I_2 = I_3 \ldots$$

b. the total resistance is the sum of the resistances of its components

$$R_{total} = R_1 + R_2 + R_3 \ldots$$

c. the potential drops across components are proportional to their resistance

$$\frac{V_1}{V_2} = \frac{R_1}{R_2}$$

d. the total potential drop is equal to the sum of the potential drops of the components

$$V_{total} = V_1 + V_2 + V_3 \ldots$$

4. Parallel circuits. Parallel circuits are those in which each component has its own current path. In a parallel circuit:

a. the total current in the circuit is equal to the sum of the currents in all of the branches of the circuit

$$I_{total} = I_1 + I_2 + I_3 \ldots$$

b. the potential difference across each branch of the circuit is the same as the potential difference of the entire circuit.

$$V_{total} = V_1 = V_2 = V_3 \ldots$$

c. ohm's law. ($I = V/R$) applies to each of the currents in equation 4a above. The current T, can be replaced by V/R to produce the equation

$$\frac{V_{total}}{R_{total}} = \frac{V_1}{R_1} + \frac{V_2}{R_2} + \frac{V_3}{R_3} \ldots$$

and since all values of V are equal, according to equation 4b,

$$\frac{1}{R_{total}} = \frac{1}{R_1} + \frac{1}{R_2} + \frac{1}{R_3} \ldots$$

Stated in words: The reciprocal of the total resistance is equal to the sum of the reciprocals of the branch resistances.

5. Electric Power. Power, was previously defined as the rate of doing work, or expending energy.

a. $P = \frac{W}{t}$

The units of electrical power are *watts*, defined as joules per second.

Since electric potential is work per unit charge, then

b. $V = \frac{W}{q}$, or $W = qV$

Substituting qV for W, we can rewrite 5a

c. $P = \frac{W}{t} = \frac{qV}{t}$

Current (I) is equal to charge per unit time. Therefore q/t in equation (5c) can be replaced with I to produce the equation

d. $P = IV$
watt = ampere × volt

This is the fundamental equation for power in an electric circuit. All other power equations may be derived from it. Using it in combination with Ohm's Law, $V = IR$ and substituting IR for V produces:

e. $P = I^2R$ (Joule's Law)

f. $P = \frac{V^2}{R}$

We should also note here that replacing amperes with coulombs/second and volts with joules/second (5a) can be written

$$\text{Watts} = \frac{\text{coulomb}}{\text{sec}} \times \frac{\text{joules}}{\text{coulomb}}$$

g. $\text{Watt} = \frac{\text{joules}}{\text{second}}$

Sample Problem

(1) An electric hair dryer draws 10. amperes on a 120 volt circuit. How much power does the hair dryer consume?

Given I = 10. amperes
V = 120 volts

Find P

Equation $P = IV$

Solution $P = (10 \text{ amperes})(120 \text{ volts})$

$\underline{\text{Answer} = P = 1200 \text{ watts}}$

(2) Find resistance in the hair dryer described.

Given $I = 10.$ amperes
$P = 1200$ watts

Find R

Equation $P = I^2R$

Solution
$$R = \frac{P}{I^2}$$
$$= \frac{1200}{100}$$
$$= 12 \text{ ohms}$$

An alternative solution for finding this resistance, uses Ohm's Law as follows:

Given $I = 10.$ amps
$V = 120$ volts

Find R

Equation $V = IR$

Solution
$$R = \frac{V}{I}$$
$$= \frac{120 \text{ volts}}{10. \text{ amps}}$$
$$= 12 \text{ ohms}$$

6. Electrical Energy and heat

a. Electrical energy is work done by electric circuit. It is equal to the product of the power (P) and the time (t) during which the circuit carries current.

$$W = Pt$$

$$\text{joule} = \text{watt.sec}$$

Useful alternate forms include $W = I^2RT = VIt$

Most lighting companies bill their customers in terms of watt-hours, or kilowatt-hours.

QUESTIONS

1. As the temperature of a metal conductor is reduced, the resistance of the conductor will (1) decrease (2) increase (3) remain the same

2. If the cross-sectional area of a fixed length of wire were decreased, the resistance of the wire would (1) decrease (2) increase (3) remain the same

3. Which graph best represents a material behaving according to Ohm's law?

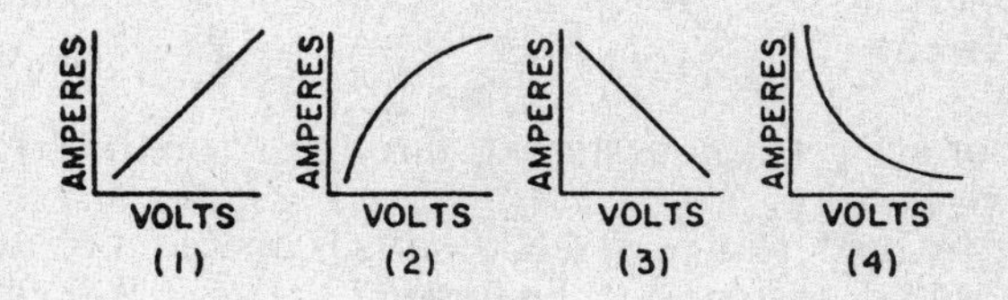

4. Three resistances of 2 ohms, 4 ohms, and 6 ohms are connected in parallel. The equivalent resistance of the three resistors is (1) less than 2 ohms (2) between 2 ohms and 4 ohms (3) between 4 ohms and 6 ohms (4) greater than 6 ohms

5. The circuit represented in the diagram below is a series circuit. The electrical energy expended in resistor *R* in 2.0 seconds is (1) 20. J (2) 40. J (3) 80. J (4) 120 J

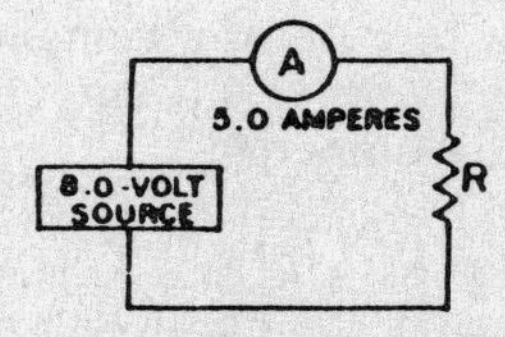

6. Electrical conductivity in liquid solutions depends on the presence of free (1) neutrons (2) protons (3) molecules (4) ions

7. What is the current in the circuit represented in the diagram below? (1) 1 A (2) 2 A (3) 3 A (4) 6 A

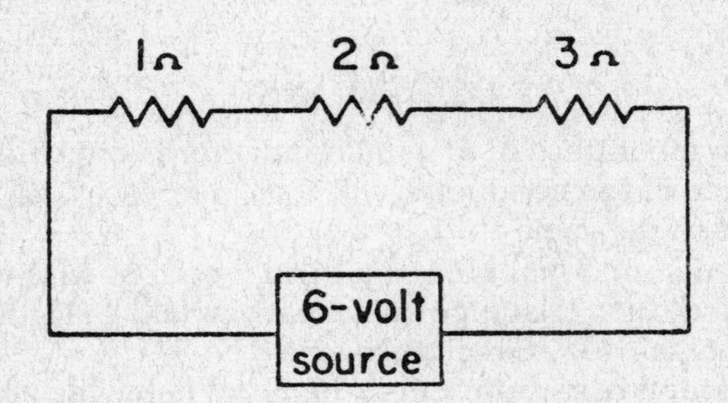

8. Which quantity must be the same for each component in any series circuit? (1) voltage (2) power (3) resistance (4) current

Base your answers to questions 9 through 13 on the information below.

An electric heater rated at 4,800 watts is operated on 120 volts.

9. What is the resistance of the heater? (1) 576,000 Ω (2) 120 Ω (3) 3.0 Ω (4) 40. Ω

10. How much energy is used by this heater in 10.0 seconds? (1) 1.15J (2) 40. J (3) 4.8×10^3 J (4) 4.8×10^4 J

11. If the heater were replaced by one having a greater resistance, the amount of heat produced each second would (1) decrease (2) increase (3) remain the same

12. If another heater is connected in parallel with the first one and both operate at 120 volts, the current in the first heater will (1) decrease (2) increase (3) remain the same

13. If the original heater were operated at less than 120 volts, the amount of heat produced would (1) decrease (2) increase (3) remain the same

14. The diagram below represents a segment of a circuit. The current in wire *X* may be (1) 1 ampere (2) 2 amperes (3) 3 amperes (4) 4 amperes

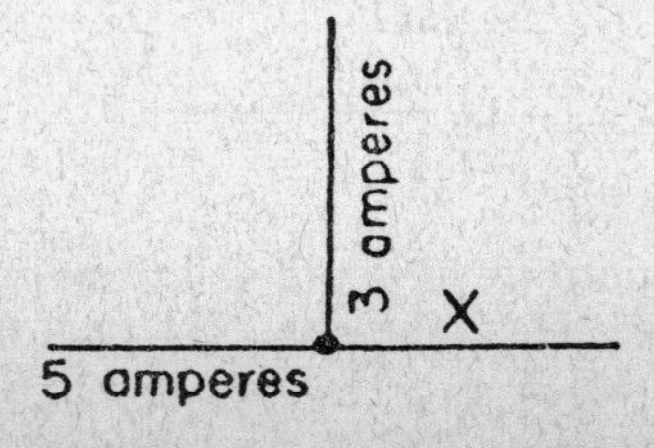

15. Which graph best represents the relationship between the current (1) and the potential difference (*V*) in a circuit in which resistance remains constant?

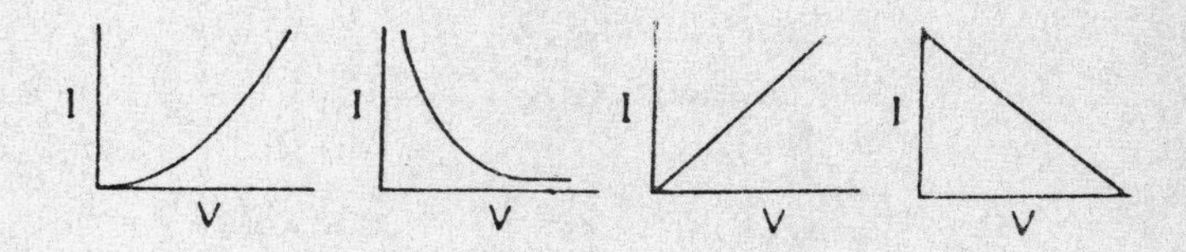

Base your answers to questions 16 through 20 on the graph below which represents data obtained by applying different potential differences to a metallic conductor at a constant temperature.

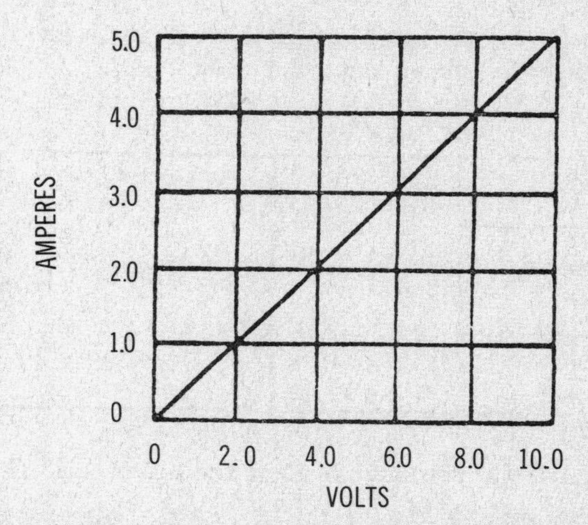

16. The resistance of the conductor is approximately
(1) 1.0 ohm (2) 2.0 ohms (3) 0.5 ohm (4) 4.0 ohms

17. At 6.0 volts, what is the rate of use of energy by the conductor?
(1) 54 watts (2) 18 watts (3) 12 watts (4) 6.0 watts

18. If the temperature of the conductor is increased, the amount of current at 10 volts would be (1) less (2) greater (3) the same

19. If the length of the conductor were increased, the amount of current at 10 volts would be (1) less (2) greater (3) the same

20. Compared to a conductor of the same material with a larger cross-sectional area, the resistance of this conductor is
(1) less (2) greater (3) the same

21. In the circuit shown in the diagram below, the rate at which electrical energy is being expended is resistor R_1 is
(1) less than in R_2 (2) greater than in R_2
(3) less than in R_3 (4) greater than in R_3

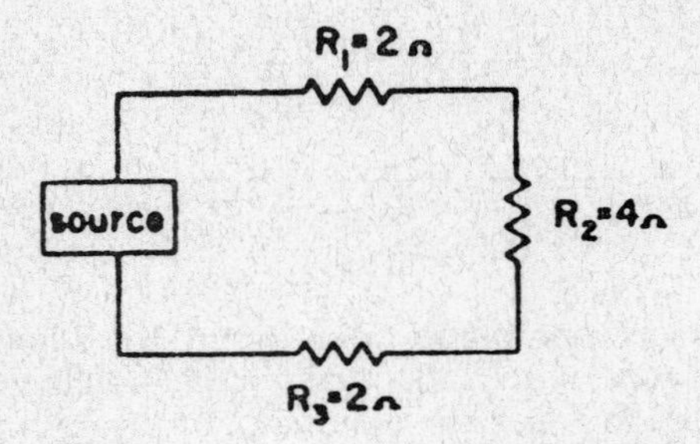

Base your answers to questions 22 through 26 on the diagram below which represents an electrical circuit.

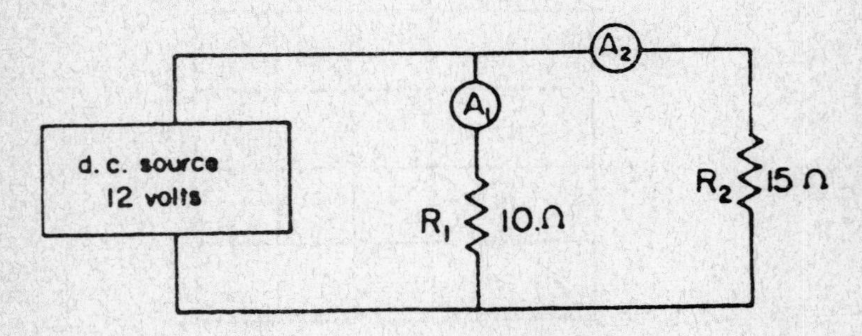

22. The equivalent resistance of the circuit is (1) 25 Ω
(2) 6.0 Ω (3) 5.0 Ω (4) 0.17 Ω

23. The potential difference across R_2 is (1) 1.0 V (2) 2.0 V
(3) 10. V (4) 12 V

24. The magnitude of the current in ammeter A_1 is
(1) 120 A (2) 2.0 A (3) 1.2 A (4) 0.83 A

25. Compared to the current in A_1, the current in A_2 is
(1) less (2) greater (3) the same

26. If another resistance were added to the circuit in parallel, the equivalent resistance of the circuit would (1) decrease
(2) increase (3) remain the same

27. Conductivity in ionized gases depends upon
(1) free electrons, only (2) positive ions, only
(3) negative ions, only (4) positive ions, negative ions, and free electrons

28. If the potential difference between points A and B in the electric circuit shown is 10 volts, what is the voltage between points B and C? (1) 5 volts (2) 10 volts (3) 20 volts (4) 30 volts

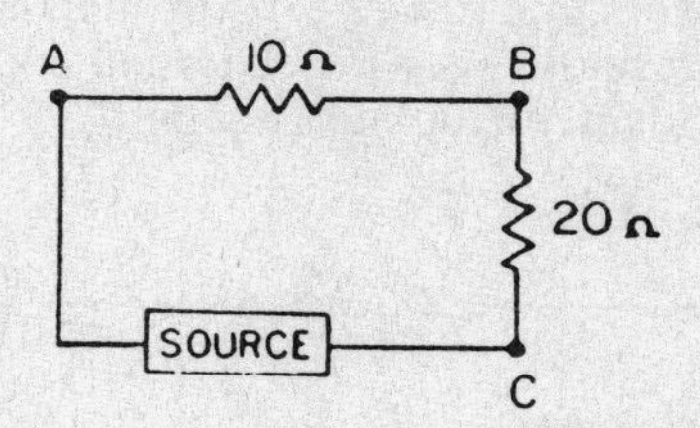

Base your answers to questions 29 through 33 on the diagram below which represents three resistors connected in parallel across 24-volt source. The ammeter reads 3.0 amperes.

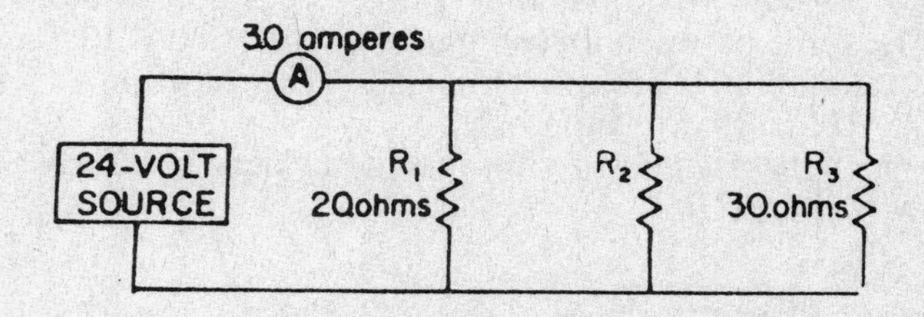

29. The equivalent resistance in the circuit is (1) 0.13 ohm (2) 8.0 ohms (3) 58 ohms (4) 72 ohms

30. The current in R_1 is (1) 0.83 ampere (2) 1.5 amperes (3) 3.0 amperes (4) 1.2 amperes

31. The potential difference across R_3 is (1) 8.0 volts (2) 24 volts (3) 48 volts (4) 72 volts

32. If the ratio of the current in R_3 to the current in R_2 is 4.5, the resistance of R_2 is (1) 5.0 ohms (2) 8.0 ohms (3) 24 ohms (4) 60. ohms

33. The power supplied to the circuit is (1) 220 watts (2) 190 watts (3) 72 watts (4) 24 watts

Base your answers to questions 34 and 35 on the diagram below which represents an electrical circuit.

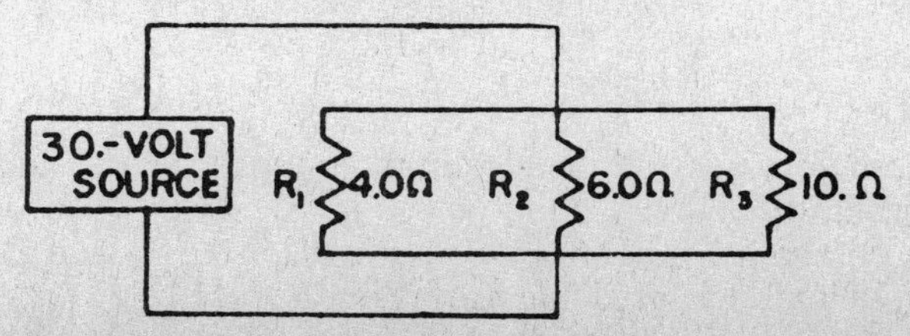

34. the equivalent resistance of R_1, R_2, and R_3 is approximately (1) 10. ohms (2) 2.0 ohms (3) 20. ohms (4) 7.0 ohms

35. The current in R_1 is (1) 3.8 amperes (2) 7.5 amperes (3) 15 amperes (4) 60. amperes

Base your answers to questions 36 through 40 on the diagram below which represents two small, charged conducting spheres, identical in size, located 2.00 meters apart.

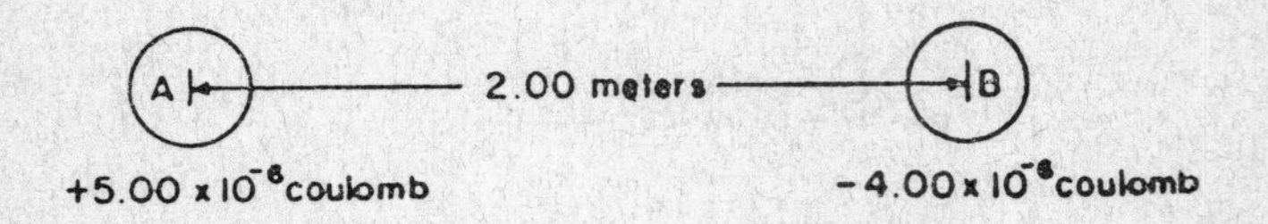

36. What is the net combined charge on both spheres?
(1) $+1.00 \times 10^{-6}$ coulomb (2) -1.00×10^{-6} coulomb
(3) $+9.00 \times 10^{-6}$ coulomb (4) -9.00×10^{-6} coulomb

37. The force between these spheres is (1) 1.80×10^{-2} newton (2) 3.60×10^{-2} newton (3) 4.50×10^{-2} newton (4) 9.00×10^{-2} newton

38. Which diagram best represents the electric field between the two spheres?

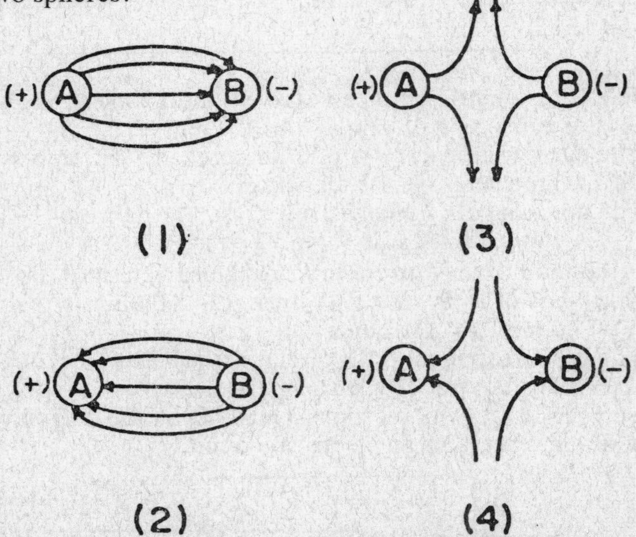

Base your answers to questions 39 and 40 on the *additional* information below.

A conductor is connected between spheres A and B and then removed after charge has transferred.

39. During the period of the conductor was attached, what was the direction of the flow of the charged particles? (1) Protons flowed from A to B, only. (2) Electrons flowed from B to A, only. (3) Protons flowed from A to B as electrons flowed from B to A. (4) Protons flowed from B to A as electrons flowed from A to B.

40. What is the net charge on each sphere? (1) Each has a charge of $+5.0 \times 10^{-7}$ coulomb. (2) Each has a charge of -5.0×10^{-7} coulomb. (3) Each has a charge of $+4.5 \times 10^{-6}$ coulomb. (4) Sphere A has a charge of $+10 \times 10^{-6}$coulomb

Base your answers to questions 41 through 44 on the diagram below which represents resistors R_1 nd R_2 connected to a constant power source of 40 volts. A_1, A_2, and A_3 represent ammeters.

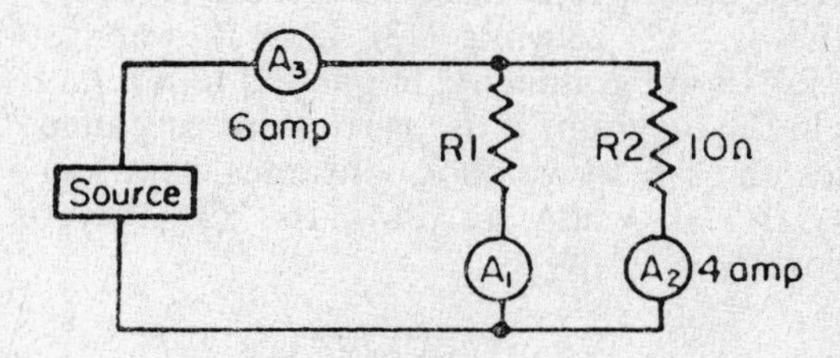

41. What is the reading of ammeter A_1? (1) 1 ampere (2) 2 amperes (3) 3 amperes (4) 4 amperes

42. The potential difference across R_1 is (1) 10 volts (2) 12 volts (3) 20 volts (4) 40 volts
The power supplied to the circuit is (1) 80 watts (2) 120 watts (3) 160 watts (4) 240 watts

43. Compared to the power supplied to R_1, the power supplied to R_2 is (1) less (2) greater (3) the same

44. If a third resistor is connected in parallel with R_2 the effective resistance of the circuit will (1) decrease (2) increase (3) remain the same

45. The diagram below represents a segment of a circuit. What is the current in ammeter A? (1) 1 ampere (2) 0 amperes (3) 3.5 amperes (4) 7 amperes

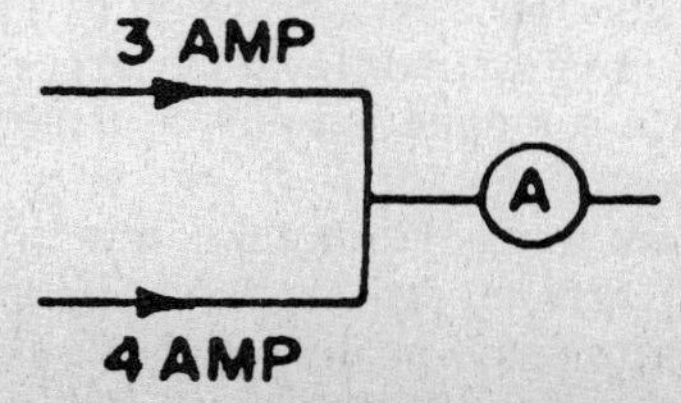

46. The diagram represents a circuit with two resistors in series. If the total resistance of R_1 and R_2 is 24 ohms, the resistance of R_2 is (1) 1.0 ohm (2) 0.50 ohm (3) 100 ohms (4) 4.0 ohms

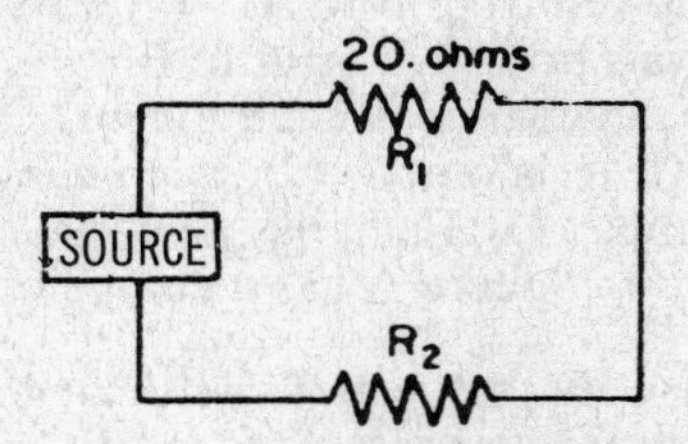

47. The potential difference across a 100.-ohm resistor is 4.0 volts. what is the power dissipated in the resistor? (1) 0.16 watt (2) 25 watts (3) 4.0×10^2 watts (4) 4.0 watts

48. Two resistors are connected in parallel to a 12-volt battery is shown in the diagram. If the current in resistance R is 3.0 amperes, the rate at which R consumes electrical energy is (1) 1.1×10^2 watts (2) 36 watts (3) 24 watts (4) 4.0 watts

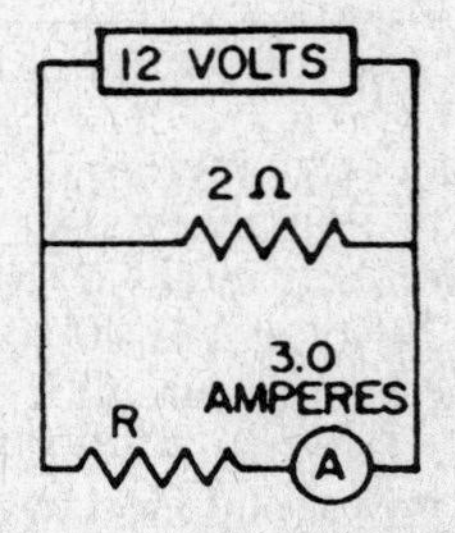

49. If energy is used in an electric circuit at the rate of 20 joules per second, then the power supplied to the circuit is (1) 5 watts (2) 20 watts (3) 25 watts (4) 100 watts

50. If 6.0 joules of work is done to move 2.0 coulombs of charge from point A to point B, what is the electric potential difference between points A and B? (1) 6.0 volts (2) .33 volt (3) 3.0 volts (4) 12 volts

51. Which condition must exist between two points in a conductor in order to maintain a flow of charge? (1) a potential difference (2) a magnetic field (3) a low resistance (4) a high resistance

52. Electric current in a solid metal conductor is caused by the movement of (1) electrons, only (2) protons, only (3) both electrons and protons (4) neutrons

53. The diagram below represents a segment of an electrical circuit. What is the current in wire *AB*? (1) 1 ampere (2) 2 amperes (3) 5 amperes (4) 6 amperes

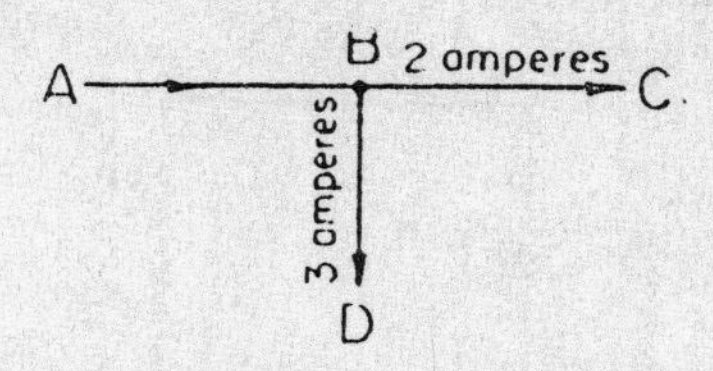

54. If the current in a wire is 2.0 amperes and the potential difference across the wire is 10. volts, what is the resistance of the wire? (1) 5.0 ohms (2) 8.0 ohms (3) 12 ohms (4) 20 ohms

55. If 4 joules of work are required to move 2 coulombs of charge through a 6-ohm resistor, the potential difference across the resistor is (1) 1 volt (2) 2 volts (3) 6 volts (4) 8 volts

Base your answers to questions 56 through 59 on the diagram of the circuit below.

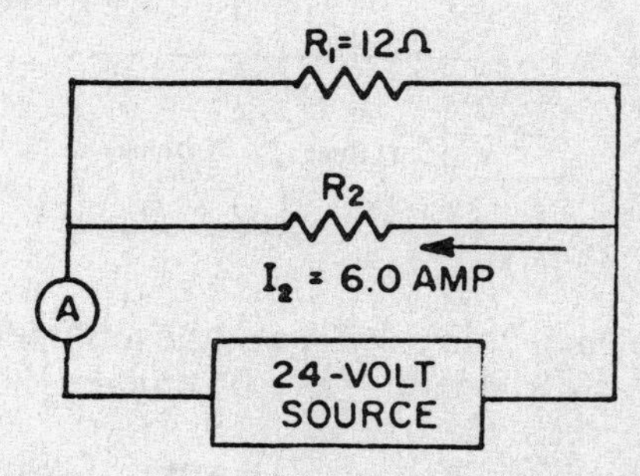

56. The current in ammeter A is (1) 1.0 ampere (2) 2.0 amperes (3) 6.0 amperes (4) 8.0 amperes

57. How much energy is used by the 12-ohm resistor in ½ hour? (1) 48 joules (2) 8.64×10^4 joules (3) 3.6×10^3 joules (4) 1.1×10^4 joules

58. If resistance R_2 were removed, the potential difference across R_1 would (1) decrease (2) increase (3) remain the same

59. If resistance R_2 were removed, the current in ammeter A would (1) decrease (2) increase (3) remain the same

60. The electric field between two parallel plates connected to a 45-volt battery is 500. volts per meter. the distance between the plates is closest to (1) 11 m (2) 22 m (3) 50. m (4) 0.090 m

61. Compared to the potential drop across the 10-ohm resistor shown in the diagram, the potential drop across the 5-ohm resistor is (1) the same (2) twice as great (3) one-half as great (4) four times as great

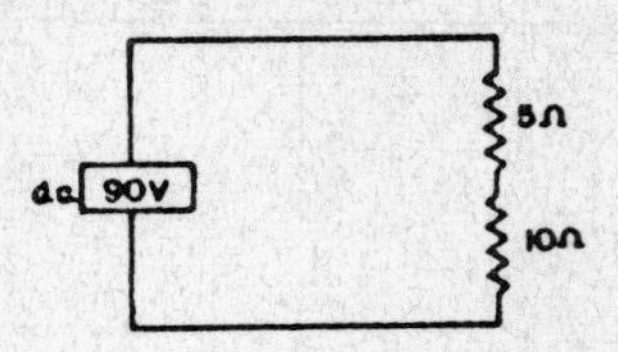

62. As two parallel conductors with currents in the same direction are moved apart, their force of (1) attraction increases (2) attraction decreases (3) repulsion increases (4) repulsion decreases

63. Compared to the current in the 10.-ohm resistance in the circuit shown below, the current in the 5.0-ohm resistance is (1) one-half as great (2) one-fourth as great (3) the same (4) twice as great

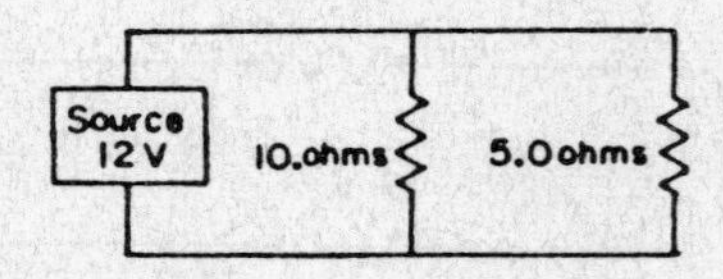

64. What is the current in a 1,200-watt heater operating on 120 volts? (1) 0.10 ampere (2) 5.0 amperes (3) 10. amperes (4) 20. amperes

65. An ampere-volt is a unit of (1) work (2) resistance (3) energy (4) power

66. A potential difference of 12 volts is applied across a circuit which has a 4.0-ohm resistance. What is the magnitude of the current in the circuit? (1) 0.33 ampere (2) 48 amperes (3) 3.0 amperes (4) 4.0 amperes

67. The resistance of a metallic wire conductor is inversely proportional to its (1) tensile strength (2) cross-sectional area (3) length (4) temperature

68. Most metals are good electrical conductors because (1) their molecules are close together (2) they have high melting points (3) they have many intermolecular spaces through which the current can flow (4) they have a large number of free electrons

69. When 20. coulombs of charge pass a given point in a conductor in 4.0 seconds, the current in the conductor is
(1) 80. amperes (2) 0.20 ampere (3) 16 amperes
(4) 5.0 amperes

70. The ratio of the potential difference across a conductor to the current in the conductor is called (1) current
(2) conductance (3) resistance (4) electric potential

71. The number of electrons that pass a certain point in a conductor in a given amount of time is the
(1) potential difference (2) charge (3) resistance
(4) electric current

72. If the length of a copper wire is reduced by half then the resistance of the wire will be (1) halved (2) doubled
(3) quartered (4) quadrupled

73. How much work is done in moving 6 electrons through a potential difference of 2.0 volts? (1) 6.0 eV (2) 2.0 eV
(3) 3.0 eV (4) 12 eV

Magnetism

The magnetic compass, long used by mariners, depends on the fact that a freely suspended magnet will align itself in a general north-south direction. By definition, the end of a suspended magnet that points toward the north pole of the earth's axis of rotation is called the *NORTH POLE* of the magnet and the opposite end is called its *SOUTH POLE.*

A. Magnetic Force

In general, magnetic forces are created when electric charges are in motion. this force is one of attraction when two sets of charges are moving along parallel lines in the same direction. The force is one of repulsion when the charges are moving in opposite directions. There is no magnetic force when the charges are moving at right angles to each other.

When two parallel conductors are 1 meter apart the force of attraction or repulsion between each meter length of the conductors can be found using the equation:

$$F_{\text{magnetic per meter}} = 2k' I_1 I_2 / r$$

F = magnetic force in newtons
I_1, I_2 = two currents in amperes
r = distance between conductors in meters
k' = magnetic constant 10^{-7}N/A^2

Because it is possible to measure the magnetic forces with precision, the ampere is now defined in terms of the magnetic forces it will create. There is an ampere of current in each of two parallel wires that are a meter apart, if the current creates a magnetic force of 2×10^{-7} newtons between the wires for each meter length of the wires.

The quantity of charge that is carried by an ampere of current as it moves past a given point in one second is now accepted as the definition of the *coulomb*.

B. Magnetic field

Just as it is convenient to speak of interaction between masses in terms of gravitational fields and between charges in terms of electric fields, it is convenient to speak of magnetic fields existing in regions where magnetic forces are in effect. Magnetic fields, therefore, exist in regions around magnets and electric currents.

1. **Direction.** The direction on the magnetic field at a given location can be determined with a compass. It is the direction in which the *north* pole of the compass points in the field.

2. **Magnetic Flux Lines.** A magnetic field is mapped by using imaginary lines of force called **magnetic flux lines**. These flux lines always form closed paths that never cross. The lines show the direction of the field. The SI unit of flux is the *weber*.

3. **Flux Density.** The magnetic field at some point in space can be measured by positioning a wire of given length l, carrying a current to I at the reference point. The wire is then oriented to achieve maximum force. The strength of the magnetic field (B) is defined by the equation:

$$B = \frac{\text{F magnetic}}{\text{Il}} = \frac{\text{newtons}}{\text{ampere meter}} \text{ or } \frac{\text{webers}}{\text{meter}^2} = \text{tesla}$$

The strength of the magnetic field (B) is referred to as the *magnetic induction* or *magnetic flux density*. The SI unit, tesla is equal to weber/m^2 The symbol for tesla is T. The unit is the cgs system is the gauss.

$$1 \text{ tesla} = \frac{1 \text{ weber}}{\text{meter}^2} = 10^4 \text{ gauss}$$

$$\text{or } 1 \text{ gauss} = 10^{-4}/\text{W/m}^2 = 10^{-4}\text{ T}$$

a. **Permeability.** Many materials have the capacity to change the flux density of a magnetic field. The property is called *permeability* and it is expressed the terms of the factor by which it changes the value. The permeability of a vacuum is one and that of air is nearly one.

4. **Magnetic Field around a Straight Conductor.** The magnetic lines of force that surround a straight conductor are concentric circles around the conductor in a plane perpendicular to the conductor.

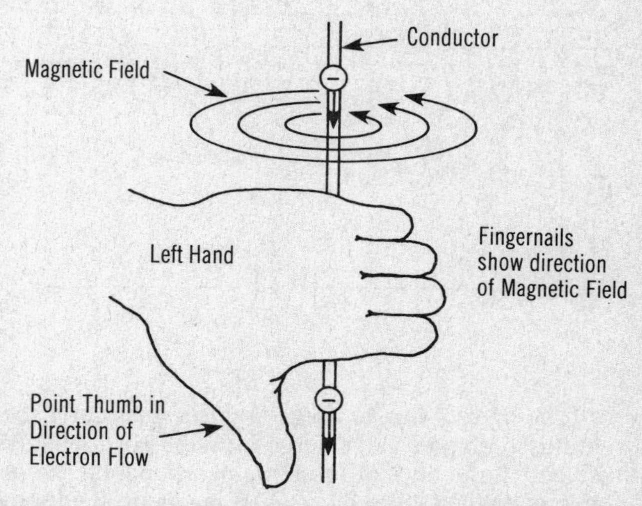

5. **Magnetic field around a Loop.** The diagram shows magnetic flux lines along a circular conductor called a loop. The fields produced by the parts of the loop are cumulative, producing a strong field inside the loop and a weaker field outside.

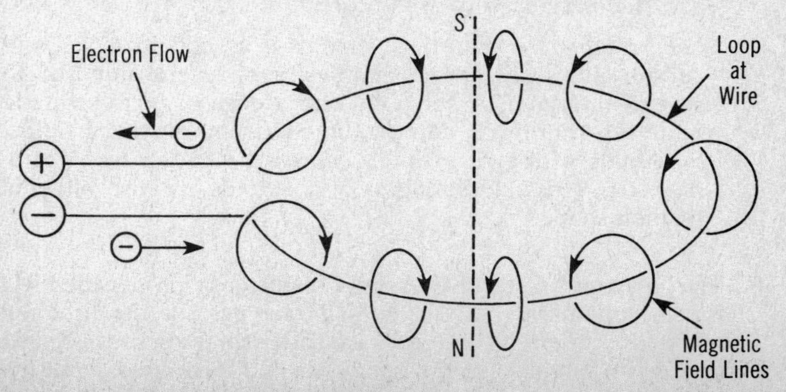

The faces of the loop show the polarity using the hand rule shown previously.

6. **Magnetic field around a Solenoid.** One can obtain a very strong uniform field by using many loops in a series. Such a series of loops produces a coil, or a cylinder known as a solenoid.

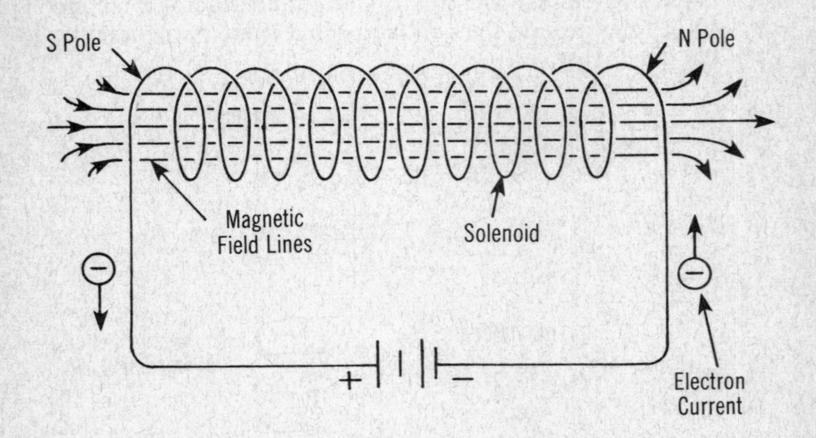

The lines of flux or magnetic field lines enter the solenoid at the south pole and emerge from the north pole. Within the solenoid, the lines of force are nearly parallel to its axis and perpendicular to its faces. The magnetic field strength of a solenoid is dependent on:

a. the number of loops
b. the current
c. the permeability of the core

Ferromagnetic materials, such as iron, cobalt, nickel and certain alloys have permeabilities from several hundred to several thousand. A solenoid with a core of ferromagnetic material, therefore, can produce extremely strong fields. Solenoids with ferromagnetic cores are known as *electromagnets*. The field strengths of solenoids are also affected by their shape.

C. **The Force on a Current-Carrying Conductor in a Magnetic Field** is perpendicular to both the field and the current. the direction of this force can be found by considering the increased flux density on one side of the conductor (additive fields) and the decrease on the other side (subtractive fields). Of course, the direction can also be found using an appropriate hand rule described below.

D. Magnetic Effects of Moving Charges

Charged particles in motion constitute a current. As such, they always produce magnetic fields and are affected by them as well.

1. The force on a moving charge is proportional to:

a. the charge
b. the flux density
c. the components of velocity perpendicular to the field

The direction of the force is perpendicular to the field and to the velocity. If the charge is negative the direction of the force can be found using the left hand as shown in the diagram.

The magnitude of this force is given by the equation

$$F_{magnetic} = Bqv$$

where:

F is the force in newtons,
B is the strength of the magnetic field in teslas, and
v is the velocity of the particle in meters per second.

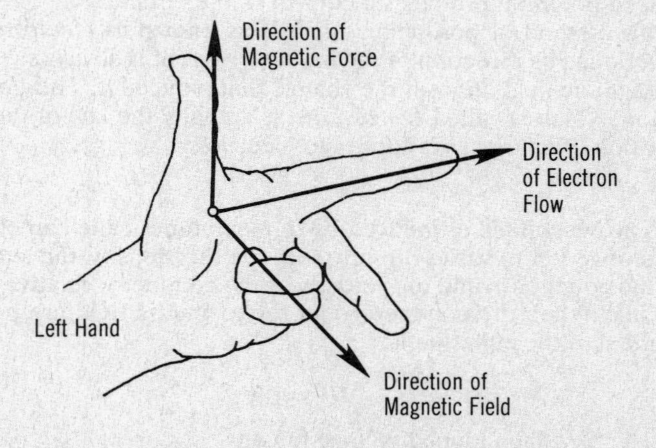

F. Magnetic Nature of Matter

All substances have some magnetic properties.

1. **Diamagnetic** substances reduce the flux density and repel magnetic materials.
2. **Paramagnetic** substances increase flux density and attract magnetic materials. In most cases, however, these forces are very weak. The exceptions are the ferromagnetic substances, which are very strongly paramagnetic.
3. **A Domain** is a clusters of atoms of magnetic substances which are grouped in such a way as to produce a single, strong magnetic field by an additive effect of the fields around the currents produced by spinning electrons.
4. **Field around a permanent magnet** consists of continuous lines of magnetic flux which enter the south pole and emerge from the north pole.

IV. Electromagnetic Induction

In 1831, Michael Faraday discovered that an electric voltage (EMF) is *induced* whenever a magnetic field *moves* with respect to a conductor. If the conductor is part of a complete circuit, the induced potential produces a current in the circuit.

This method of producing an EMF is known as *electromotive induction*. The direction of the induced current is always such that its magnetic field opposes the change that induced it. This generalization, which is called *Lenz's Law*, is actually the law of the conservation of energy as it applies to electricity.

A. The Magnitude of Induced EMF

The Magnitude of Induced EMF (sometimes called an electromotive force) varies directly with the flux density, the length of the conductor, and the velocity of the conductor relative to the flux. When the conductor and the magnetic field are perpendicular, the equation is:

$$E = Blv$$

E is the induced voltage in volts,
B is the field intensity in teslas (or webers/meter^2)
l is the length in meters
v is the velocity in meters/second.

QUESTIONS

Base your answers to questions 1 through 4 on the circuit diagram below which represents a solenoid in series with a variable resistor and a voltage source.

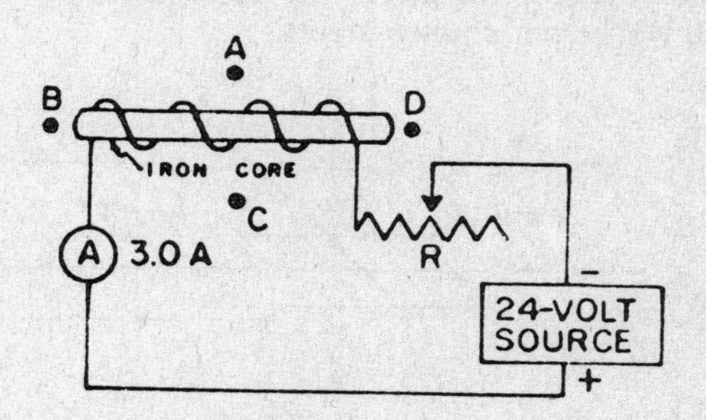

1. The resistance of the circuit is (1) 72 Ω (2) 24 Ω (3) 12 Ω (4) 8.0 Ω

2. The direction of the magnetic field inside the iron core is toward point (1) A (2) B (3) C (4) D

3. If the resistance of the variable resistor is decreased, the magnetic field strength of the solenoid will (1) decrease (2) increase (3) remain the same

4. If the iron core is removed from the solenoid, the magnetic field strength of the solenoid will (1) decrease (2) increase (3) remain the same

5. Each diagram below represents a cross section of a long, straight, current-carrying wire with the electron flow into the page. Which diagram best represents the magnetic field near the wire?

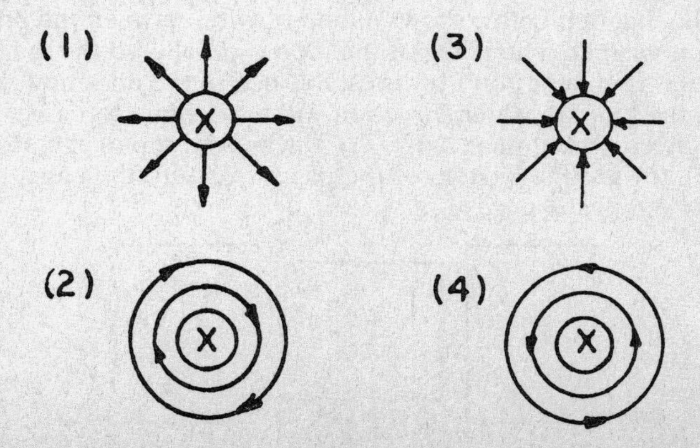

Base your answers to questions 6 through 10 on the diagram below which represents a helium ion with a charge of +2 elementary charges moving toward point A with a constant speed (v) of 2.0 meters per second perpendicular to a uniform magnetic field between the poles of a magnet. The strength of the magnetic field is 0.10 weber per square meter.

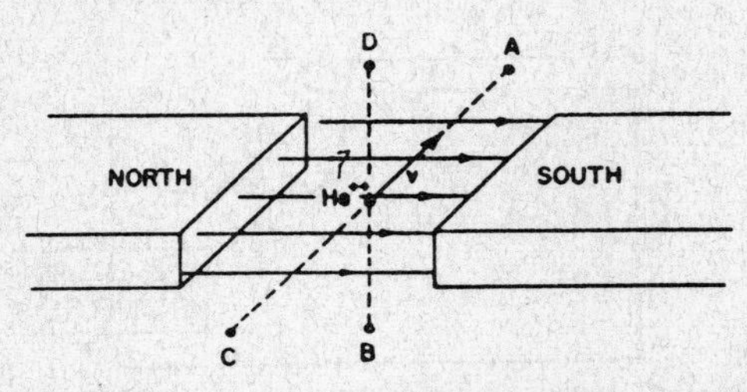

6. The direction of the magnetic force on the helium ion is toward point (1) A (2) B (3) C (4) D
7. The magnitude of the magnetic force exerted on the helium ion is (1) 3.2×10^{-20} N (2) 6.4×10^{-20} N (3) 0.10 N (4) 0.20 N
8. If the strength of the magnetic field and the speed of the helium ion are both doubled, the force on the helium ion will be (1) halved (2) doubled (3) the same (4) quadrupled
9. If the polarity of the magnet is reversed, the magnitude of the magnetic force on the helium ion will (1) decrease (2) increase (3) remain the same
10. The helium ion is replaced by an electron moving a the same speed. Compared to the magnitude of the force on the helium ion, the magnitude of the force on the electron is (1) less (2) greater (3) the same
11. The diagram below shows a loop of wire between the poles of a magnet. The plane of the loop is parallel to the magnetic field. If an electron flow is established in the direction shown in the loop, in which direction will a magnetic force be exerted on segment AB? (1) toward the top of the page (2) toward the bottom of the page (3) into the page (4) out of the page

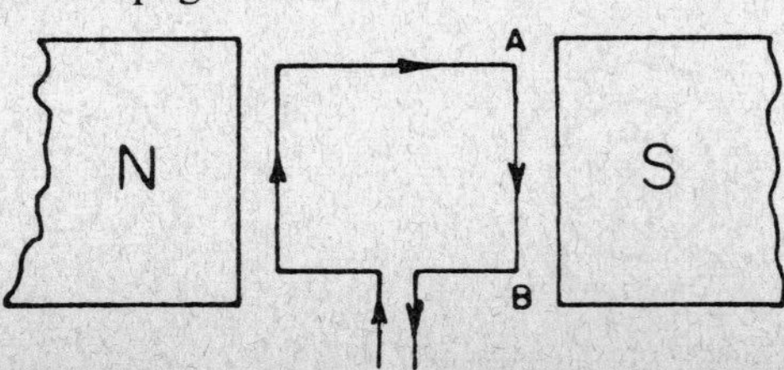

12. Which diagram best represents a magnetic field between two magnetic poles?

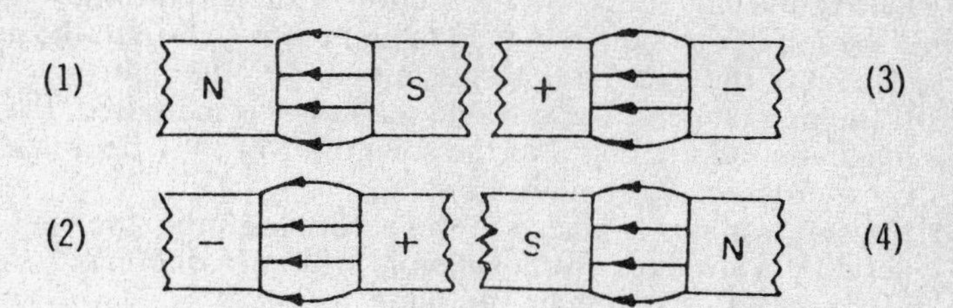

13. The diagram below represents a straight conductor in a uniform magnetic field. The field is directed into the page and the electron flow in the conductor is to the right. What is the direction of the magnetic force on the wire? (1) toward the left (2) toward the right (3) toward the top of the page (4) toward the bottom of the page

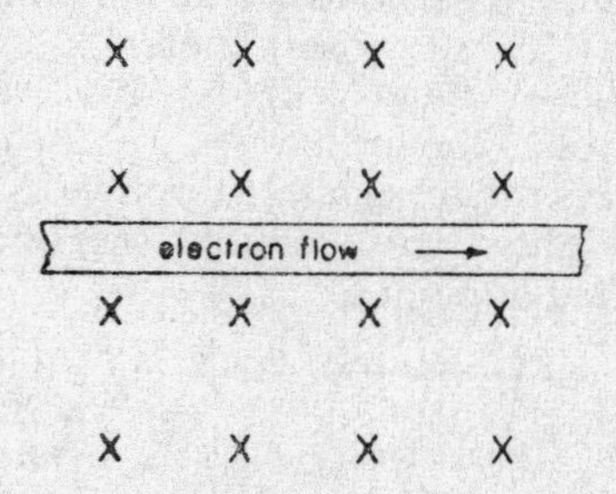

14. If a charged particle moving through a magnetic field experiences a magnetic force, the angle between the magnetic field and the force exerted on the particle is (1) 0° (2) 45° (3) 90° (4) 180°

Base your answers to questions 15 through 19 on the diagram below which represents a circuit containing a solenoid on a cardboard tube; a variable resistor *R*, and a source of potential difference.

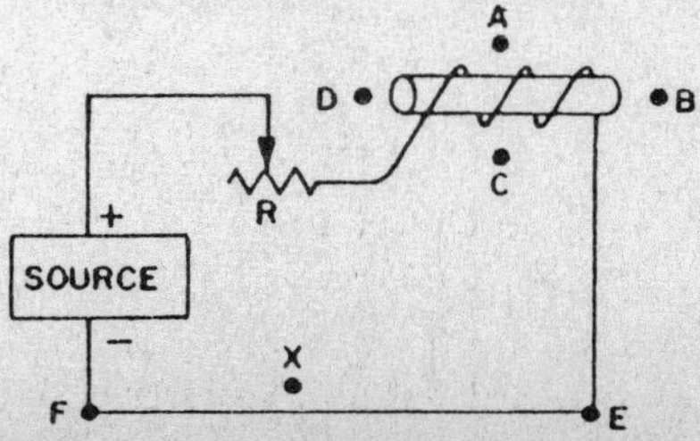

15. The north pole of the solenoid is nearest to point
(1) A (2) B (3) C (4) D

16. Due to the current in the *FE* section of the circuit, the direction of the magnetic field at point *X* is (1) into the page (2) out of the page (3) to the left (4) to the right

17. If the resistance of the variable resistor *R* is increased, the magnetic field strength of the solenoid will (1) decrease (2) increase (3) remain the same

18. If a soft iron rod is placed in the cardboard tube, the magnetic field strength of the solenoid will (1) decrease (2) increase (3) remain the same

19. If the number of turns in the solenoid is increased and the current is kept constant, the magnetic field strength of the solenoid will (1) decrease (2) increase (3) remain the same

20. A particle with a charge of 2×10^{-6} coulomb crosses a uniform magnetic field perpendicularly. The particle experiences a force of 1×10^{-3} newton. If the particle has a speed of 1×10^{6} meters per second, the magnitude of the field strength is (1) 2×10^{3} webers/meter^2
(2) 2×10^{-3} weber/meter^2 (3) 5×10^{-1} weber/meter^2
(4) 5×10^{-4} weber/meter^2

Base your answers to questions 21 through 25 on the diagram below which represents a cross section of an operating solenoid. A compass is located at point C.

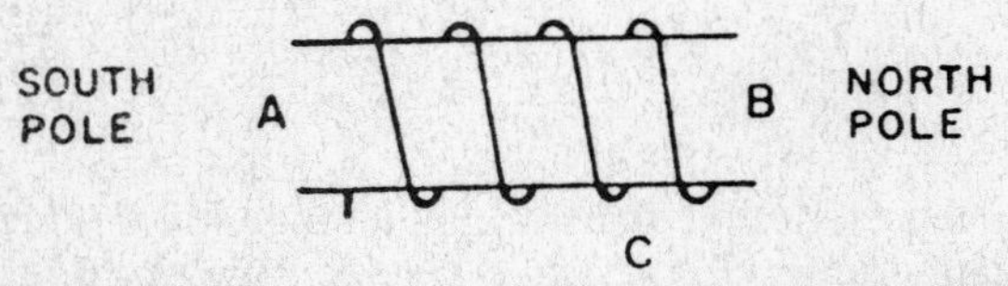

21. Which diagram best represents the shape of the magnetic field around the solenoid?

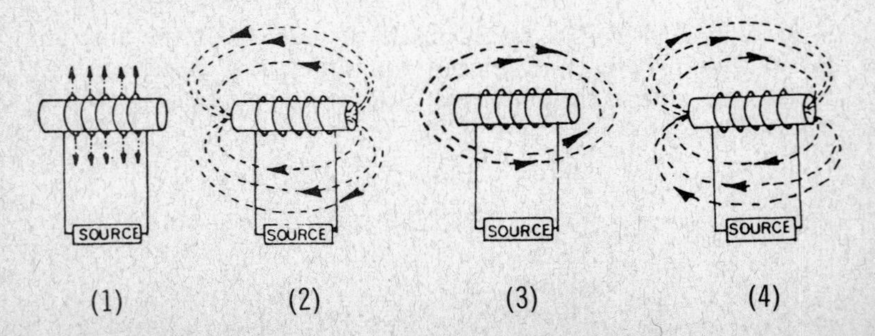

22. Which shows the direction of the compass needle at point *C*?

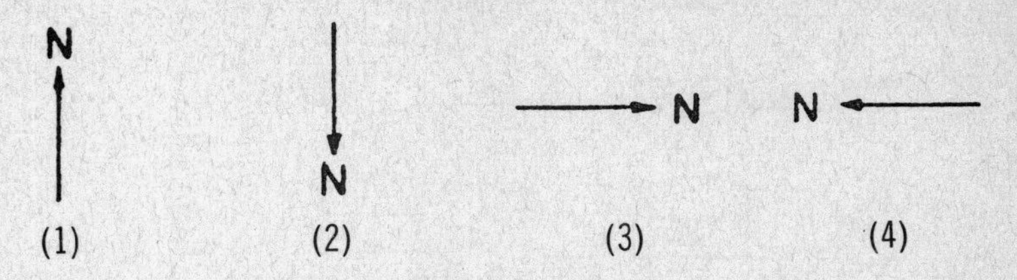

23. If *B* is the north pole of the solenoid, which diagram best represents the direction of electron flow in one of the wire loops?

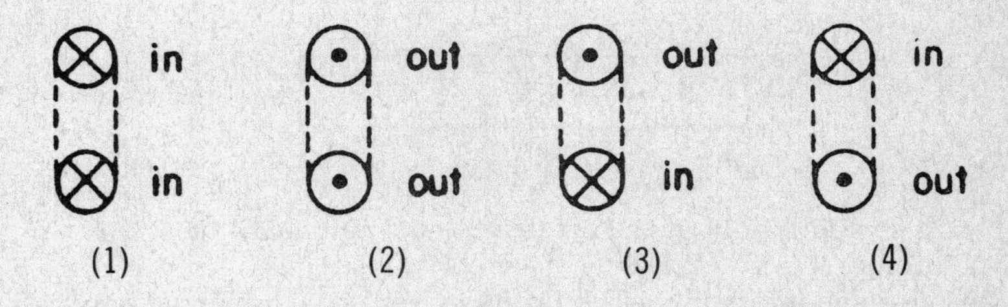

24. If the current in the solenoid is doubled and the number of turns halved, the magnetic field strength of the solenoid will (1) decrease (2) increase (3) remain the same

25. If a highly permeable iron rod were inserted into the solenoid, the strength of the magnetic field inside the solenoid would (1) decrease (2) increase (3) remain the same

26. In the diagram below, *A, B, C*, and *D* are points in the magnetic field near a current-carrying loop. At which points is the direction of the magnetic field into the page?
(1) A and B (2) B and C (3) C and D (4) A and D

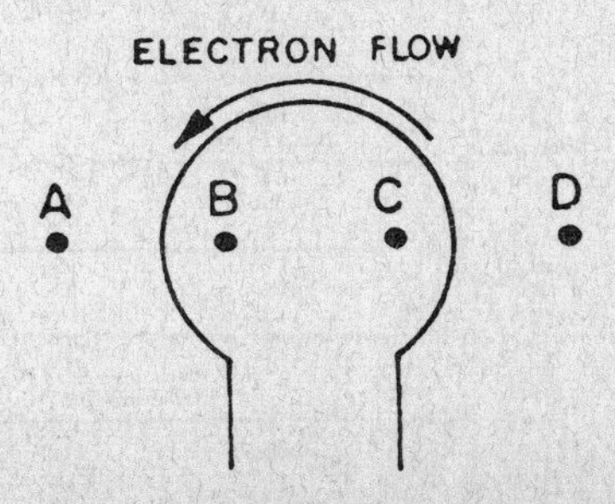

27. Which diagram below best represents a magnetic field?

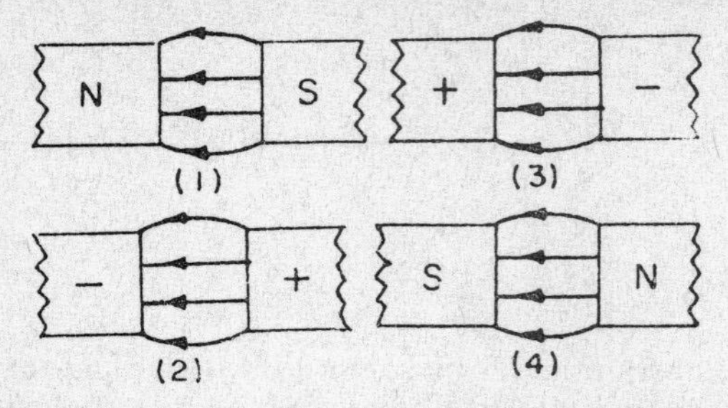

28. The arrows in the diagram below indicate the direction of the electron flow.

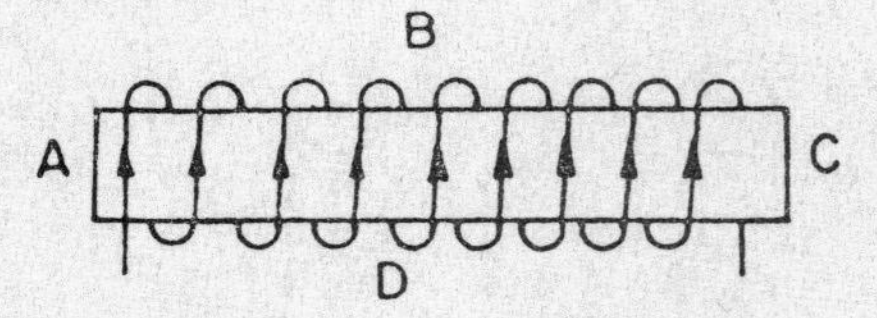

The south pole of the electromagnet is located closest to point (1) A (2) B (3) C (4) D

29. A magnetic field will be produced by (1) moving electrons (2) moving neutrons (3) stationary protons (4) stationary ions

30. The field around a permanent magnet is caused by the motions of (1) nucleons (2) protons (3) neutrons (4) electrons

31. Magnetic flux density may be measured in (1) newtons (2) teslas (3) coulombs (4) joules

32. Which diagram best represents the magnetic field around a current-bearing conductor?

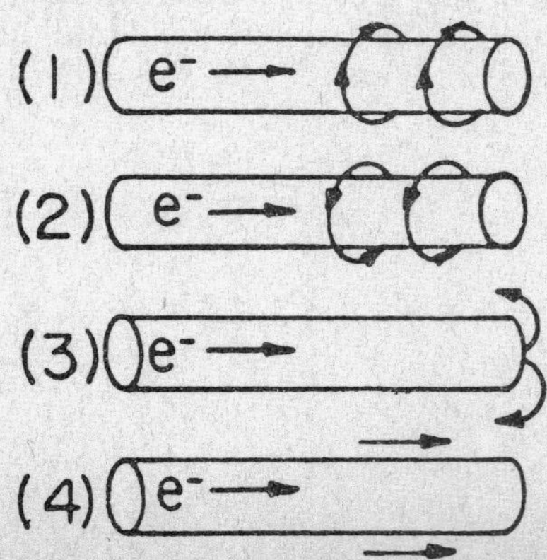

33. In the diagram below, what is the direction of the magnetic field at point A? (1) to the left (2) to the right (3) toward the top of the page (4) toward the bottom of the page

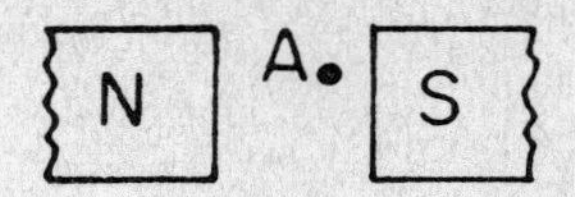

34. Electrons flow in a loop of wire as shown in the diagram. What is the direction of the magnetic field at point A? (1) into the paper (2) out of the paper (3) toward the left (4) toward the right

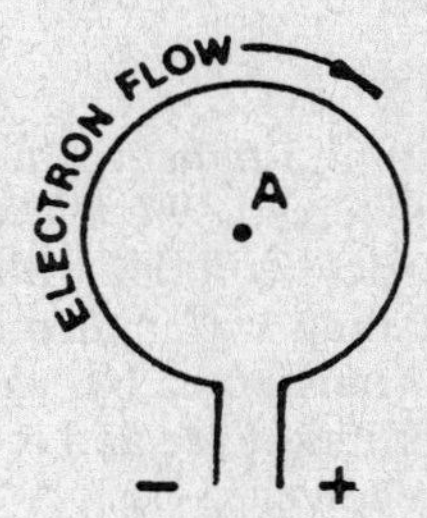

35. In the diagram, in which direction is the magnetic field at point X? (1)toward A (2) toward B (3) toward C (4) toward D

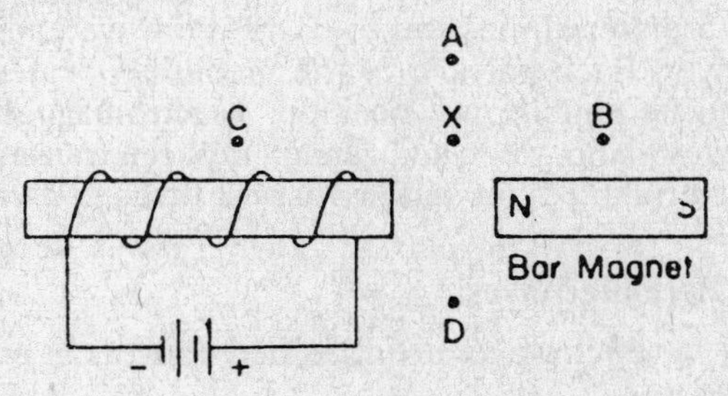

UNIT IV. WAVE PHENOMENA

I. Introduction to Waves

In order to study sound and light it is helpful to see how these forms of energy are transferred. We know that energy is transferred when a moving particle collides with other particles. Winds turning a windmill and falling water turning a turbine are examples. Energy can also be transferred by *wave motion*. The energy originates as a vibration of its source and is transferred by waves to create a vibration of its receiver.

A. Transfer of Energy

Mechanical waves such as sound waves, water waves or waves in a rope require a material medium to carry the wave energy from source to receiver. Each particle of the medium vibrates a small distance around its equilibrium position. **Electromagnetic** waves such as light waves, radio waves and gamma rays can transmit energy without any material medium and travel best through a vacuum.

B. Pulses and Periodic Waves

Another way in which waves are classified consists of pulses and periodic waves.

1. Pulses in a material medium

When the compressed portion of a spring is released, its energy will travel to the other end. Such a single disturbance is called a **pulse**.

a. Speed

Although the spring itself doesn't move, the compression pulse can be seen to travel from one end to the other. Its velocity depends mostly on the nature of the medium and to a lesser degree on the nature of the impulse.

b. **Reflection and transmission**

When a pulse reaches the end of its medium, its energy is either transformed into some other form of energy or the pulse is reflected or transmitted. Usually part of the pulse will be reflected and part will be transmitted to the new medium. The reflection can be observed by sending a pulse through a rope. If the far end of the rope is hanging free, a large displacement occurs here and the pulse returns with no change of its phase. If the far end of the rope is fastened to a rigid wall, the pulse is reflected and returned as an inverted pulse.

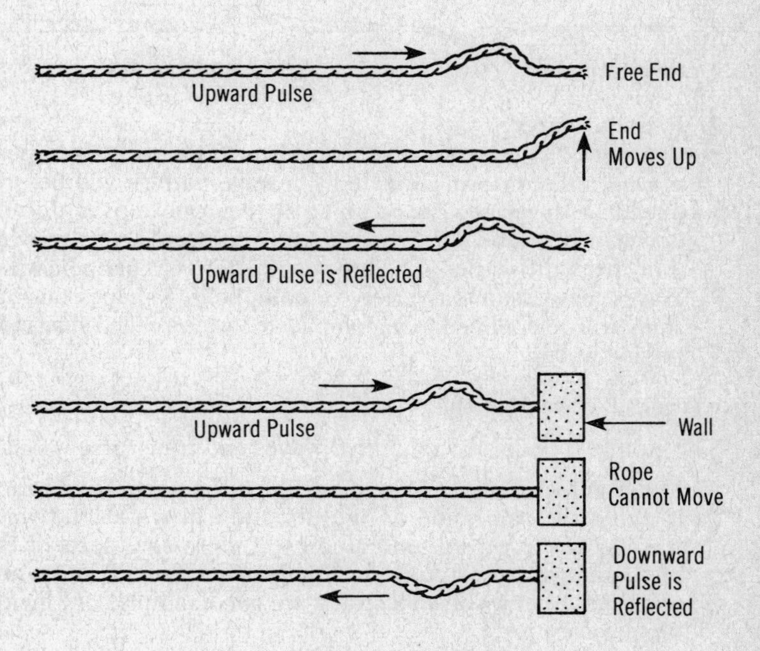

If the far end of the rope is attached to a light cord part of the pulse is reflected back along the rope and part is transmitted to the cord. No inversion occurs for either the reflected or the transmitted portions of the pulse.

If the other end is attached to a rope of greater density, the reflected pulse is inverted.

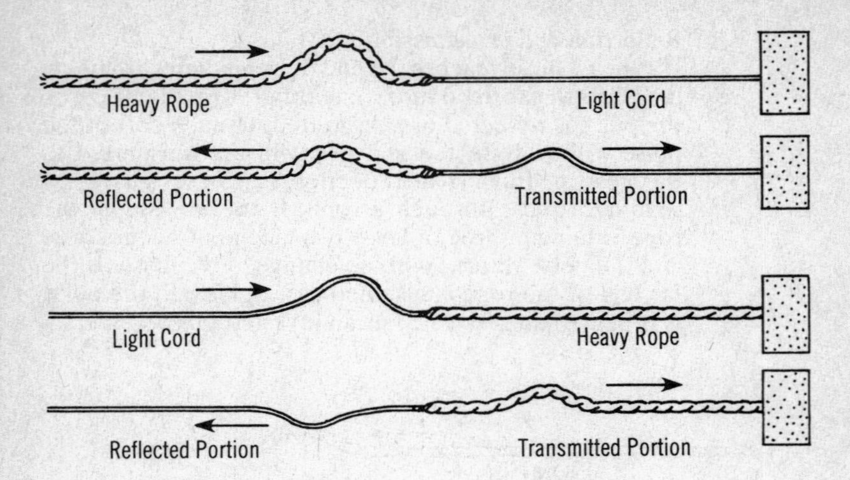

2. **Periodic waves**

A periodic wave is a series of pulses. If the end of a rope is moved up and down repeatedly, a wave pattern will be produced in the rope. Every point of the rope moves up and down in turn and the whole pattern appears to be moving away from the source. The traveling pattern is a **periodic wave**. Waves may assume a variety of complicated shapes but our study will be confined to uncomplicated waves called **sinusoidal** or **sine** waves.

C. Types of Wave Motion

Two simple types are longitudinal waves and transverse waves.

1. **Longitudinal waves** are those in which the particle vibration is along the same line as the direction in which the wave travels. These waves generated by compressing parts of the medium, are called compressional waves. Sound waves and compressional waves in a spring are two examples of longitudinal waves.

2. **Transverse waves** are those in which particles of the medium vibrate right angles to the direction in which the wave travels. Waves in a rope, and electromagnetic waves, such as light waves are examples of transverse waves. The vibration may be in any plane, but it is always perpendicular to the direction in which the wave moves.

   a. **Polarization**

   A transverse wave may be polarized so that all its vibrations occur in only one plane. The up and down movement of a vibrating rope is an example of a polarized wave. If the rope is moved in a circular motion, vibrations in all transverse directions would be produced and the wave would not be polarized. This wave could be polarized by passing it through a vertical slit. The slit would eliminate all those vibrations which are not vertical.

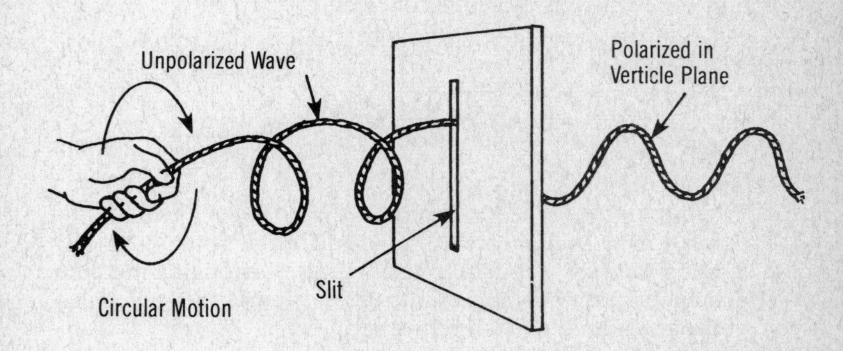

Longitudinal waves cannot be polarized; since the particles of the medium vibrate along same direction as the wave travels.

3. Other types of waves include **torsional waves**. When the end of long, thin metal rod is twisted and then released, a torsion pulse will travel along the length of the rod. Regular repetition of these pulses produces torsional waves.

   Some waves are combinations of longitudinal and transverse forms and others may include torsional components as well. Ocean waves are interesting examples of complex waves.

QUESTIONS

1. A wave is generated in a rope which is represented by the solid line in the diagram below. As the wave moves to the right, point P on the rope is moving toward which position?
   (1) A (2) B (3) C (4) D

2. Which will generally occur when a pulse reaches a boundary between two different media? (1) The entire pulse will be reflected. (2) The entire pulse will be absorbed. (3) The entire pulse will be transmitted. (4) Part of the pulse will be transmitted and part will be reflected.

The diagram below represents a pulse traveling from left to right in a stretched heavy rope. The heavy rope is attached to light rope which is attached to a wall.

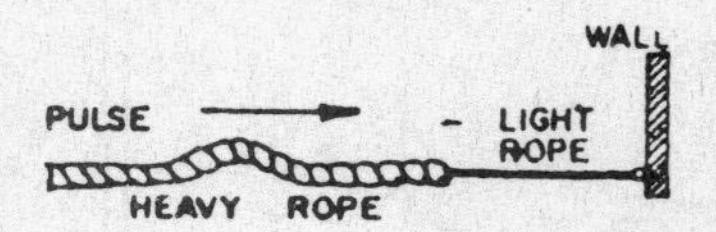

3. When the pulse reaches the light rope, its speed will (1) decrease (2) increase (3) remain the same

4. A transverse wave moves to the right (→) through a medium. Which diagram best represents the motion of the molecules of the medium due to the wave motion?

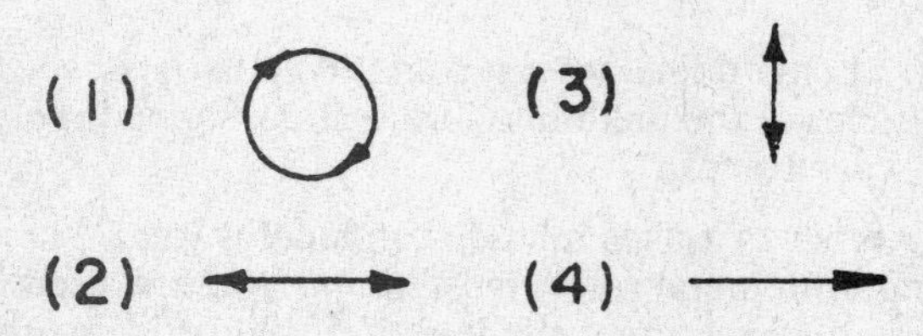

5. A sound wave can *not* be (1) reflected (2) refracted (3) diffracted (4) polarized

6. Which pair of pulses represented below, moving through the same medium, will produce the pulse shown at the right when they are superimposed?

(1) (3)

(2) (4)

7. In which type of wave is the disturbance parallel to the direction of wave travel? (1) torsional (2) longitudinal (3) transverse (4) circular

8. If the wave properties of a particle are difficult to observe, it is probably due to the particle's (1) small size (2) large mass (3) low momentum (4) high charge

9. If the frequency of a sound wave in air at STP remains constant, its energy can be varied by changing its (1) amplitude (2) speed (3) wavelength (4) period

10. Which characterizes a polarized wave? (1) transverse and vibrating in one plane (2) transverse and vibrating in all directions (3) circular and vibrating at random (4) longitudinal and vibrating at random

11. A single vibratory disturbance which moves from point to point in a material medium is known as a (1) phase (2) pulse (3) distortion (4) wavelet

12. A longitudinal wave can *not* be (1) polarized (2) diffracted (3) refracted (4) reflected

13. Whether or not a wave is longitudinal or transverse may be determined by its ability to be (1) diffracted (2) reflected (3) polarized (4) refracted

14. Wave motion in a medium transfers (1) energy, only (2) mass, only (3) both mass and energy (4) neither mass nor energy

15. Longitudinal waves are involved in the transmission of (1) light (2) radar (3) sound (4) photons

II. Common Characteristics of Periodic Waves

A. Frequency

The frequency (f) is the number of complete waves passing a given point in a second. Since a complete vibration ends in the same position in which it begins, it has become known as a "cycle". Frequency is measured in cycles per second (sec^{-1}) which are called hertz, with the symbol Hz.

B. Period

The period (T) of a wave is the time it takes to complete one cycle. It can also be defined as the time required for a wave to pass a given point. It should be obvious that the period is the reciprocal of the frequency:

$$T = \frac{1}{f} \text{ or } f = \frac{1}{T}$$

If an object is vibrating at a frequency of 5Hz, then $f = 5\ \text{sec}^{-1}$ or 5/s

$$\text{and } T = \frac{1}{5/s} \text{ or } \frac{1}{5} \text{ second.}$$

C. Amplitude

The amplitude of a wave is the maximum displacement of a particle from its rest position.

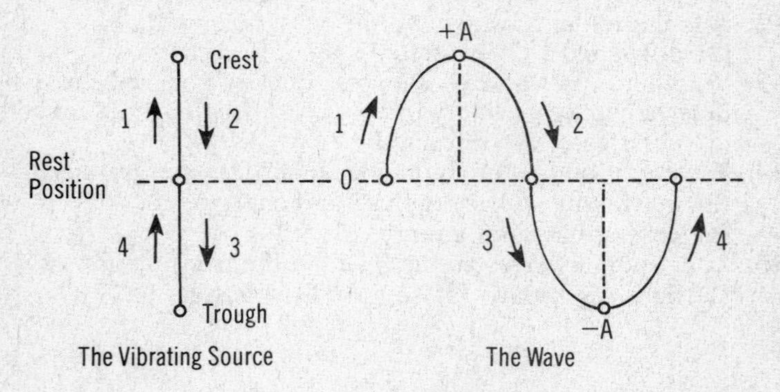

Note that the vibrating source, in generating each wave begins a cycle at 0, and proceeds to a maximum displacement (+A) and returns to 0. Then to −A and back to 0.

D. Phase

The term phase is a useful concept for describing waves. Since a wave repeats itself each cycle, it can be divided like a circle into 360 degrees. When two points on a periodic wave are at the same displacement from their position of equilibrium and moving in the same direction, they are said to be "in phase". Two points are in phase if they reach the same position at exactly the same time. If one is at a crest when the other is at a trough they are 180 degrees out of phase, or in phase opposition.

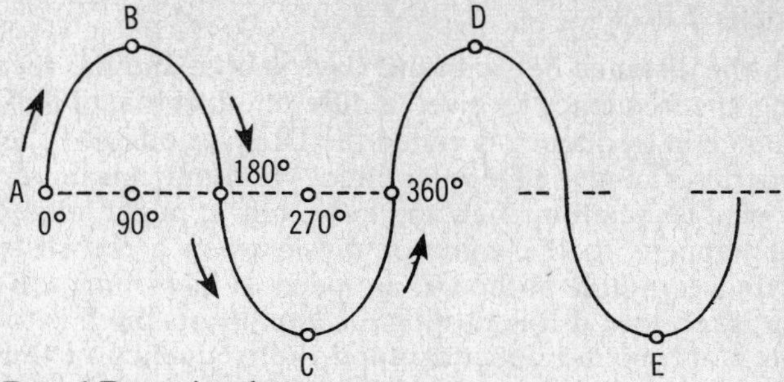

Points B and D are in phase.
Points B and C are in phase opposition.

E. Wave length

Wave length is the distance between two corresponding points on successive waves. In terms of phase, wave length is the distance between two consecutive points in phase. For example, the distance between two successive crests is one wave length. The symbol for wavelength is the Greek letter "lambda", written λ. The wavelength is directly is directly proportional to the speed of the wave, and inversely proportional to its frequency.

$$\lambda = \frac{v}{f}$$

F. Speed

The speed of a wave depends on the properties of the medium and the nature of the wave. If the frequency and wavelength are known, then speed, v, can be calculated using the relationship:

$$v = f\,\lambda$$

This equation is applicable to all waves.

Sample Problem

A 128 Hz tuning fork produces a sound wave whose wavelength is 2.64m. What is the speed at which the wave travels?

Given $f = 128\ \text{sec}^{-1}$
$\lambda = 2.64\ \text{m}$

Find v

Equation $v = f\,\lambda$

Solution $v = (128\ \text{sec}^{-1})\ (2.68\ \text{m})$

Answer = v = 343 m/sec

G. Doppler Effect

When the distance between the wave source and the receiver is changing, the frequency received is different than that of the source. This change in frequency is called the Doppler effect.

Before describing the Doppler effect on sound, it is important to review what is meant by high and low pitch. It might help to think of what happens to the voice or music when a record is being played on a turntable at the wrong speed. If its turning too slowly, its pitch is too low, if it is turning too rapidly, its pitch is too high. The pitch of a sound depends upon the frequency of the sound wave, but because pitch is subjective, frequency and pitch are *not* identical.

Next time you're waiting at a railroad crossing as a train passes try to listen carefully. If the train sounds is horn as passes the crossing, the horn will seem higher pitched while the train is approaching and suddenly drop to a lower pitch when it is leaving.

As the train approaches the frequency you hear is greater than that produced by the horn. This is because more pulses arrive per unit time. As the train moves away from you, the lower pitch is heard because less pulses arrive per unit time. Now, if that train is traveling at a constant speed, you'll notice a constant high pitch as it approaches, a sudden change as it passes, and a constant low pitch as it moves away. The Doppler effect may be due to a motion of the receiver, the source, or both together.

H. Wave Propagation

1. Wave fronts

When a pebble is dropped into a pond of still water, it creates a series of circles moving out from the point where the pebble entered the water. These circles represent what are known as circular waves fronts. A wave front is the locus of adjacent points of a wave which are in phase. The distance between two successive wave fronts is one wavelength.

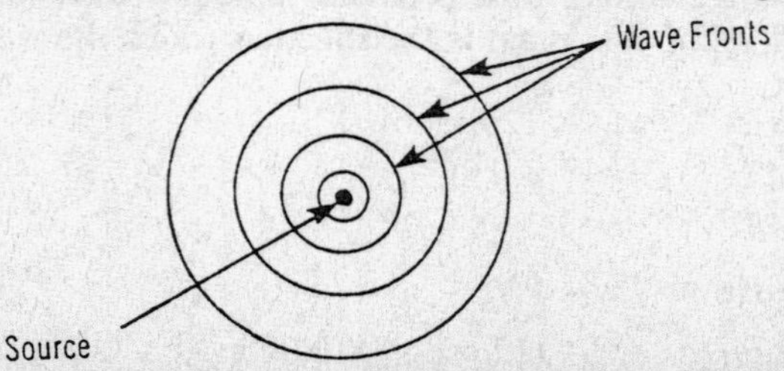

In three dimensional space, of course, a wave front may be viewed as the surface of a sphere whose radius is the distance from the source. As the distance from the source becomes extremely large, a portion of the wave front appears to be a straight line as in linear or plane wave fronts. Linear wave fronts can be produced in a ripple tank, using a flat ruler to create the disturbance.

2. Huygens' Principle

A logical way to predict the future movements of any wave front was described by Huygens. It is known as Huygens' Principle and it states that every point on a wave front may be considered a source of new wavelets with the same speed. As these wavelets move away from their source they intersect to form the pattern of the next wave front. In this way each wave front can be treated as the source of the succeeding one.

QUESTIONS

1. Which distance on the diagram below identifies the amplitude of the given wave? (1) AE (2) AB (3) AC (4) AD

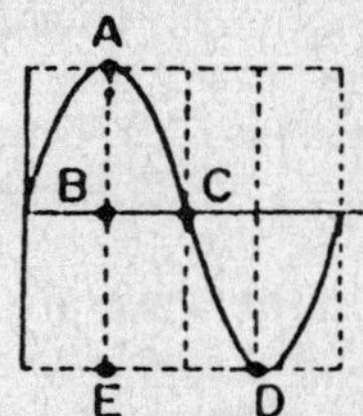

2. As the frequency of the wave generated by a radio transmitter is increased, the wavelength (1) decreases (2) increases (3) remains the same

3. Compared to the speed of light in a vacuum, the speed of light in a dispersive medium is (1) less (2) greater (3) the same

4. As the frequency of a wave increases, the period of that wave (1) decreases (2) increases (3) remains the same

5. The diagram below represents a wave traveling in a uniform medium. Which characteristic of the wave is constant? (1) amplitude (2) frequency (3) period (4) wavelength

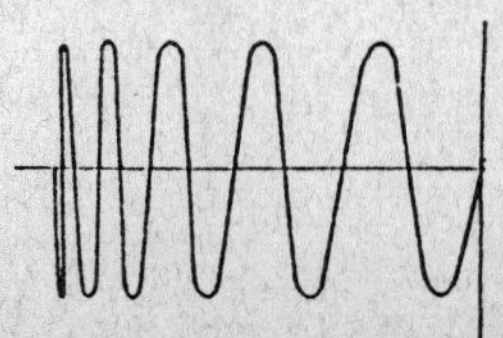

6. If the frequency of a wave is increased, its wavelength will (1) decrease (2) increase (3) remain the same
7. According to the theory of matter waves, as the momentum of a particle increases, its wavelength (1) decreases (2) increases (3) remains the same

Base your answers to questions 8 through 12 on the diagram below which represents a segment of a periodic wave traveling to the right in a steel spring.

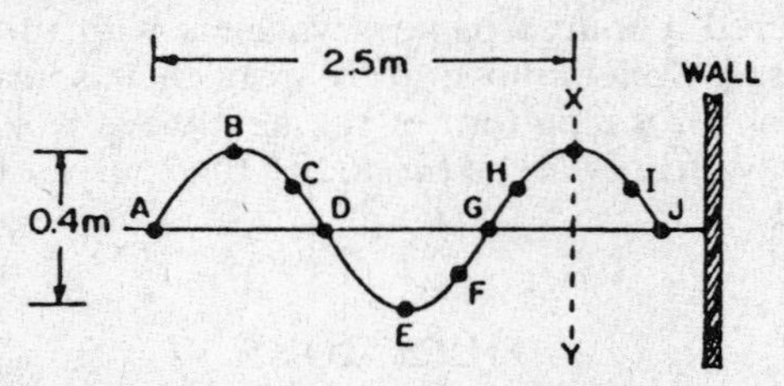

8. What is the amplitude of the wave? (1) 2.5 m (2) 2.0 m (3) 0.2 m (4) 0.4 m
9. What is the wavelength of the wave? (1) 1.0 m (2) 2.0 m (3) 2.5 m (4) 0.4 m
10. If a wave crest passes line XY every 0.40 second, the frequency of the wave is (1) 1.0 Hz (2) 2.5 Hz (3) 5.0 Hz (4) 0.4 Hz
11. Which two points on the wave are in phase? (1) A and D (2) B and E (3) C and I (4) C and H
12. What type of wave is illustrated by the diagram? (1) torsional (2) longitudinal (3) elliptical (4) transverse
13. The vibrating tuning fork shown in the diagram below produces a constant frequency. The tuning fork is moving to the right at a constant speed, and observers are located at points A, B, C, and D. Which observer hears the *lowest* frequency? (1) A (2) B (3) C (4) D

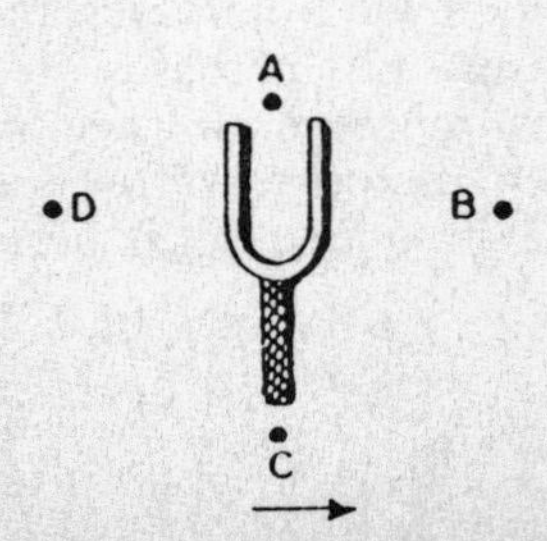

14. As the phase difference between two interfering waves changes from 0° to 180°, the amplitude of the resultant wave (1) decreases (2) increases (3) remains the same

15. If the frequency of a sound wave is 440. cycles per second, its period is closest to (1) 2.27×10^{-3} second/cycle (2) 0.75 second/cycle (3) 1.33 seconds/cycle (4) 3.31×10^2 seconds/cycle

16. If the period of a wave is doubled, its wavelength will be (1) halved (2) doubled (3) unchanged (4) quartered

Base your answers to questions 17 through 21 on the diagram below which represents a transverse wave.

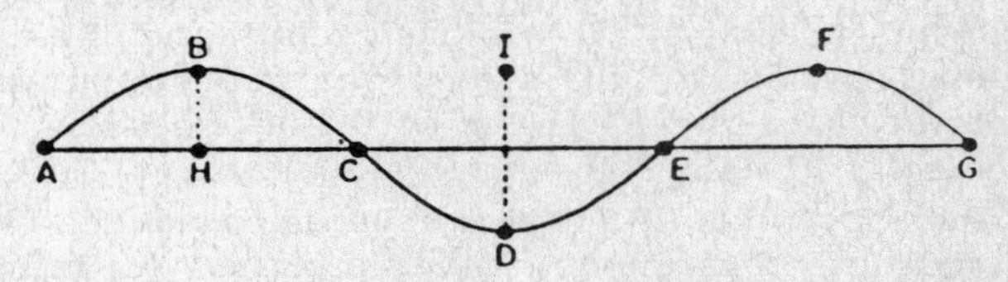

17. Which two points are in phase? (1) A and C (2) B and D (3) C and E (4) B and F

18. The amplitude of the wave is the distance between (1) A and C (2) A and E (3) B and H (4) I and D

19. How many cycles are shown in the diagram? (1) 1 (2) 2 (3) 3 (4) 1.5

20. A wavelength is the distance between points (1) A and C (2) A and E (3) B and H (4) I and D

21. If the period of the wave is 2 seconds, its frequency is (1) 0.5 cycle/sec (2) 2.5 cycles/sec (3) 3.0 cycles/sec (4) 1.5 cycles/sec

22. A medium in which waves of different frequencies travel at different speeds and may be separated is called (1) a dispersive medium (2) a nondispersive medium (3) an inelastic medium (4) a coherent medium

23. Which point on the wave shown in the diagram is 180° out of phase with point P? (1) 1 (2) 2 (3) 3 (4) 4

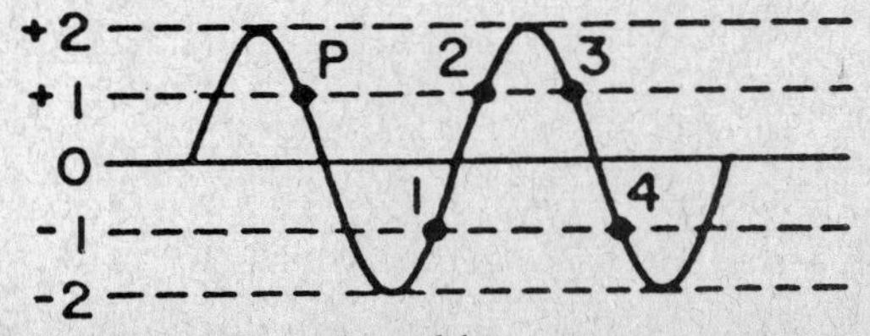

24. Periodic waves are produced by a wave generator at the rate of one wave every 0.50 second. The period of the wave is (1) 1.0 s (2) 2.0 s (3) 0.25 s (4) 0.50 s

25. The amplitude of the wave shown below is represented by the distance between points (1) A and B (2) A and C (3) A and D (4) E and D

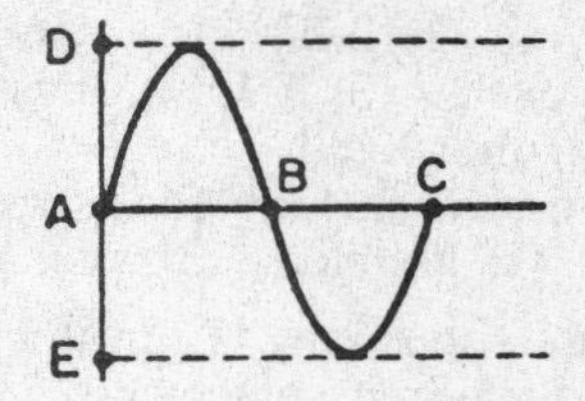

26. A wave traveling at 5.0×10^4 meters per second has a wavelength of 2.5×10^1 meters. What is the frequency of the wave? (1) 1.25×10^6 Hz (2) 2.0×10^3 Hz (3) 5.0×10^{-4} Hz (4) 5.0×10^3 Hz

27. If the speed of a wave in a medium is dependent on the frequency, the medium is (1) transparent (2) refractive (3) translucent (4) dispersive

28. As a wave enters a medium, there may be a change in the wave's (1) frequency (2) speed (3) period (4) phase

29. What total distance will a sound wave travel in air in 3.00 seconds at STP (1) 3.31×10^2 m (2) 6.62×10^2 m (3) 9.93×10^2 m (4) 9.00×10^8 m

30. The driver of a car hears the siren of an ambulance which is moving away from her. If the actual frequency of the siren is 2,000 hertz, the frequency heard by the driver may be (1) 1,900 Hz (2) 2,000 Hz (3) 2,100 Hz (4) 4,000 Hz

31. What is the period of a wave with a frequency of 250 hertz? (1) 1.2×10^{-3} s (2) 2.5×10^{-3} s (3) 9.0×10^{-3} s (4) 4.0×10^{-3} s

Base your answers to questions 32 through 36 on the diagram below which represents a vibrating string with a periodic wave originating at A and moving to G, a distance of 6.0 meters.

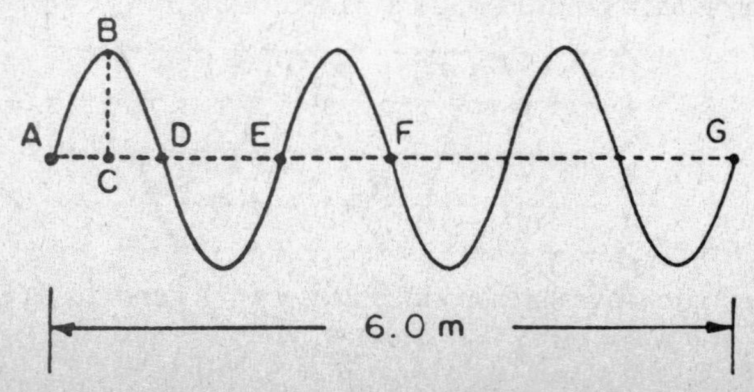

32. What type of wave is represented by the diagram? (1) elliptical (2) longitudinal (3) torsional (4) transverse

33. What is the wavelength of this wave? (1) 1.0 m (2) 2.0 m (3) 3.0 m (4) 6.0 m

34. Which phenomenon would occur if the waves were reflected at G and returned back to A through the oncoming waves? (1) diffraction (2) dispersion (3) standing waves (4) Doppler effect

35. As the wave moves toward G, point E on the string will move (1) to the left and then to the right (2) vertically down and then vertically up (3) diagonally down and then diagonally up (4) diagonally up and then diagonally down

36. If the waves were produced at a faster rate, the distance between points D and E would (1) decrease (2) increase (3) remain the same

37. Compared to the period of a wave of red light, the period of a wave of blue light is (1) less (2) greater (3) the same

38. A car's horn is blowing as the car moves at constant speed toward an observer. Compared to the frequency of the sound wave emitted by the horn, the observed frequency is (1) constant, and lower (2) constant, and higher (3) the same (4) varying and lower

39. In the diagram below, a train of waves is moving along a string. What is the wavelength? (1) 1 m (2) 2 m (3) 3 m (4) 6 m

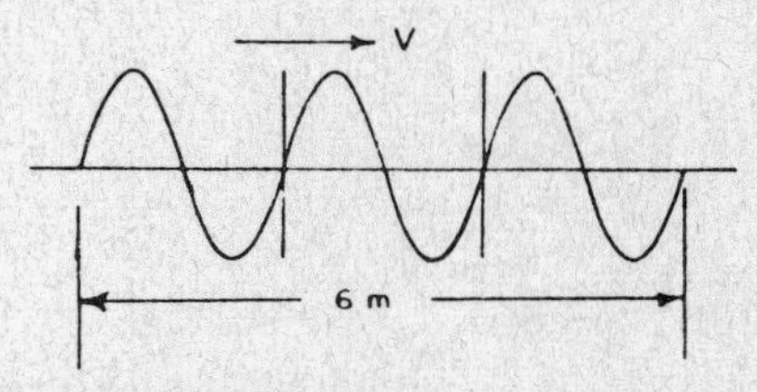

III. Periodic Wave Phenomena

In describing periodic waves the direction in which the wave is travelling is shown by an arrow called a **ray**. Rays are always drawn perpendicular to the wave front. The rays of a spherical wave are the radii of the sphere with the source at its center. Rays of a plane wave are a series of parallel lines.

Interference

When two waves reach the same point simultaneously, either constructive or destructive interference will occur at this point.

1. Superposition

The disturbance experienced where two waves meet is the algebraic sum of the individual disturbances that each of the two waves would have caused acting singly. If the component disturbances are vectors, the resultant is a vector sum. This fundamental principle applies to all forms of wave motion and is called the principle of **superposition**.

a. **Constructive interference**

Maximum constructive interfering occurs at points where the two waves are in phase. The net displacement is greater than that produced by either one alone.

b. **Destructive Interference**

Maximum destructive interference occurs where the phase difference between the waves is 180°. The waves cancel each other if their amplitudes are the same.

2. Two sources in phase

The ripple tank provides us with a good demonstration of the results when two sources produce equal waves crossing each other in a ripple tank.

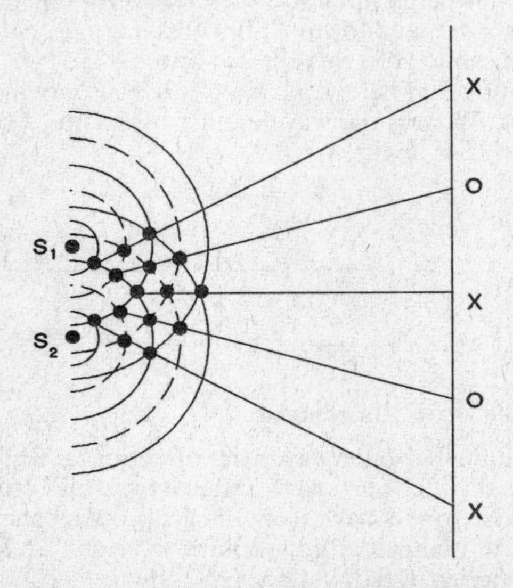

There will be many places, marked "o" in the diagram where the crest of one wave and the trough of another arrive at the same time. There is a complete destructive interference at these points. Since there is no vibration, they are referred to as **nodes**.

At the points marked x, two crests or two troughs come together and produce crests or troughs of **increased** amplitude called **antinodes.** If both wave fronts are straight, the nodes and antinodes alternate in a series of parallel lines called nodal or antinodal lines. These nodal lines represent the loci of points where the path difference to the source is $n\lambda-\lambda/2$ where n is any integer and lambda (λ) is the wavelength. Nodal lines will form where the path difference is any **odd** number of wave lengths. **Antinodal** lines will form where the path difference is any **even** number of wavelengths.

Standing waves

Whenever two waves of the same frequency and amplitude pass through each other in opposite directions they produce a **standing wave**.

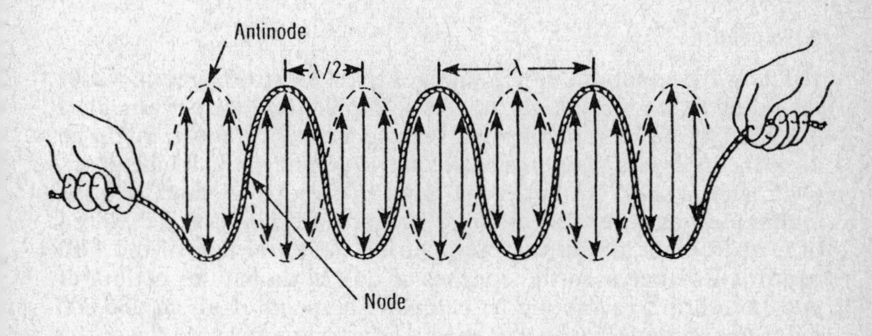

All points in a standing wave are vibrating but the amplitude varies from the maximum at the antinodes to the minimum at the nodes. Nodes and antinodes alternate every quarter wavelength. In the diagram, the two hands are vibrating in phase. Crests from both sources arrive at the antinodes at the same time. The two troughs will arrive at the antinodes a half cycle later. The distance between any two adjacent nodes or any two antinodes is a half wavelength. Standing waves are most commonly produced by reflections at a fixed boundary of the medium. Since the incident and reflected waves have the same frequency and amplitude, they will produce standing waves wherever they meet. It is relatively easy to measure the distance between adjacent nodes of water waves, sound waves or radio waves. Since the distance between adjacent nodes is equal to half a wavelength, measuring the distance between nodes provides a convenient way to determine wavelengths.

Resonance

Every object has a natural frequency at which it vibrates when disturbed. This frequency can be altered by changing the size or shape of the object or the materials that it contains. You can observe this by striking water glasses filled to different levels and listening to the tones that each makes.

Adding small amounts of energy in time with the natural frequency can cause a build-up of very strong vibrations by a process called *resonance*. For example, when a child on a swing is given small pushes in time with the natural frequency of the swing, very large vibrations are produced. Also, soldiers are commanded to break step when crossing a bridge to avoid a possible resonance that could result in disaster.

IV. Light

Light is an electromagnetic disturbance that produces the sensation of sight.

A. Speed

The wave equation, $v = f\lambda$, applies to all electromagnetic waves including light waves. Light travels so rapidly that measurement of its speed has been very difficult. Galileo described how it might be done by uncovering lanterns between two people about a mile apart. Over such a short distance, the time involved is so short that his technique was never successful. The first determination is credited of Olaf Roemer, an astronomer. During the latter half of the 17th century, Roemer used the eclipses of one of the moons of Jupiter and the earth's revolution to calculate a speed of about 186,000 miles per second. Fizeau was the first person to measure the speed of light on the surface of the earth. He used an intense source of light, mirrors, and a toothed wheel to alternately cover and uncover the light. About the same time (mid nineteenth century), Foucault used a rotating mirror instead of a toothed wheel. Both these values were in agreement. Albert A. Michelson, the first American to win a Nobel Prize, spent most of his life studying light and optics. He made many improvements to the basic idea of the rotating mirror, which produced an extremely accurate measurement. The accepted value for the speed of light, to three significant digits, is 3.00×10^8 meters per second (very close to 186,000 miles per second). The conventional symbol for the speed of light constant is "c".

$$c = f\lambda$$

1. **In space.** The speed of light in space is one of the most important concepts when studying natural phenomena. Einstein was bold enough to state clearly that the speed of light in space is

constant, regardless of the motion of its source or its receiver. This required and continues to require a reexamination of all classical physics. New assumptions concerning length, time, mass and energy must be made and new transformations from one inertial system to another must be formulated.

All electromagnetic radiation travels at the same velocity, c, in a vacuum. According to the principles of relativity, no objects can ever attain this speed.

2. **In a material medium.** The speed of light in a material medium is always less than its speed in a vacuum. It varies with frequency and the optical density of the medium. The speed of light in air is very close to its speed in a vacuum. For most practical purposes, it may be considered to be the same.

B. Reflection

Light falling on an object may be absorbed, transmitted, or reflected. **Reflection** occurs when light falls on an object and is turned back. The laws of reflection are the same for all wave phenomena.

1. **Laws of Reflection**
   a. When light falls on any reflecting surface, the incident ray, the reflected ray, and the normal all lie in the same plane.
   b. The angle of reflection is always equal to the angle of incidence: $\theta_i = \theta_r$
2. **Regular reflection.** The law of reflection applies at each point on a reflecting surface. When the surface is smooth and flat, such as a plane mirror, the image is a perfect replica of the light source. The image is virtual (meaning that it *can't* be projected on a screen), it is erect, and image distance and object distance are equal. This is known as **regular**, or **specular** reflection.
3. **Diffuse reflection** takes place when the reflecting surface is **irregular**. The laws of reflection still hold but the irregular surface makes normals to the surface that are not parallel. Therefore, the reflected light is scattered.

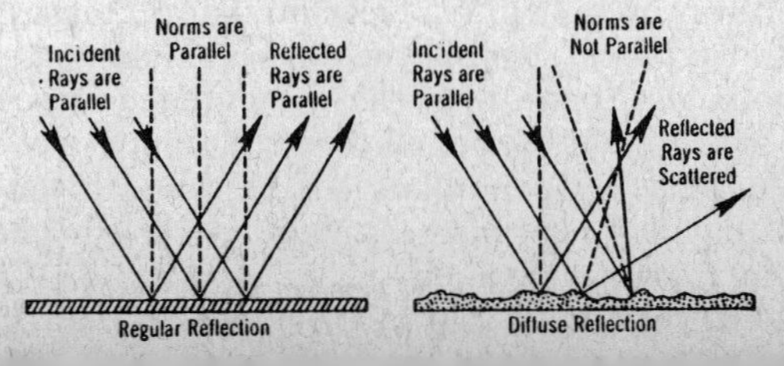

C. Refraction

When a wave crosses the boundary into another medium with different properties, some of the energy will be reflected. The remainder will be absorbed or transmitted through the new medium at a different speed. This causes any ray except those parallel with the normal to change direction in the new medium. This change in direction is called **refraction**. The expression, "oblique rays," refer to any rays that are not perpendicular to the boundary of the medium.

1. The **index of refraction (n)** of a medium is the ratio of the speed of light in a vacuum (c) to its speed in the material medium (v)

$$(2) \qquad n = \frac{c}{v}$$

a. Snell's Law

Snell found that the ratio of the sine of the angle of incidence in air to the sine of the angle of refraction in a given medium was a constant for that given medium. This constant value (n) is called the index of refraction for that medium. A more generalized statement for refractions that occur between any two media is expressed by the equation:

$$n_1 \sin \theta_1 = n_2 \sin \theta_2$$

where θ_1 is the angle of the incident ray with the normal (the angle of incidence) in medium 1,

θ_2 is the angle of the refracted ray with the normal (the angle of refraction) in medium 2,

n_1 is the index of refraction of medium 1,

n_2 is the index of refraction of medium 2,

Written in a more useful form for problem solving: $n_1 \sin\theta_1 = n_2 \sin\theta_2$. These relationships assume that the incident ray, the refracted ray and the normal are all in the same plane.

b. Refraction and speed

Waves moving into a medium that **decreases** their speed, are refracted (bent) toward the normal, making the angle of refraction is less than the angle of incidence. Since the frequency (f) of the wave doesn't change and since $\lambda = v/f$, it follows that a decrease in speed causes a decrease in wavelength. Waves moving into a medium that **increases** their speed are refracted away from the normal.

2. **The Critical angle** (θ_c) is that angle of incidence for which the angle of refraction is 90°. This can only happen when the velocity of the light increases causing the angle of refraction to be greater than the angle of incidence. By measuring the critical angle of material, its index of refraction can be calculated using the equation:

$$\sin \theta_c = \frac{1}{n}$$

3. **Total internal reflection** is observed whenever the angle of incidence in an optically dense medium is greater than the critical angle. In these cases, no light passes into the less dense medium and all the light is reflected back making the boundary act like a perfect mirror.

4. **Dispersion** is the separation of light of different frequencies (called polychromatic light) into its component waves. In dispersive media waves of different frequencies have different speeds. They will each have a different index of refraction in the new medium and will be refracted at different angles. Differences in frequency of light waves affect the eye as differences in color. Polychromatic light, such as sunlight, can be separated into its component frequencies when it passes through a dispersive medium such as a prism.

a. Effect of medium

Differences in density and elasticity of media will result in differences in the speed at which waves travel. These differences have no effect on frequency. When a wave passes from one medium to another, its speed changes and its new wavelength can be calculated by dividing its new speed by its frequency.

b. Dispersive medium

A medium in which the speed of the waves depends on frequency is called a **dispersive medium**. The name comes from the word dispersion which means to separate. In such media, waves of high frequency travel more slowly than those of low frequency. The separation of visible light into its component colors by a glass prism is an example of dispersion. Glass is a dispersive medium for light.

c. Nondispersive medium

A medium in which speed is the same for all frequencies is called a **nondispersive medium**. All elastic media, including air and water are nondispersive for sound waves of low amplitude.

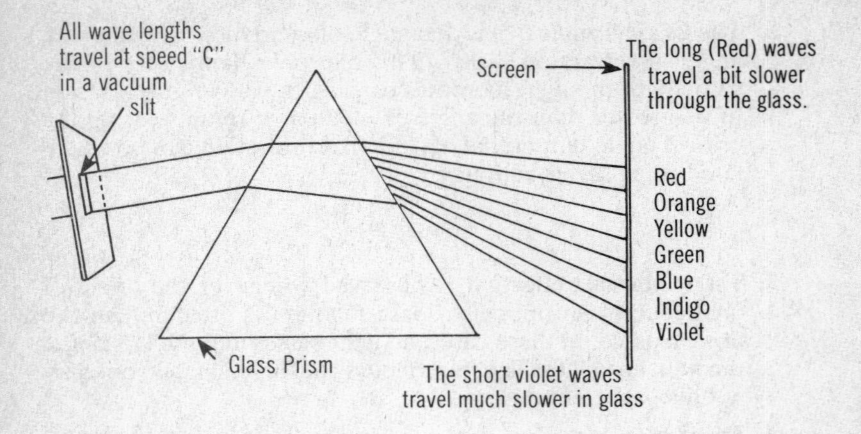

d. Wave Nature of Light

During the 17th century, two different theories evolved in attempts to explain the nature of light. Basically, they differed in that one held that light is a stream of tiny particles; the other, that light travels as waves. Since particles and waves obey different laws, the theories conflicted. Particles travel in straight lines, while waves bend around corners (diffraction). When particles collide they rebound, while waves produce standing wave patterns (interference).

Much of the behavior of light can be interpreted in terms of wave phenomena.

1. **Interference of light.** Interference phenomena can be produced only by waves. Since it has been shown that light produces interference patterns, it must be concluded that light has wave properties.

   a. **Coherent sources** are sources that produce waves with a constant phase relation. Early experiments used weak monochromatic light sources passing through a barrier containing two slits to provide coherent light. More recently lasers are used as sources of coherent light.

   b. **Double slit.** Light from two coherent pint sources produce a stationary interference pattern. This is a pattern of alternating nodal and antinodal lines. Antinodal lines are regions of maximum light energy and will form bright spots on a screen. Using Young's double slit arrangement the antinodal "lines" are planes, rather than lines.

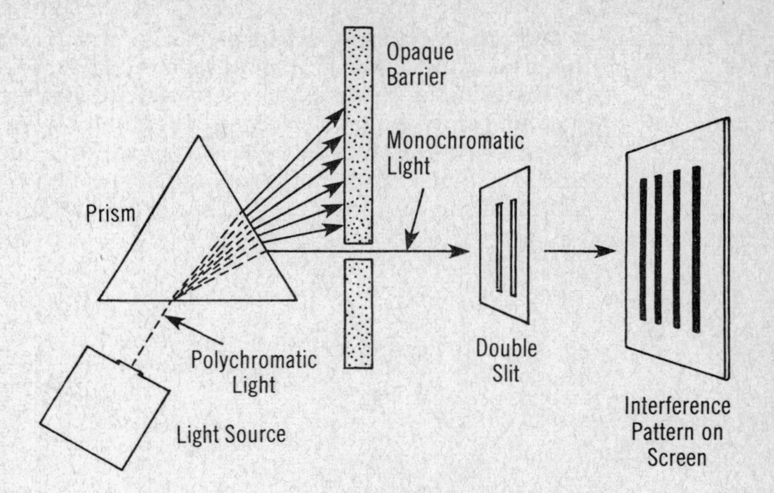

Each plane produces a vertical bar of light on the screen. The alternate light and dark bars on the screen can be explained only as the result of diffraction at the slits and interference on the screen. Young's experiments and others like them seemed to have settled that it was waves over particles in the 19th century. But the question has been reopened in the 20th century.

The relationship between wavelength and color can be demonstrated with a source of white light and a double slit arrangement. Provided the slits are narrow enough, each of the interference bars, except the central one, has a different color. Starting from the side of the bar nearest the center, they range from violet, through blue, green, yellow, orange, and end in red. The effect of the slits is to diffract longer wavelengths more than shorter wavelengths. The combination of colors produced from a given source is referred to as the **spectrum** of that source.

(1) **Diffraction gratings.** In order to measure wavelengths, we must use a very narrow slit. This reduces the amount of light passing through. To overcome this problem, the diffraction grating which uses as many as 20,000 slits to the inch was devised. Slits are formed by ruling scratches in glass or some other transparent material.

(2) **A diffraction grating spectrometer** is an instrument that uses a diffraction grating to separate and measure wavelengths. Using the geometry described in the diagram below, we can derive the equation for wavelength (λ) in terms of the slit spacing (d), the distance the light is diffracted from the center line (x) and the distance from the slit to spectrum line (L).

$$\frac{\lambda}{d} = \frac{x}{L}$$

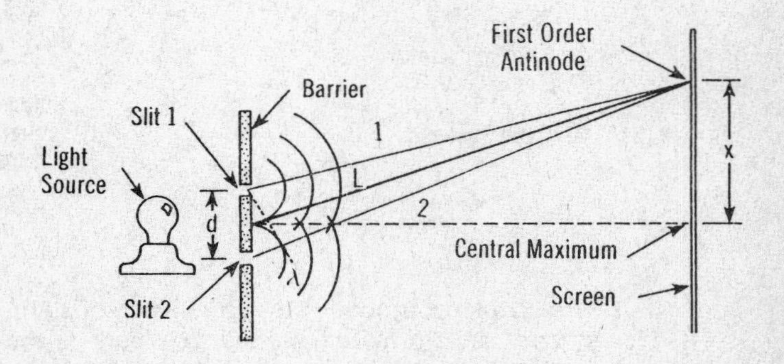

Plane waves from a distant source arrive at slits 1 and 2 simultaneously. After diffraction by these narrow slits, constructive interference occurs halfway between the slits along the central antinode. Constructive interference will also occur at the first order antinode at the point where path length 2 is exactly one wavelength longer than path length 1. Note that the small triangle formed by sides d, λ, and the short dotted line is similar to the large triangle formed by side x,L, and the long dotted line. Since corresponding sides of similar triangles are proportional,

$$\lambda/d = x/L$$

$$\text{and } \lambda = dx/L$$

2. Transverse Nature of Light. Remember that waves may be longitudinal or transverse. The strongest evidence that light is a transverse wave is that it can be polarized. If non-polarized light is passed through a material which permits only light vibrating in one plane to pass through, only light vibrating in that plane emerges. If we cause this light to pass through a second polarization "filter", the intensity of the transmitted light is observed to change from a maximum when the two filters have parallel axes to a minimum (practically zero, or darkness) when the axes are perpendicular.

E. Electromagnetic Radiation

Electromagnetic radiations are transverse wave disturbances propagated through space at the speed of 3.00×10^8 meters per second (c). this is best known as the speed of light. It is generally accepted that these radiations originate from field disturbances produced by accelerating charged particles.

1. **Electromagnetic spectrum.** The electromagnetic spectrum includes radio waves and infrared waves, which are longer than visible light; and ultraviolet rays, X-rays and gamma rays, which have wavelengths shorter than those of visible light. Names were derived from sources and the divisions *do* overlap, but the following points are clearly defined:
   a. Frequency (f) and wavelength (λ) are inversely proportional, and velocity in a vacuum (c) is constant at 3.0×10^8 m/sec:
   $$\lambda = c/f \text{ or } f = c/\lambda$$
   b. Visible light is only a very small portion of the electromagnetic spectrum.
   c. The different effects on receivers are due to differences in frequency.
   d. As frequency increases, the wave nature becomes less apparent.

2. **Doppler Effect.** The Doppler effect is exhibited by electromagnetic radiations as well as sound waves. The Doppler effect may be summarized as follows:
   a. There is an increase in the observed frequency when the distance between the source and the receiver is decreasing.
   b. There is a decrease in the observed frequency when the distance between the source and the receiver is increasing.

   Since speed is constant in space, these changes in frequency can be used to measure the velocity of a moving object that is emitting waves at a constant frequency.

   Investigations depending upon, or related to the Doppler effect include, among many others:

a. certain types of radar that measure speed.

b. radial velocity of stars.

c. speed of earth satellites, based on shifts in frequency of radio waves transmitted by them.

d. determining whether objects in distant space are moving towards us or away from us.

e. radar waves are reflected at increased frequencies from objects that are approaching.

f. random motions of molecules in a gas discharge tube result in a spreading of each observed spectral line due to a Doppler shift.

QUESTIONS

1. Which is *not* in the electromagnetic spectrum? (1) light waves (2) radio waves (3) sound waves (4) X-rays

2. As the wavelength of light in a vacuum increases, its speed (1) increases (2) decreases (3) become the same

Base your answers to questions 3 through 5 on the diagram below. The diagram shows two lights rays originating from sources S in medium y. The dashed line represents a normal to each surface.

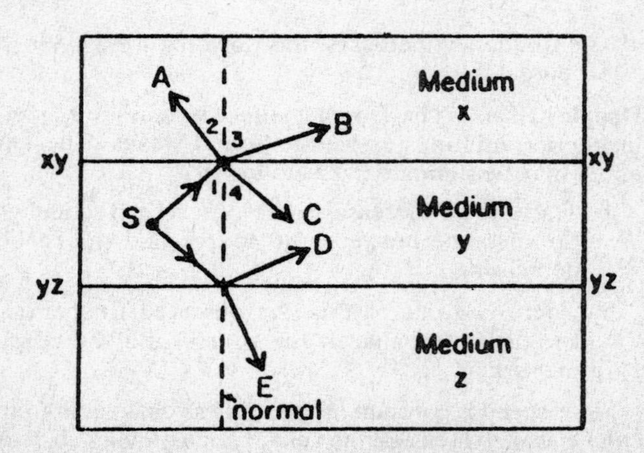

3. Which light ray would *not* be produced in this situation? (1) A (2) B (3) C (4) E

4. A reflected light ray is ray (1) A (2) B (3) C (4) E

5. Which two angles must be equal? (1) 1 and 2 (2) 2 and 3 (3) 3 and 4 (4) 1 and 4

6. Which will most likely occur when light passes through a double slit? (1) refraction (2) diffraction, only (3) interference, only (4) diffraction and interference

7. Only coherent wave sources produce waves that (1) are the same in frequency (2) have the same speed (3) have a constant phase relation (4) are polarized in the same plane

Base your answers to questions 8 through 12 on the diagram below which represents monochromatic light incident upon a double slit in barrier A, producing an interference pattern on screen B.

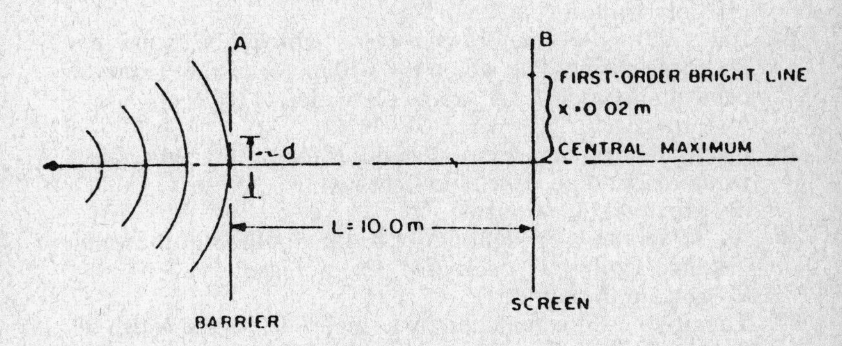

8. The observed phenomenon in this experiment is produced by (1) refraction and reflection (2) polarization and reflection (3) diffraction and interference (4) diffraction and polarization

9. If x = 0.02 meter, L = 10.0 meters, and the wavelength of the incident light is 5.0×10^{-7} meter, the distance d between the slits is (1) 2.5×10^{-4} m (2) 2.0×10^{-2} m (3) 2.5×10^{-2} m (4) 4.0×10^{-5}m

10. If the distance between the slits is increased, the distance x will (1) decrease (2) increase (3) remain the same

11. If the distance from the slits to the screen is increased, the distance x will (1) decrease (2) increase (3) remain the same

12. If the wavelength of the incident light were increased, the distance x would (1) decrease (2) increase (3) remain the same

Base your answers to questions 13 through 17 on the diagram below. An interference pattern is produced on screen B when light passes through the double slit in barrier A.

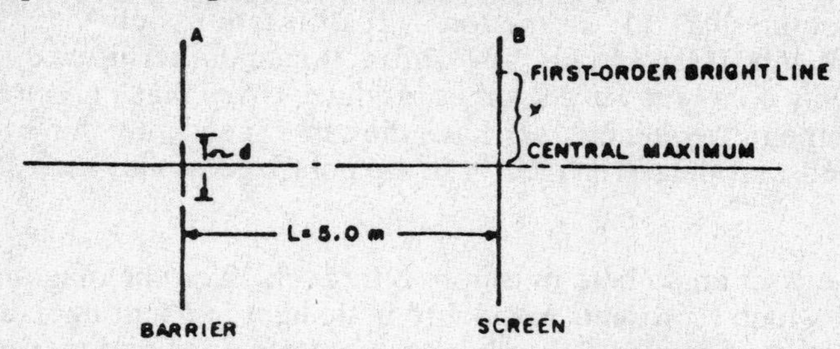

13. The pattern produced on the screen is the result of interference and (1) reflection (2) refraction (3) diffraction (4) polarization

14. If d = 3.0×10^{-4} meter, what color light will be found in a first-order bright line which is 0.010 meter distant from the central maximum? (1) red (2) violet (3) green (4) orange

15. Compared to the value of distance x for yellow light, the value of distance x for blue light will be (1) less (2) greater (3) the same

16. As the screen is brought closer to the double slit, the value of distance x will (1) decrease (2) increase (3) remain the same

17. The double slit is replaced by a single slit. As the width of the single slit is decreased, the width of the central maximum will (1) decrease (2) increase (3) remain the same

18. In a vacuum, all electromagnetic waves have the same (1) frequency (2) wavelength (3) speed (4) energy

Base your answers to questions 19 through 23 on the diagram below which represents a side view of part of an interference pattern produced on a screen by light passing through a double slit. The distance (d) between the centers of the slits is 1.0×10^{-4} meter.

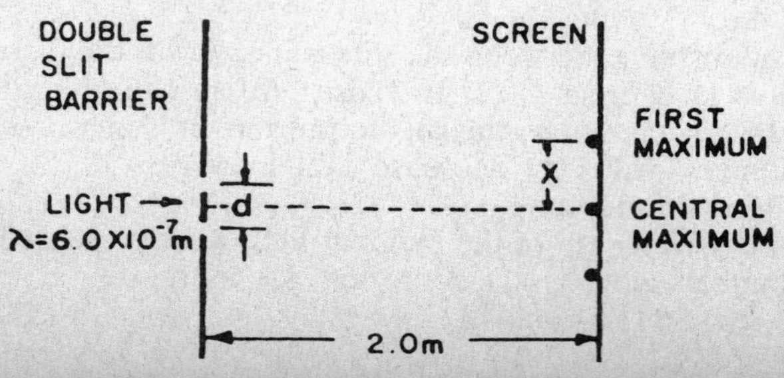

19. What is the distance x from the central maximum to the first maximum in the interference pattern? (1) 1.0×10^{-4} m (2) 6.0×10^{-7} m (3) 1.2×10^{-3} m (4) 1.2×10^{-2} m

20. The difference between the distances from each of the slits to the first maximum is (1) 1 wavelength (2) ¾ wavelength (3) ½ wavelength (4) ¼ wavelength

21. If the distance between the double slits were increased, distance x would (1) decrease (2) increase (3) remain the same

22. If light of a longer wavelength were used, distance x would (1) decrease (2) increase (3) remain the same

23. Compared to the width of the central maximum produced by the two slits, the width of the central maximum produced by just one of the two slits would be (1) less (2) greater (3) the same

24. As the wavelength of a visible lightbeam is increased from violet to red, the speed of the light in a vacuum (1) decreases (2) increases (3) remains the same

Base you answers to questions 25 through 28 on the diagram and the information below.

Red light passing through a double slit is producing a stationary interference pattern on a screen as shown on the diagram.

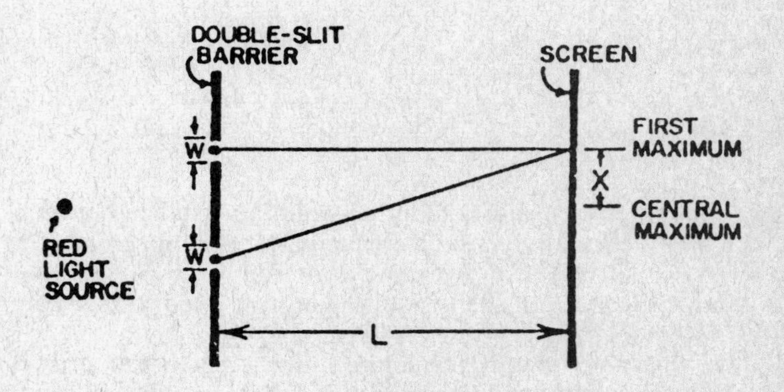

25. The interference pattern is reduced because the light passing through the two slits is (1) dispersed (2) polarized (3) diffracted (4) refracted

26. If the distance L from the slits to the screen were increased, the distance X between the bright lines of the pattern on the screen would (1) decrease (2) increase (3) remain the same

27. If blue light were substituted for the red light source, the distance X between the bright lines of the pattern on the screen would (1) decrease (2) increase (3) remain the same
28. If a single slit with the same width (W) as one of the double slits were used, the width of the central maximum of the interference pattern on the screen would (1) decrease (2) increase (3) remain the same
29. The pattern of bright and dark bands observed when monochromatic light passes through two narrow slits is due to (1) polarization (2) reflection (3) refraction (4) interference

Base your answers to questions 30 through 32 on the diagram below which represents monochromatic light incident upon a double slit, producing an interference pattern on a screen. The distance between the slits is 1.0×10^{-4} meter. The distance from the slits to the screen is 1.0 meter, and the distance between the central maximum and the first-order maximum is 5.5×10^{-3} meter

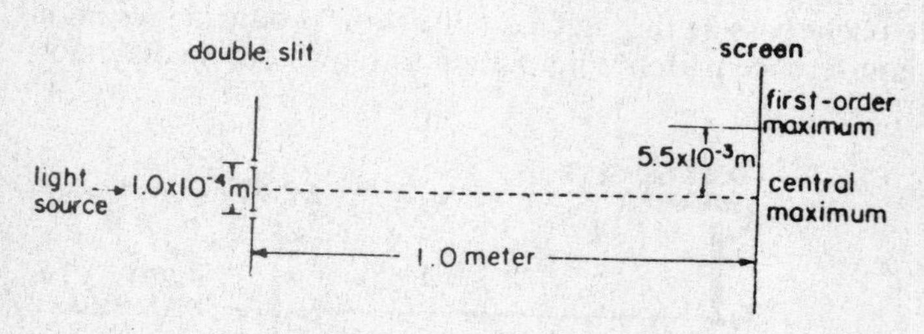

30. For which color of incident light would the distance between the bright bands in the interference pattern be the greatest? (1) blue (2) green (3) orange (4) red
31. What is the color of the incident light? (1) red (2) green (3) violet (4) yellow
32. If the distance between the slits and the screen is reduced and the slit width held constant, the distance between the central maximum and the first-order maximum would be (1) less (2) greater (3) the same
33. Compared to a double-slit diffraction pattern, a single-slit diffraction pattern can be readily identified by the (1) color of the bright bands (2) intensity of the dark bands (3) evenly spaced bright and dark bands (4) relatively wider central bright band

Base your answers to questions 34 and 35 on the diagram below which shows light from a monochromatic source incident on a screen after passing through a double slit.

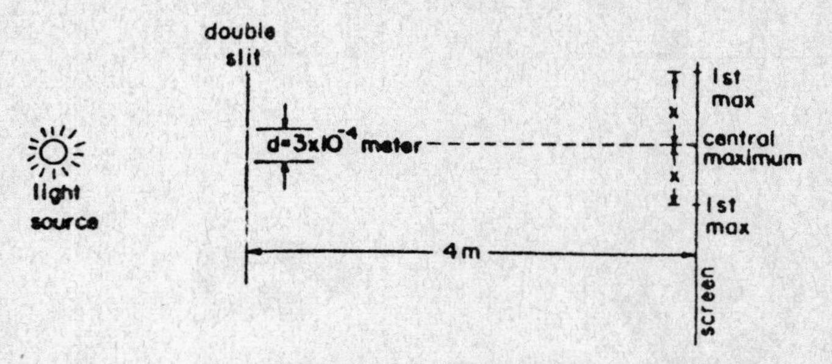

34. What is the wavelength of the light source if the distance between the central maximum and the first maximum is 0.01 meter? (1) 6.7×10^5 m (2) 8.3 m (3) 3.3×10^1 m (4) 7.5×10^{-7} m

35. If the distance d between the slits is decreased, the distance x between the central maximum and the first maximum will (1) decrease (2) increase (3) remain the same

36. The patterns of light produced by the interaction of light with a thin film are a result of (1) diffusion (2) polarization (3) diffraction (4) interference

Base your answers to questions 37 through 41 on the diagram which represents red light incident upon a double-slit barrier, producing an interference pattern on a screen. The wavelength of the red light is 6.6×10^{-7} meter, the distance (d) between the slits is 2.0×10^{-3} meter, and X is 3.3×10^{-4} meter.

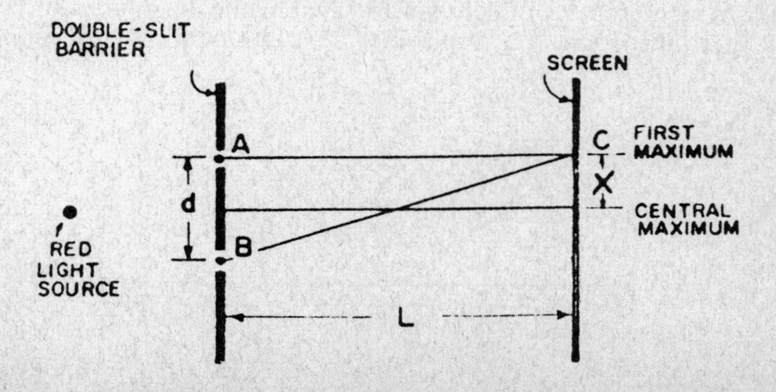

37. Which diagram best describes the pattern observed on the screen?

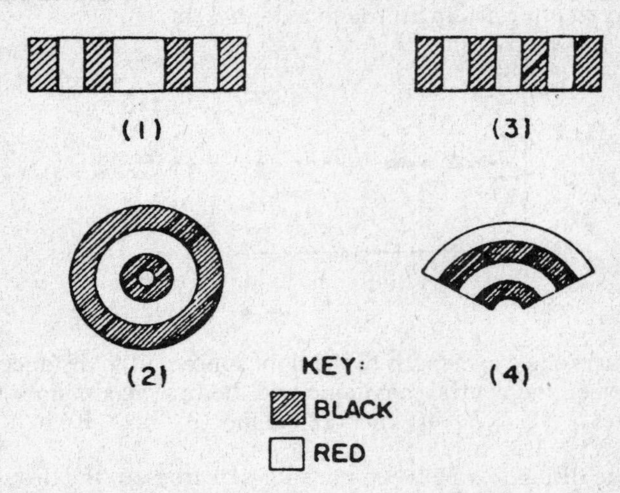

38. The distance (L) from the slits to the screen is (1) 1.0 m (2) 6.6×10^{-7} m (3) 3.0×10^{3} m (4) 4.0×10^{-6} m

39. The path difference between BC and AC is (1) λ (2) 2λ (3) $\lambda/2$ (4) $3/2\ \lambda$

40. The red light source is replaced with a blue light source. Compared to distance X when the red light source is used, what would distance X be when the blue light source is used? (1) less (2) greater (3) the same

41. If the blue light source is accelerated to the left, away from the slits, the Doppler shift would cause the distance X on the screen to (1) decrease (2) increase (3) remain the same

42. Which property of light is illustrated by the diagram below? (1) diffraction (2) dispersion (3) refraction (4) reflection

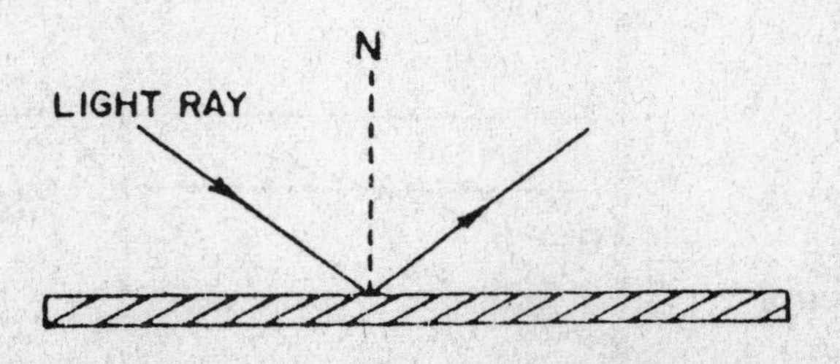

43. Which diagram best represents the phenomenon of diffraction?

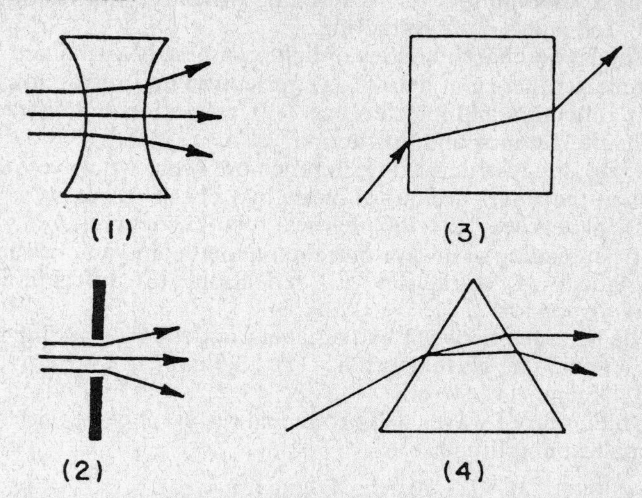

44. Which phenomenon is best explained by the wave theory?
(1) reflection (2) illumination (3) interference
(4) the photoelectric effect

45. Standing waves are produced by two waves traveling in opposite directions in the same medium. These two waves must have (1) the same amplitude and the same frequency
(2) the same amplitude and different frequencies
(3) different amplitudes and the same frequency
(4) different amplitudes and different frequencies

46. A wave spreads into the region behind a barrier. This phenomenon is called (1) diffraction (2) reflection
(3) refraction (4) interference

47. Standing waves are produced by the interference of two waves with the same (1) frequency and amplitude, but opposite directions (2) frequency and direction, but different amplitudes (3) amplitude and direction, but different frequencies (4) frequency, amplitude, and direction

48. The diagram at the right represents waves passing through a small opening in a barrier. This is an example of
(1) reflection (2) refraction
(3) polarization
(4) diffraction

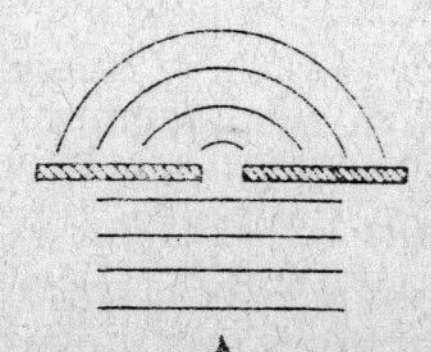

49. Which wave phenomenon could *not* be demonstrated with a single wave pulse? (1) a standing wave (2) diffraction (3) reflection (4) refraction

50. Which two characteristics of light can best be explained by the wave theory of light? (1) reflection and refraction (2) reflection and interference (3) refraction and diffraction (4) interference and diffraction

51. Maximum destructive interference between two waves occurs when the waves are out of phase by (1) 45 degrees (2) 90 degrees (3) 180 degrees (4) 360 degrees

52. The spreading of a wave into the region behind an obstruction is called (1) refraction (2) reflection (3) diffraction (4) dispersion

53. The distance between two adjacent nodes of a standing wave is 45 cm. The wavelength is (1) 23.5 cm (2) 4.5 cm (3) 45 cm (4) 90. cm

54. Which pair of waves will produce a resultant wave with the smallest amplitude

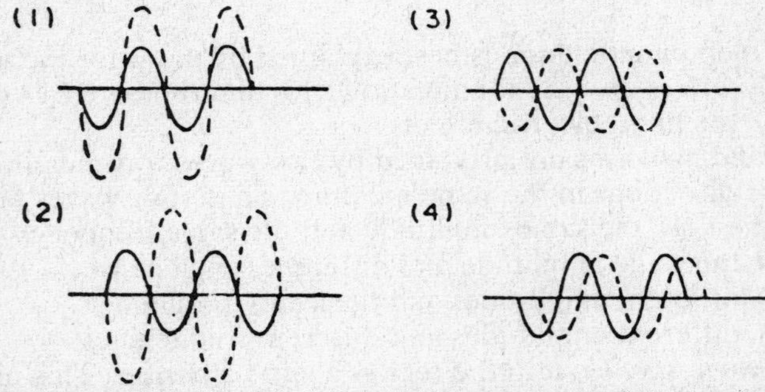

55. The Doppler Effect helps explain observed changes of light (1) reflection (2) refraction (3) frequency (4) speed

In the diagram below, a wave of 1.00-meter amplitude is moving from left to right in an elastic cord.

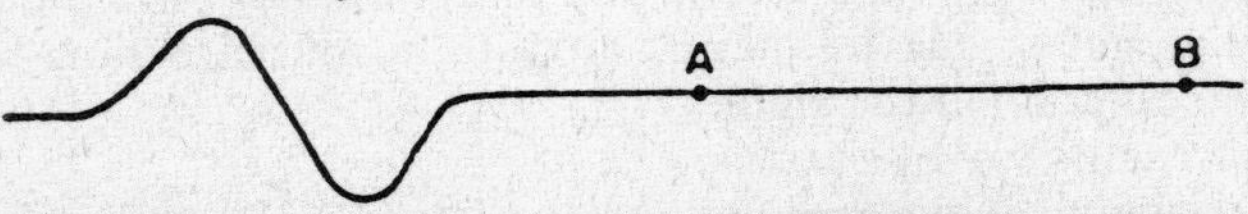

56. The displacement of point A after the wave has passed point B will be (1) 1.00 meter up (2) 1.0 meter down (3) 0.500 meter up or down (4) 0.00 meter

57. As a wave is refracted, which characteristic of the wave will remain unchanged? (1) velocity (2) wavelength (3) frequency (4) direction

58. Which diagram best illustrates the diffraction of waves?

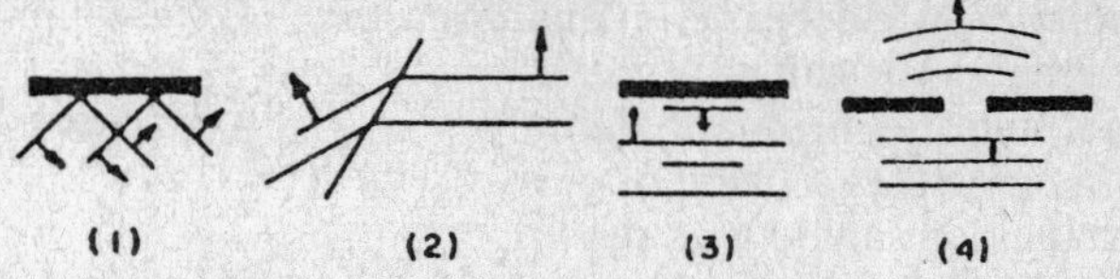

59. Which pair of moving pulses in a rope will produce destructive interference?

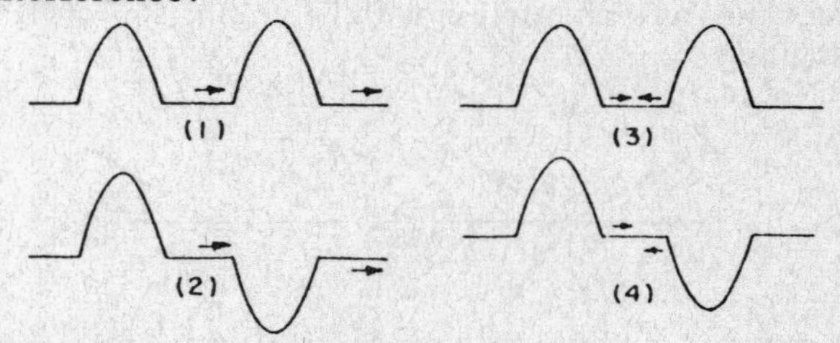

60. The diagram at the right shows two pulses traveling in opposite directions through a transmitting medium. Which diagram best represents the resulting pulse when the pulses combine at point P?

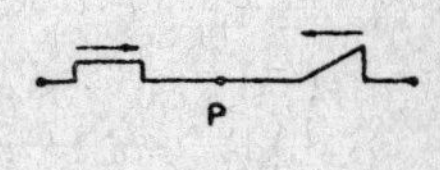

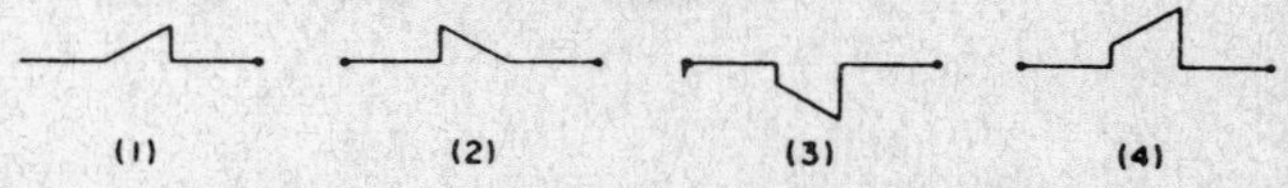

61. A beam of monochromatic red light passes obliquely from air into water. Which characteristic of the light does not change? (1) direction (2) velocity (3) frequency (4) wavelength

62. Which formula represents a constant for light waves of different frequencies in a vacuum? (1) $f\lambda$ (2) f/λ (3) λ/f (4) $f + \lambda$

63. If the frequency of a light wave in a vacuum is increased, its wavelength (1) decreases (2) increases (3) remains the same

64. A ray is reflected from a surface as shown in the diagram below. Which letter represents the angle of incidence? (1) A (2) B (3) C (4) D

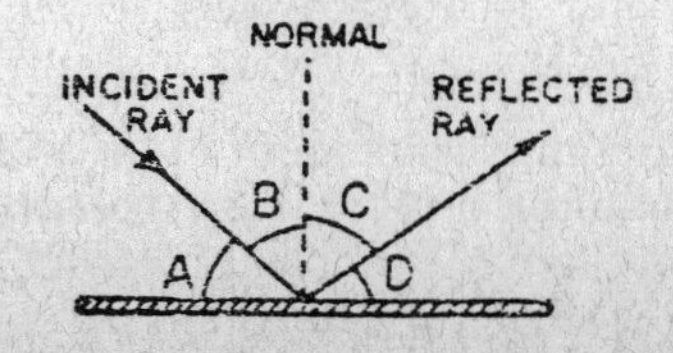

65. When a wave enters a medium of higher refractive index, its velocity (1) decreases (2) increases (3) remains the same

66. A ray of monochromatic light AB in air strikes a piece of glass at an incident angle θ as shown in the diagram at the right. Which diagram below best illustrates the ray's interaction with the glass?

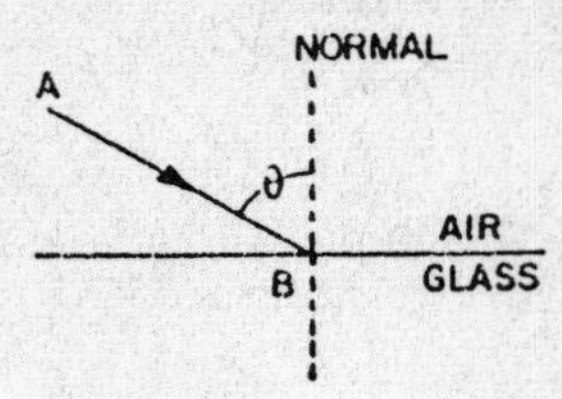

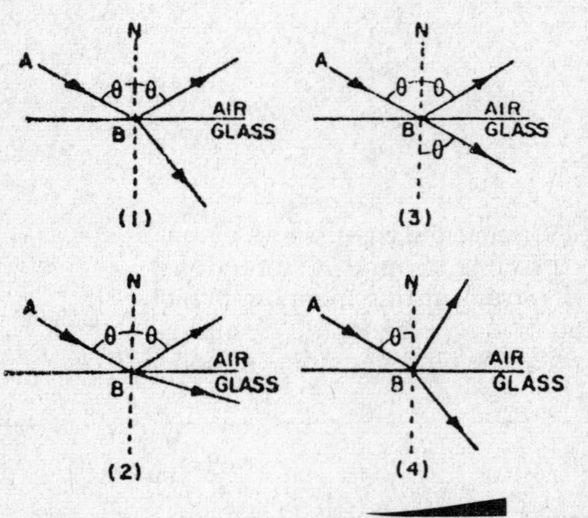

67. Which phenomenon of light is illustrated by the diagram below? (1) refraction (2) dispersion (3) regular reflection (4) diffuse reflection

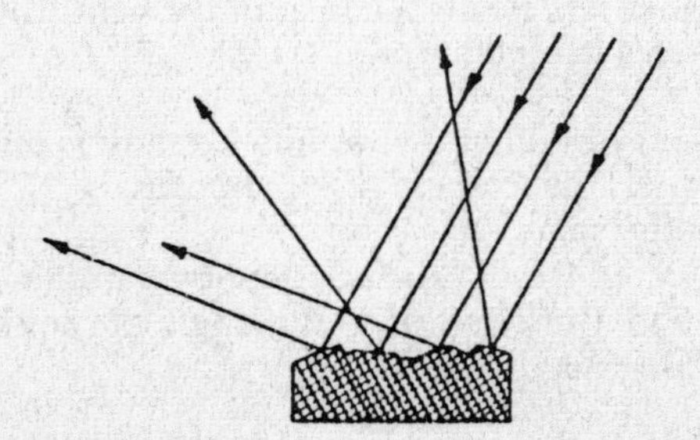

68. If a ray of light in glass is incident upon an air surface at an angle greater than the critical angle, the ray will (1) reflect, only (2) refract, only (3) partly refract and partly reflect (4) partly refract and partly diffract

69. In the diagram below, ray AB is incident on surface XY at point B. If medium 2 has a lower index of refraction than medium 1, through which point will the ray most likely pass? (1) E (2) F (3) C (4) D

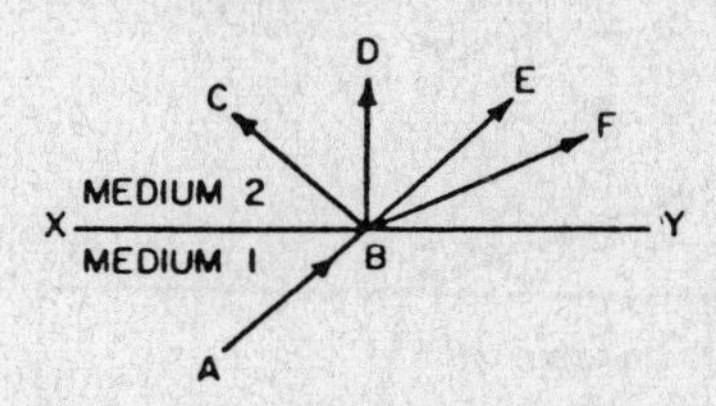

70. The diagram shows a ray of light (R) incident upon a surface at an angle greater than the critical angle. Through which point is the ray most likely to pass? (1) A (2) B (3) C (4) D

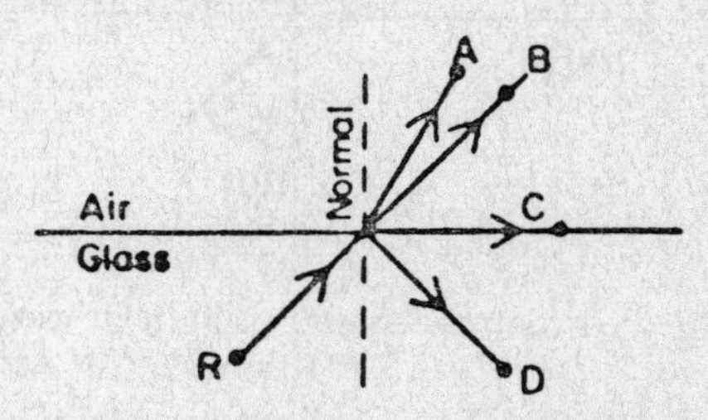

71. As monochromatic light passes from air into glass, its speed (1) decreases (2)increases (3) remains the same

The diagram below represents a light ray traveling from crown glass into air. The position of the light source is changed to vary the angle θ.

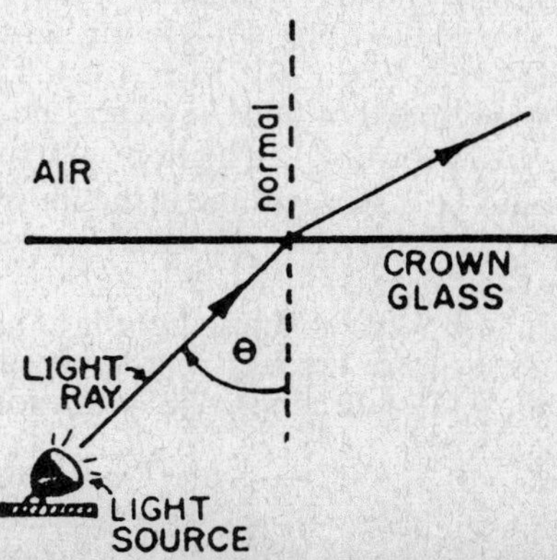

72. As θ approaches the critical angle, the angle of refraction approaches (1) 0° (2) 41° (3) 90° (4) 180°

Base your answers to questions 23 through 26 on the diagram below which represents two media with parallel surfaces in air and a ray of light passing through them.

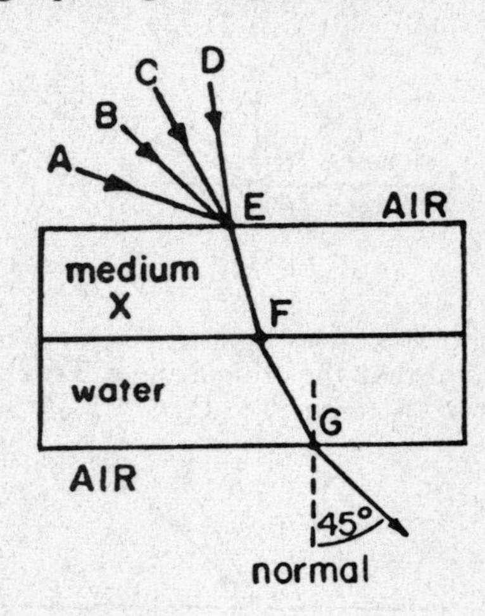

73. What is the approximate speed of light in water?
(1) 4.4×10^7 m/s (2) 2.3×10^8 m/s (3) 3.0×10^8 m/s
(4) 4.0×10^8 m/s

74. If the angle of refraction in air is 45 degrees, what is the sine of the angle of incidence in water? (1) 0.53 (2) 0.71
(3) 0.94 (4) 1.33

75. Which line best represents the incident ray in the air?
(1) AE (2) BE (3) CE (4) DE

76. What is the sine of the critical angle for light passing from water into air? (1) 0.50 (2) 0.75 (3) 0.87 (4) 1.33

77. Compared to the speed of light in water, the speed of light in medium X is (1) lower (2) higher (3) the same

78. Ray EFG would be a straight line if the index of refraction for medium X were (1) less than 1.33 (2) greater than 1.33
(3) equal to 1.33

79. The separating of polychromatic light into its component frequencies as it passes through a prism is called
(1) inteference (2) diffraction (3) diffusion (4) dispersion

80. Which diagram best represents the reflection of an object O by plane mirror M?

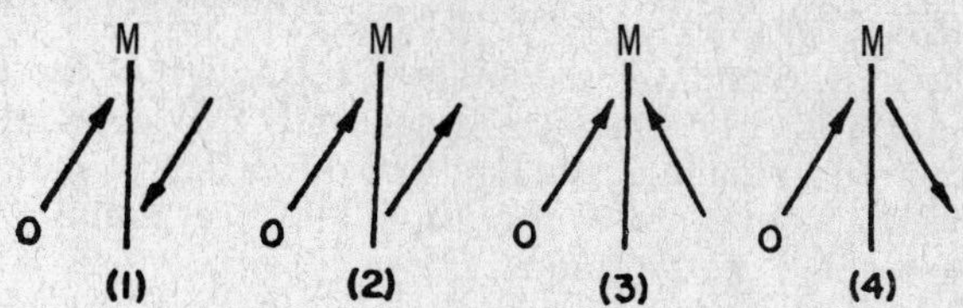

Base your answers to questions 81 through 87 on the accompanying diagram which represents a ray of monochromatic light incident upon the surface of plate X. The values of n in the diagram represent absolute indices of refraction.

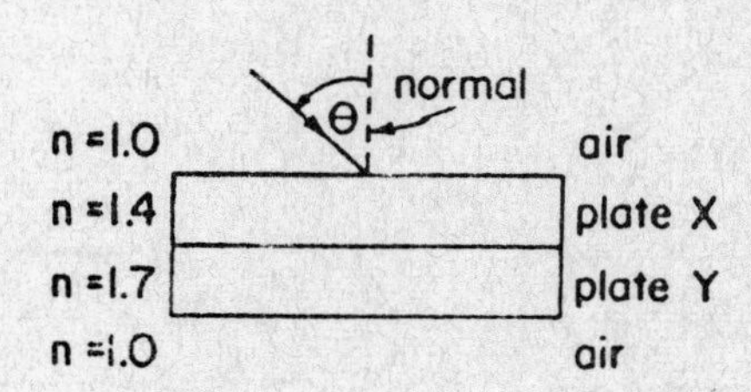

81. What is the relative index of refraction of the light going from plate X to plate Y? (1) 1.0/1.7 (2) 1.0/1.4 (3) 1.7/1.4 (4) 1.4/1.7

82. The speed of the light ray in plate X is approximately (1) 1.8×10^8 m/sec (2) 2.1×10^8 m/sec (3) 2.5×10^8 m/sec (4) 2.9×10^8 m/sec

83. How long will it take a light wave to travel a distance of 100. meters? (1) 3.00×10^{10} s (2) 3.00×10^8 s (3) 3.33×10^{-7} s (4) 3.33×10^7 s

84. Which phenomenon of light is illustrated by the diagram at the right? (1) regular reflection (2) diffuse reflection (3) diffraction (4) refraction

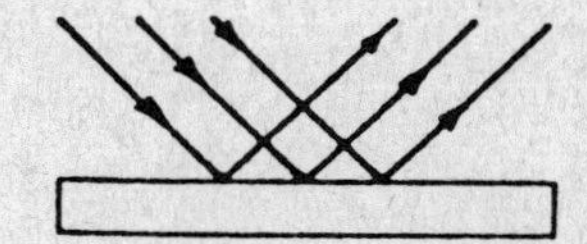

85. What is the sine of the critical-angle for the light ray at the boundary between plate Y and the air? (1) 1.0 (2) 0.83 (3) 0.71 (4) 0.59

86. Compared to angle θ, the angle of refraction of the light ray in plate X is (1) smaller (2) greater (3) the same

87. Compared to angle θ, the angle of refraction of the ray emerging from plate Y into air will be (1) smaller (2) greater (3) the same

88. Which arrow best represents the path that a monochromatic ray of light will travel as it passes through air, benzene, Lucite, and back into air? (Benzene: N = 1.50)

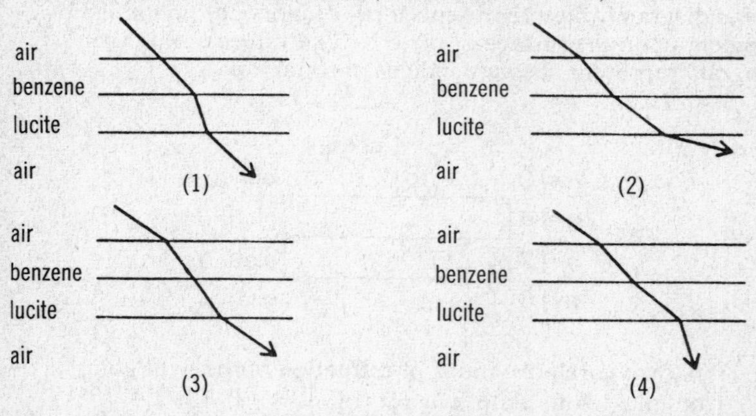

89. As a ray of monochromatic light enters a block of flint glass from air, its speed (1) decreases (2) increases (3) remains the same

90. Compared to the speed of light in a material medium, the speed of light in a vacuum is (1) less (2) greater (3) the same

91. The image formed by a diverging lens is (1) enlarged (2) inverted (3) real (4) virtual

92. The diagram below represents a wave traveling from medium 1 to medium 2. The relative index of refraction may be determined by calculating the ratio of (1) θ_1/θ_2 (2) $\sin\theta_2/\sin\theta_1$ (3) $\sin\theta_1/\sin\theta_2$ (4) n_1/n_2

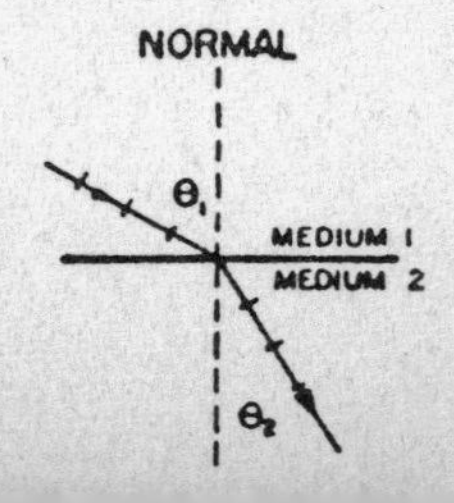

93. The speed of light in Glycerol is the same as the speed of light in (1) diamond (2) flint glass (3) air (4) corn oil

94. In which medium will the wavelength of red light be the *shortest*? (1) flint glass (2) crown glass (3) glycerol (4) diamond

95. Four identically shaped converging lenses are made of crown glass, flint glass, Lucite, and fused quartz. Which lens would have the *shortest* focal length? (1) crown glass (2) flint glass (3) Lucite (4) quartz

96. Which of the diagrams representing light rays reflecting from a surface illustrates diffuse reflection?

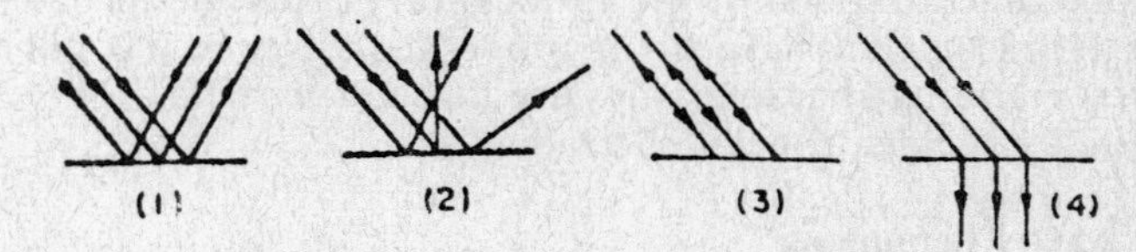

97. In which diagram is angle θ a critical angle?

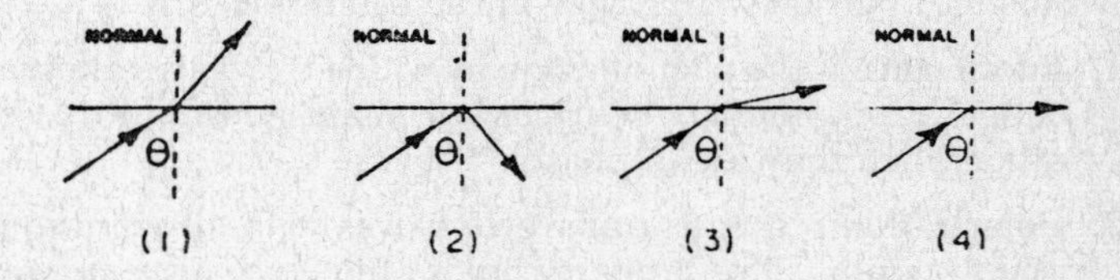

98. The diagram at the right represents light rays approaching a diverging lens parallel to the principal axis. Which diagram below best represents the light rays after they have passed through the diverging lens?

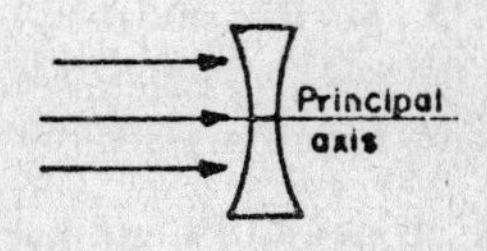

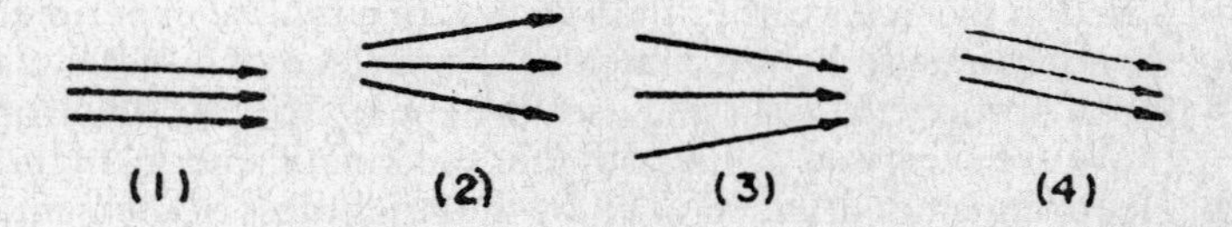

99. In which medium does light travel at the slowest speed? (1) water (2) corn oil (3) alcohol (4) air

100. If the speed of light in a medium is 2.0×10^8 meters per second, the index of refraction for the medium is (1) 1.0 (2) 2.0 (3) 1.5 (4) 0.67

UNIT V. MODERN PHYSICS

I. Dual Nature of Light

Soon after Newton formulated his theory that light consisted of particles, Huygens introduced his wave theory. This was in the 17th century and, although by the 19th century experiments had shown that light seemed to have all the properties of waves, it is still agreed that some of its characteristics are those of particles. This duality is true for all electromagnetic radiations.

A. Wave Phenomena

Interference, polarization and diffraction are properties exhibited by waves, not particles. To review these properties briefly:

1. **Interference** is the phenomenon by which two waves alternately cancel each other by being out of phase and reinforce each other when they are in phase.
2. **Polarization** refers to transverse waves with all vibrations in one direction. This property holds only for transverse waves. Light can be polarized.
3. **Diffraction** refers to the spreading of waves in all directions at its edges. Light tends to spread around corners without casting sharp shadows.

B. Particle Phenomena

1. Photo-electric effect experiments indicate that light is composed of particles rather than waves. In a typical experiment, monochromatic light is aimed at the surface of a metal. The metal absorbs the light and uses the energy to eject electrons.

   We are surprised to find out that the kinetic energy of these electrons is not affected at all by the brightness of the light. If light were composed of waves, a bright light would consist of high amplitude waves that carry a great deal of energy. It is reasonable to expect that this energy would be transferred to any electrons that are released.

   According to particle theory, light is made of small particles rather than waves. The brighter the light, the more particles it contains. Each particle of light striking the metal releases an electron. Thus, a bright light which contains many particles would release more electrons than a dim light. A simple experiment can be performed to show that this is exactly what happens.

2. The maximum kinetic energy of the emitted electrons (called photoelectrons) depends on the *frequency* of the incident *radiation,* and the nature of the emitting surface. The emission of photoelectrons is a random phenomenon.

II. The Quantum Theory

The introduction of the wave theory of light provided a convenient explanation for characteristics of light moving through space, or through substances. Difficulties in the wave model were encountered when trying to explain those phenomena which involved interactions between light and matter. More specifically those questions dealing with emission or absorption of radiation needed some alternative theory. The *quantum theory* was developed for that purpose around 1900.

A. The *quantum* is a discrete packet of energy emitted or absorbed by atomic oscillators. Max Planck announced the quantum theory as his explanation for the distribution of black body radiation.

1. Planck's Constant

Planck began his work in 1900 with one of his assumptions being that the energy per quantum increases as the frequency of the radiation increases. In order to express the relationship as an equation, a constant had to be inserted:

$$E = hf$$

E is the energy per quantum
f is the frequency of the emitted radiation
h is Planck's constant

When E is in joules and F in hertz, then $h = 6.63 \times 10^{-34}$ joule sec.

B. Explanation of Photo Electric Effect

Five years after Planck advanced his theory, in 1905 Einstein proposed an explanation for the photo electric effect based on the assumption that *all* electromagnetic radiation was quantized, even when traveling in space. He suggested that these packets of energy of which light was composed could be calculated from Planck's equation and that each electron is emitted as a result of absorbing the energy of one of the packets.

1. A Photon is a quantum of light energy. Each photon acts on an individual basis. Since its energy varies directly with frequency it follows that it varies inversely with its wavelength.

Velocity of electromagnetic radiation (c) is the product of wavelength (λ) and frequency (f) so c/λ can be substituted for f in the expression $E = hf$, or

$$E = \frac{hc}{\lambda}$$

2. Photo electric equation

The above equations should reveal the kinetic energy of the emitted electron, but as the electron frees itself from the surface of the emitter, it loses some of its energy in doing so. The energy it loses in this task is called the *work function* (w) and the maximum kinetic energy of the released electrons is a linear function of the frequency of photons.

$$KE_{max} = hf - w_o$$

A graph showing kinetic energy of ejected electrons versus frequency of incident radiation would reveal:

a. the slope of Planck's constant

b. the frequency needed to overcome the work function will be the intercept on the frequency axis (f_o)

$$\text{work function} = hf_o$$

3. The threshold frequency

The threshold frequency is the minimum photon frequency needed to free the electron. The energy of the photon required to free the electron is the *work function* of the emitting material.

C. Photon-Particle Collisions

According to Einstein's mass-energy relationship, all investigations should demonstrate that both energy and momentum are conserved in photon-particle collisions. Arthur Compton studied such collisions in 1922, using X-rays.

1. Photon momentum

Einstein predicted that the momentum of a photon would be inversely proportional to its wavelength. This was tested by Compton who directed X-rays at graphite targets and compared the wavelengths before and after the collisions. He found that some of the scattered rays had a longer wavelength than the original ones. In later experiments he found that he could account for the increased wavelength by assuming that the X-ray photons lost some of their energy and momentum to electrons during photon-electron collisions. He confirmed that the energy and momentum gained by the electrons just equals that lost by the X-rays. The momentum (p) of a photon is given by

$$p = \frac{E}{C} = \frac{hf}{C} = \frac{h}{\lambda}$$

A photon moves with the speed of light in any reference frame. It cannot have a rest mass but it does carry *momentum* and it *can* exert a *force*.

D. Matter Waves

Electromagnetic waves have some properties that would be expected to exist only in particles. In 1923-1924 de Broglie proposed that moving particles should also have some of the properties of waves. The wavelength of a particle is given by

$$\lambda = \frac{h}{mv} \text{ or } \lambda = \frac{h}{p}$$

The wave nature of matter under ordinary circumstances is not observed because the wavelengths are very small. Behavior such as diffraction and interference is difficult to demonstrate with particles. However, Davisson and Germer sent a beam of electrons through crystal which acted like a diffraction grating and produced interference patterns. The observed wavelengths were equal to h/p.

III. Models of the Atom

Dalton's model of atomic structure pictured atoms as solid spheres, indivisible and complete in themselves. Thomson's discovery of the electron required a revision so that the atom would include both negative and positive parts. Thomson suggested that atoms were solid, positively charged spheres, in which the negatively charged electrons were embedded.

A. The Rutherford Model

In the 19th century, Ernest Rutherford conducted a series of experiments to investigate atomic structure by bombarding thin sheets of gold foil with alpha particles from a radioactive source. The source was put into a lead chamber with a hole in it. The particles passing through the hole produced a narrow beam which he directed toward the foil. Those particles that passed through were projected on a fluorescent screen producing a flash of light. As much as 99% of the alpha particles passed straight through the gold foil onto the screen. This suggested that most of the foil consisted of empty space. However, the remaining particles were seen to scatter in all directions. The distribution of the particles as a function of their scattering angles was investigated. The evidence led Rutherford to propose a model of the atom in which the positive charge was concentrated in an extremely dense core, with minute, negatively charged electrons widely separated from this nucleus by empty space.

1. The alpha particle

The alpha particle is a nucleus of a helium atom composed of two protons and two neutrons. Many naturally radioactive substances emit alpha decay. The energies associated with the alpha particles of the Rutherford experiments were in the range of 4.5 to 9.0 MeV. These energies require velocities from 1.5×10^7 m/s to 2.0×10^7 m/s.

2. Alpha particle scattering

Rutherford and his group reported that the most surprising result of their alpha particle scattering experiments was that some particles were deflected through extremely large angles. Some, almost 180 degrees. Analysis of these large angles could be explained only if the alpha particles were experiencing coulomb forces of repulsion. This evidence provided the basis for his proposed atomic model with mass and positive charge concentrated in the nucleus.

a. **The scattering angle**

The scattering angle, θ (theta), is the angle through which the alpha particle is deflected after encountering the positively charged nucleus of the target atom.

b. **The impact parameter**

The impact parameter (P) is the distance between the path leading to a head-on collision with the nucleus and the original path actually taken by the alpha particle. As P gets smaller, θ gets larger. Head-on collisions, where P is .0 and θ is 180°, are *not* very likely to occur.

3. Trajectories of alpha particles

The deflected particles follow hyperbolic paths resulting from the repulsion coulomb forces between the alpha particle and the protons in the nuclei of atoms in the foil.

4. Scattering and atomic number

Assuming they all have the same energy, the number of particles scattered beyond a given angle will depend on the charge (and therefore, atomic number) on the target nucleus.

5. Dimensions of atomic nuclei

Evidence from the alpha scattering experiments was very convincing that the radius of nucleus is extremely small compared to that of the atom. The approximate size of the nuclear radius can be found by finding the shortest distance between the center of the particle and the center of the nucleus in a head-on collision. In such a collision, the particle converts all of its kinetic energy to potential energy.

This technique cannot be applied for particles with enough energy to penetrate the nucleus. The radii of atomic nuclei are in the order of 10^{-14} meter. The radius of atoms, including electrons is about 10^{-10} meter which is ten thousand times larger. The atom, therefore, appears to be mainly empty space.

B. The Bohr Model of the Hydrogen Atom

In 1913, Neils Bohr proposed a model of the hydrogen atom in which the electron revolves around the nucleus in a circular orbit. The Bohr model provided a satisfactory model for the gydrogen atom, but not for atoms containing many electrons. For this reason, it has been replaced by a wave mechanical model. It did involve a number of assumptions departing from traditional theory.

1. Bohr's assumptions

a. Even though an orbitting electron accelerates toward the nucleus, Bohr had to assume that it does not emit electromagnetic energy.

b. Permitted orbits for electrons are limited, with each one being associated with a specific energy state. The angular momentum of the electron in each of these orbits must be a whole number multiple of the quantity $h/2\pi$; h being Planck's constant.

$$mvr = n\ h/2\pi$$

(mvr, mass x velocity $\times$ radius = angular momentum)

C. Emission of radiation of specific energies when hydrogen gas is exposed to high voltages can be explained by assuming them to be the result of electrons changing from higher to lower energy states. This change is expressed

$$hf = E_2 - E_1$$

h is Planck's constant
f is the frequency of the emitted radiation
E_2 is the higher energy state
E_1 is the lower energy state

2. Energy levels

a. **Excitation**

The experiments of J. Frank and G. Hertz provided supporting evidence for the assumption of fixed energy states when they bombarded gas molecules with high energy electrons in 1914. They found that the molecules accepted energy from the electrons only in discrete amounts. This process of raising energy of particles is called excitation.

Summarizing the additional findings:

1) Excitation energies are different for different gases.
2) Excited atoms subsequently release the energy as photons
3) The potential energy needed to change the energy of an atom to a higher state is called the *excitation* or resonance potential.

Atoms can be excited by other methods, such as electrical discharge; applying large amounts of heat (thermal excitation); and electromagnetic excitation.

b. **Ground state**

Ground state is the lowest possible energy state of an atom of atomic particle.

c. **Ionization potential**

Ionization potential is the amount of energy necessary for an electron to break away and leave its atom. Since the electron moves from ground state to infinity the atom loses a negative charge and becomes a positively charged ion. The ionization potential for hydrogen is 13.6 electron volts (eV).

3. Standing Waves

An alternate way of explaining discrete energy levels is by applying the ideas of the standing or stationary waves to the atom. The possible orbits in the atom become described as those whose circumference $2\pi r$ is an integral number of wavelengths.

$$2\pi r = n\lambda$$

The waves of course, describe the *probability* of finding the electron at a particular position. Substituting the de Braglie wavelength, h/mv for λ and rearranging the terms:

$$mvr = nh/2\pi$$

This is exactly what Bohr's second assumption demanded with respect to angular momentum.

IV. Atomic Spectra

A. Excitation and emission

Spectral lines are produced as excited electrons return to ground state they produce bright line spectra that are characteristic for each element. these spectra can be used to identify elements. Electrons may be elevated several energy levels during excitation. As each of them returns to its ground state it is possible, that several photons will be emitted with energies equal to difference between different pairs of energy states.

$$E_{photon} = E_i - E_f$$

Where: E photon is the energy of the photon that is created
E_i is the initial energy of an energized electron
E_f is the final energy of the electron

B. Absorption spectra

A gas that is too cool to emit light will absorb light. If white light is sent through the gas and then through a prism the normally continuous spectrum of white light will have dark lines in it. The dark regions represent photons absorbed by the gas. The set of wavelength absorbed constitute an *absorption spectrum*. It has been determined that atoms absorb photons whose energies are the same as the energies of the photons they emit when subjected to excitation. Incident radiation must be exactly the right energy to be absorbed. If it's energy is high enough it may cause the atom to be ionized by loss of electrons. In other cases, incident radiation may simply pass through or be scattered.

C. The Hydrogen spectrum

The hydrogen atoms are excited, emission spectral lines occur in groups known as series. The lines are in a definite order. They are widely spaced at the lower frequencies and crowded close together at higher frequencies

1. Balmer series

The spectral lines in the group known as the Balmer series result from excited electrons at each of the high energy levels returning to the second (n = 2) principal energy level. Some of these lines are visible.

2. Other series

Letting $n_1 = 1$ produces a high frequency series in the ultraviolet range and letting $n_1 = 3$ produces a series in the infrared range.

The highest frequency spectral lines are created when electrons at higher energy levels return to the ground state ($n_1 = 1$). The light produced has very short wavelengths in the ultra-violet range; which are invisible.

Similarly when electrons at higher energy levels return to the n = 3 level, a series of lines are created in the infrared range, which are also invisible.

QUESTIONS

1. How much energy is needed to raise a hydrogen atom from the n = 2 energy level to the n = 4 energy level? (1) 10.2 eV (2) 2.55 eV (3) 1.90 eV (4) 0.65 eV
2. If the momentum of a neutron increases, the wavelength of the neutron will (1) decrease (2) increase (3) remain the same

3. Targets made up of nuclei of successively larger atomic numbers will cause the amount of scattering of a beam of alpha particles to (1) decrease (2) increase (3) remain the same

4. The threshold frequency for a photoemissive surface is 6.4×10^{14} hertz. Which color light, if incident upon the surface, may produce photoelectrons? (1) blue (2) green (3) yellow (4) red

5. In returning to the ground state from $n = 2$ state, the maximum number of photons a hydrogen atom can emit is (1) 1 (2) 2 (3) 3 (4) 4

6. What is the minimum energy needed to ionize a hydrogen atom when it is in the $n = 2$ state? (1) 1.9 eV (2) 3.4 eV (3) 12.2 eV (4) 13.6 eV

7. A hydrogen atom emits a photon with an energy of 1.63×10^{-18} joule as it changes to the ground state. The radiation emitted by the atom would be classified as (1) infrared (2) ultraviolet (3) blue light (4) red light

8. The Rutherford scattering experiments suggested that the mass of the atom is composed mostly of (1) electrons (2) positrons (3) nucleons (4) alpha particles

9. What is the *minimum* amount of energy needed to ionize a hydrogen atom at the $n = 3$ level? (1) 1.51 eV (2) 1.90 eV (3) 2.89 eV (4) 12.1 eV

10. In alpha particle scattering, the nucleus produces an effect on the scattering angles. This is primarily due to the fact that the nucleus (1) has a small total charge (2) has a mass close to that of the alpha particles (3) exerts coulomb forces (4) is widely dispersed throughout the atom

11. Which phenomenon can be explained *only* in terms of the particle model of light? (1) reflection (2) refraction (3) photoelectric effect (4) diffraction

12. Which phenomenon provides evidence that they hydrogen atom has discrete energy levels? (1) emission spectra (2) photoelectric effect (3) alpha particle scattering (4) natural radioactive decay

13. Rutherford's model of the atom showed that most of the volume of the atom is composed of (1) protons (2) electrons (3) neutrons (4) empty space

Base your answers to questions 14 through 16 on the Rutherford scattering experiments.

14. In head-on collisions, alpha particles were scattered at angles of (1) 0° (2) 90° (3) 120° (4) 180°

15. An alpha particle is a (1) positron (2) deuteron (3) gold nucleus (4) helium nucleus

16. Which diagram best represents the path of a neutron which passes close to the nucleus of an atom?

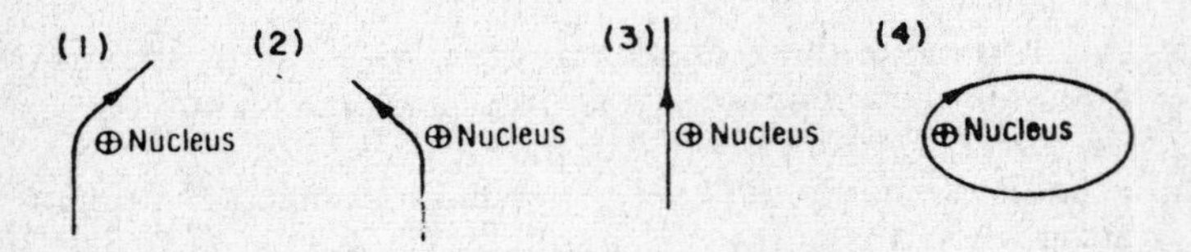

17. Which isotope is used in defining the atomic mass unit? (1) $^{1}_{1}H$ (2) $^{238}_{92}U$ (3) $^{16}_{8}O$ (4) $^{12}_{6}C$

18. The work function of a metal is 4.2 eV. If photons with an energy of 5.0 eV strike the metal, the maximum kinetic energy of the emitted photoelectrons will be (1) 0 eV (2) 0.80 eV (3) 3.8 eV (4) 9.2 eV

19. Which diagram best represents the path of an alpha particle as it passes near the nucleus of an atom?

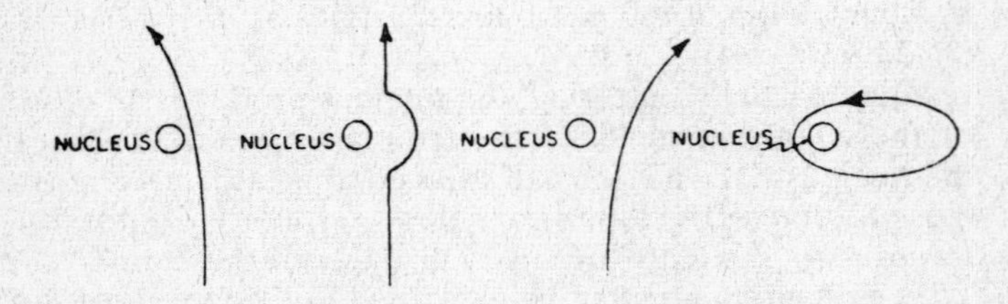

Base your answers to questions 20 through 24 on the diagram below which represents monochromatic light incident upon photoemissive surface *A*. Each photon has 8.0×10^{-19} joule of energy. *B* represents the particle emitted when a photon strikes surface *A*.

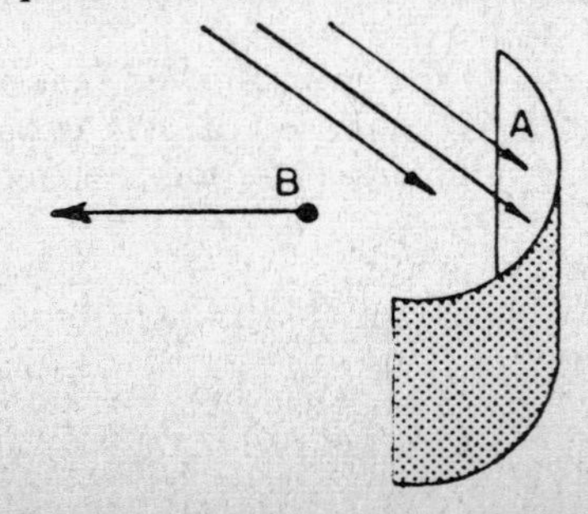

20. What is particle B? (1) an alpha particle (2) an electron (3) a neutron (4) a proton

21. If the work function of metal A is 3.2×10^{-19} joule, the energy of particle B is (1) 3.0×10^{-19} joule (2) 4.8×10^{-19} joule (3) 8.0×10^{-19} joule (4) 11×10^{-19} joule

22. The frequency of the incident light is approximately (1) 1.2×10^{15} cycles/sec (2) 5.3×10^{-15} cycles/sec (3) 3.7×10^{-15} cycle/sec (4) 8.3×10^{-16} cycle/sec

23. If only the frequency of the incident light is increased, the rate of particle emission will (1) decrease (2) increase (3) remain the same

24. If only the intensity of the incident light is increased, the energy of each emitted particle will (1) decrease (2) increase (3) remain the same

25. A hydrogen atom is excited to the $n = 3$ state. In returning to the ground state, the atom could *not* emit a photon with an energy of (1) 1.9 eV (2) 10.2 eV (3) 12.1 eV (4) 12.75 eV

26. What is the minimum amount of energy required to ionize a mercury atom in the $n = \mathrm{d}$ state? (1) 4.95 eV (2) 5.43 eV (3) 10.38 eV (4) 12.75 eV

27. In the Rutherford scattering experiment, metal foils were bombarded with (1) alpha particles (2) beta particles (3) protons (4) neutrons

28. The threshold frequency of a metal surface is in the violet light region. What type of radiation will cause photoelectrons to be emitted from the metal's surface? (1) infrared light (2) red light (3) ultraviolet light (4) radio waves

29. What is the minimum amount of energy needed to ionize a hydrogen atom in the ground state? (1) 0.0 eV (2) 10.2 eV (3) 12.75 eV (4) 13.6 eV

30. The work function of a photoelectric material can be found by determining the minimum frequency of light that will cause electron emission and then (1) adding it to the velocity of light (2) multiplying it by the velocity of light (3) adding it to Planck's constant (4) multiplying it by Planck's constant

Base your answers to questions 31 through 35 on the information below.

Incident photons with an energy of 6.0×10^{-19} joule per photon cause electrons to be ejected from a surface. The work function of the surface is 3.0×10^{-19} joule.

31. What is the frequency of the incident photons? (1) 9.0×10^{29} hertz (2) 9.0×10^{14} hertz (3) 1.1×10^{-15} hertz (4) 4.0×10^{-37} hertz

32. Compared to the original incident photons, photons with a higher energy would have a (1) longer wavelength (2) higher intensity (3) higher frequency (4) greater speed

33. What is the photoelectric threshold frequency of the surface?(1) 1.9×10^{14} hertz (2) 9.0×10^{14} hertz (3) 2.2×10^{15} hertz (4) 4.5×10^{15} hertz

34. What is the maximum kinetic energy possible for an ejected electron? (1) 3.0×10^{-38} joule (2) 3.0×10^{-19} joule (3) 3.0 joules (4) 9.0 joules

35. If the surface were replaced by a surface with a higher threshold frequency, the maximum energy possible for an ejected electron would be (1) less (2) greater (3) the same

Base your answers to questions 36 through 40 on the information below.

A beam of monochromatic light is incident upon a photoemissive surface, resulting in the emission of photoelectrons with a maximum kinetic energy of 0.200 eV. The energy of each incident photon is 2.60 eV.

36. The maximum kinetic energy of the photoelectrons is (1) 3.2×10^{-20}J (2) 2.0×10^{-19}J (3) 3.2×10^{-12}J (4) 4.2×10^{-19}J

37. What is the frequency of the incident light? (1) 1.6×10^{-19} Hz (2) 4.16×10^{-19}Hz (3) 6.27×10^{14}Hz (4) 1.59×10^{15}Hz

38. What is the work function of the photoemissive surface? (1) 0.2 eV (2) 2.4 eV (3) 2.6 eV (4) 2.8 eV

39. The original photoemissive surface is replaced by a new one with a greater work function. Compared to the original threshold frequency, the threshold frequency of the new photoemissive surface will be (1) less (2) greater (3) the same

40. If the intensity of the incident light is increased, the kinetic energy of each photoelectron will (1) decrease (2) increase (3) remain the same

Base your answers to questions 41 through 43 on the Energy Levels for Hydrogen chart in the *Physics Reference Tables.*

41. A hydrogen atom changes from the $n = 1$ energy state to the $n = 3$ energy state. This change could be caused by a single photon which has an energy of (1) 1.5 eV (2) 10.2 eV (3) 12.1 eV (4) 13.6 eV

42. Which photon could be absorbed by a hydrogen atom in the ground state? (1) 11.0-eV photon (2) 10.2-eV photon (3) 3.4-eV photon (4) 0.54-eV photon

43. Which energy level jump would show as a bright line in the *visible* spectrum of hydrogen? (1) 1 to 2 (2) 2 to 3 (3) 3 to 2 (4) 4 to 7

Base your answers to questions 44 through 48 on the graph below which represents the maximum kinetic energy of photoelectrons as a function of incident electromagnetic frequencies for two different photoemissive metals, *A* and *B*.

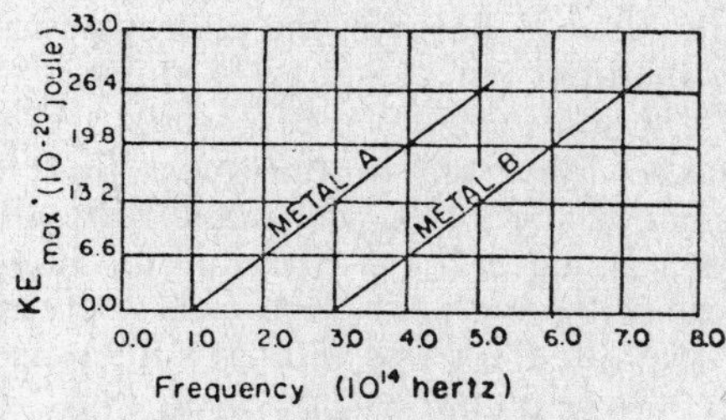

44. The slope of each line is known as (1) Bohr's constant (2) the photoelectric constant (3) Compton's constant (4) Planck's constant

45. The threshold frequency for metal *A* is (1) 1.0×10^{14} Hz (2) 2.0×10^{14} Hz (3) 3.0×10^{14} Hz (4) 0.0 Hz

46. The work function for metal *B* is closest to (1) 0.0 joules (2) 2.0×10^{-19} joule (3) 3.0×10^{-19} joule (4) 1.5×10^{-14} joule

47. Compared to the work function for metal *B*, the work function for metal *A* is (1) less (2) greater (3) the same

48. Monochromatic light with a period of 2.0×10^{-15} second is incident on both of the metals. Compared to the energy of the photoelectrons emitted by metal A, the energy of the photoelectrons emitted by metal *B* is (1) less (2) greater (3) the same

49. An atom changing from an energy state of −0.54 eV to an energy state of −0.85 eV will emit a photon whose energy is (1) 0.31 eV (2) 0.54 eV (3) 0.85 eV (4) 1.39 eV

50. Compared to the amount of energy required to excite an atom, the amount of energy released by the atom when it returns to the ground state is (1) less (2) greater (3) the same

51. In the photoelectric effect, the speed of emitted electrons may be increased by (1) increasing the frequency of the light (2) decreasing the frequency of the light (3) increasing the intensity of illumination (4) decreasing the intensity of illumination

Base your answers to questions 52 through 56 on the information below.

Photons of wavelength 2×10^{-7} meter are incident upon a photoemissive surface whose work function is 6.6×10^{-19} joule.

52. The speed of the incident photons is approximately (1) 2.0×10^{-7} m/sec (2) 6.6×10^{-19} m/sec (3) 1.3×10^{-25} m/sec (4) 3.0×10^{8} m/sec

53. The maximum kinetic energy of the photoelectrons is approximately (1) 0 joules (2) 3.3×10^{-19} joule (3) 6.6×10^{-19} joule (4) 9.9×10^{-19} joule

54. If the frequency of the incident photons is increased, the kinetic energy of the emitted photoelectrons will (1) decrease (2) increase (3) remain the same

55. If the intensity of the incident photons is decreased, the rate of emission of photoelectrons will (1) decrease (2) increase (3) remain the same

56. Photons of the same wavelength are incident upon a photoemissive surface with a lower work function. Compared to the original situation, the maximum kinetic energy of the photoelectrons emitted from the new surface would be (1) less (2) greater (3) the same

57. When alpha particles are scattered by thin metal foils, which observation indicates a very high percentage of space in atoms? (1) thicker foils scatter more. (2) The paths are hyperbolic. (3) Most pass through with little or no deflection. (4) The scattering angle is related to the atomic number.

58. Radiations such as radio, light, and gamma are propagated by the interchange of energy between (1) magnetic fields, only (2) electric fields, only (3) electric and gravitational field (4) electric and magnetic fields

59. All of the following particles are traveling at the same speed. Which has the greatest wavelength? (1) proton (2) alpha particle (3) neutron (4) electron

60. Which is conserved when a photon collides with an electron? (1) velocity (2) momentum, only (3) energy, only (4) momentum and energy

61. Which graph best represents the relationship between the energy of a photon and its wavelength?

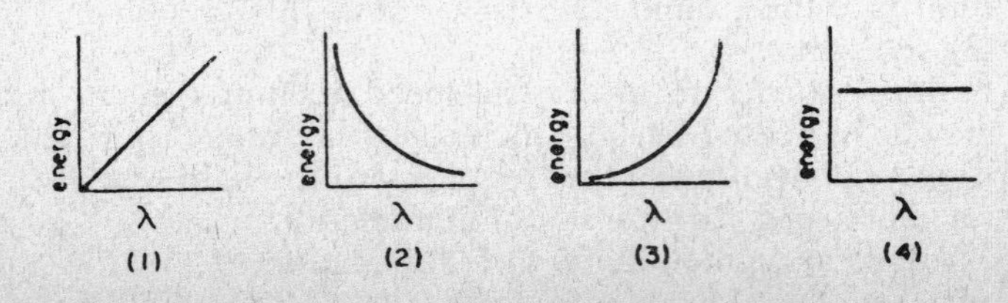

Base your answers to questions 62 through 66 on the information below.

Photons with an energy of 3.0 eV strike a metal surface and eject electrons with a maximum energy of 2.0 eV.

62. The work function of the metal is (1) 1.0 eV (2) 2.0 eV (3) 3.0 eV (4) 5.0 eV

63. If the photons had a higher frequency, what would remain constant? (1) the energy of the photons (2) the speed of the photons (3) the energy of the electrons (4) the speed of the electrons

64. If the photon intensity were decreased, there would be (1) an increase in the energy of the photons (2) a decrease in the energy of the photons (3) an increase in the rate of electron emission (4) a decrease in the rate of electron emission

65. Compared to the frequency of the 3.0-eV photons, the threshold frequency for the metal is (1) lower (2) higher (3) the same

66. If a metal with a greater work function were used and the photon energy remained constant, the maximum energy of the ejected electrons would (1) decrease (2) increase (3) remain the same

67. What is the maximum scattering angle for alpha particles incident upon a thin gold foil? (1) 0° (2) 90° (3) 180° (4) 270°

68. When electrical charges are accelerated in a vacuum, they may generate (1) sound waves (2) water waves (3) light waves (4) torsional waves

69. When incident on a given photoemissive surface, which color of light will produce photoelectrons with the greatest energy? (1) red (2) orange (3) violet (4) green

UNIT VI. MOTION IN A PLANE

This unit is an extension of kinematics in Unit I.

I. Two Dimensional Motion and Trajectories

The path of a batted baseball lies in a two dimensional plane. Its motion at any instant is a vector that may be resolved into two components: a horizontal velocity in the x direction; and a vertical velocity in the y direction. Once the ball leaves the bat, if its initial velocity (speed and direction) is known, its subsequent path may be predicted. The calculations are relatively simple if we assume that there is no wind or air resistance.

A. A Projectile Fired Horizontally

When a projectile is fired horizontally, its initial velocity in the y direction (v_{iy}) is zero. Under the influence of gravity, it will accelerate vertically toward the earth at an acceleration of 9.8 m/s^2.

At the same time that it is accelerating downward it is moving with a constant horizontal velocity (v_x). Thus at any given instant after its release, its resultant velocity can be determined. To do this, first calculate its vertical velocity using the equation $v_y = gt$. Then add vertical and horizontal velocities, as vectors at right angles, to find the resultant velocity.

Throughout the flight of the projectile, its velocity in the vertical direction is unaffected by any horizontal velocity. Thus if a ball is dropped from rest at a height of 20 meters, it will take about 2 seconds to reach the ground. If the ball is thrown horizontally from the same height, its horizontal velocity has no effect and it will also take 2 seconds to reach the ground.

Sample Problem

A baseball is thrown horizontally from a grandstand 20 meters above the ground at a speed of 10 m/s.

a. How long does it remain in flight before hitting the ground?
b. What is its maximum range (horizontal displacement) before it starts to roll along the ground?

Given $v_x = 10$ m/s
$v_{iy} = 0$
$S_y = 20$ m
$g = 9.8$ m/s²

Find t and s_x

Equations (1) $s_y = ½ (gt^2)$ (2) $s_x = v_x t$

Solution a. Solving equation (1) for t:
$t^2 = 2s_y/g$
$= 2\,(20\text{ m})/9.8\text{ m/s}^2$
$= 4\text{ s}^2$ Answer t = 2 seconds

b. $s_x = v_x t$
$= 10\text{ m/s }(2\text{s})$ Answer $s_x = 20$ meters

B. A Projectile Fired at an Angle

When a projectile is fired upward at an angle, the vertical component of its initial velocity is equal to the initial velocity multiplied by the sine of the angle. The horizontal component is equal to the initial velocity multiplied by the cosine of the angle. Treat projectile vector problems as separate linear problems remembering that the time of flight must be the same in each case.

If he projectile is launched and then lands at the same height, it may have covered a considerable vertical distance but its vertical displacement is zero.

Because the earth's gravity has no effect on a horizontal motion, it will have no effect on the horizontal velocity. The horizontal displacement, called the "range", can be found by multiplying the horizontal velocity by the time in flight.

Sample Problem

An athlete doing a running jump leaves the ground at an angle of 25 degrees and a speed of 10 m/s. How far did he jump?

Given $v = 10 \text{ m/s}$
$\theta = 25°$
$g = 9.8 \text{ m/s}^2$

Find s_x

Equation $v_y = v \sin\theta - gt$
$s_x = v \cos\theta\, t$

Solution $v_y = v \sin\theta - gt$

At the top of the jump, half of the total time (t) has elapsed. At this instant the elapsed time is 0.5t and v_y is zero.

$$v_y = v \sin\theta - gt$$
$$\theta = 10 \text{ m/s} \sin 25° - (9.8 \text{ m/s}^2)\, 0.5t$$
$$t = 0.84 \text{ s}$$
$$s_x = v \cos\theta\, t$$
$$= 10 \text{ m/s} \cos 25°\; 0.84 \text{ s}$$

Answer $s_x = 7.6 \text{ m}$

QUESTIONS

1. Projectile A is fired horizontally at an initial velocity of 100 m/s over a lake. Projectile B is fired horizontally from the same height but at an initial speed of 200 m/s. The elapsed time required for B to hit the water (1) is twice that of A, (2) half that of A (3) one quarter that of A (4) the same as that of A.

2. A baseball leaves the bat at an angle of 30° with the ground and a speed of 50 m/s. The maximum height the ball will reach is closest to (1) 30 m (2) 60 m (3) 90 m (4) 120 m

3. Projectiles are fired from a ship at the same speed but at different angles with the water. As the angle increases from 45 to 85 degrees, the range of the projectile (1) increases (2) decreases (3) increases and then decreases (4) decreases then increases

4. A rock is thrown from the edge of a cliff with an initial speed of 20 m/s and an angle of 30 degrees below the horizontal. If the rock strikes the ground below in 2 seconds, the height of the cliff is closest to: (1) 10 m (2) 20 m (3) 30 m (4) 40 m

5. A golf ball is hit for a long drive and reaches its maximum height 2 seconds later. If the initial horizontal velocity of the ball was 40 m/s, the maximum range of its flight was (1) 20 m (2) 40 m (3) 80 m (4) 160 m

II. Uniform Circular Motion

According to the first law of motion, an object remains in uniform motion unless acted upon by an unbalanced force. To change straight line motion into circular motion, an outside force must constantly pull the object toward the center of its circular path. In this case, the applied force changes (direction) of the velocity, not the speed. The result is called *uniform circular motion.* It is defined as the motion of an object at constant speed along a circular path.

1. **Centripetal acceleration** (a_c) is an acceleration directed toward the center of curvature. Its magnitude is directly proportional to the square of the speed and inversely proportional to the radius of the path

$$a_c = \frac{v^2}{r}$$

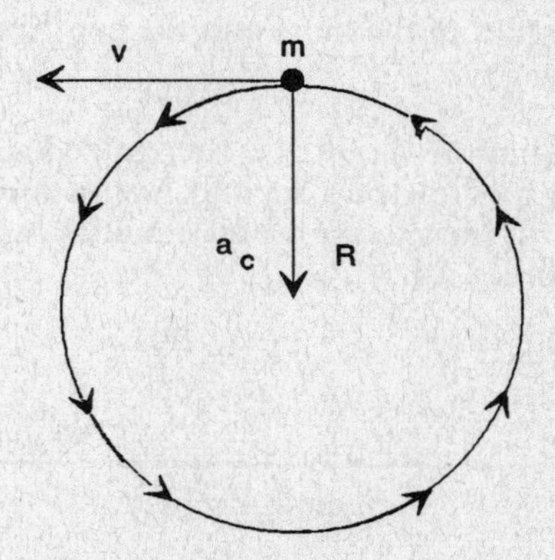

As mass m moves in a circle of radius R, its instantaneous velocity is tangent to the circle and its centripetal acceleration is directed toward the center.

Sample Problem

Calculate the speed an earth satellite must have to enter a circular orbit at an altitude of 200 km. Use 9.2 m/sec² for the acceleration due to gravity at that altitude and 6400 km for the radius of the earth.

Given $r = 200\text{ km} + 6400\text{ km} = \underline{6600\text{ km}}$
$a = 9.2\text{ m/s}^2$

Find v

Equation $a = \dfrac{v^2}{r}$ and $v = \sqrt{ar}$

Solution $9.2\dfrac{\text{m}}{\text{s}^2} \times \dfrac{10^{-3}\text{ km}}{1\text{ m}} = 9.2 \times 10^{-3}\text{ km/s}^2$

$v = \sqrt{(9.2 \times 10^{-3}\text{ km/s}^2)(6.6 \times 10^3\text{ km/s}^2)}$

$v = 7.8\text{ km/s}$

2. **Centripetal force** F_c is a vector quantity directed toward the center of curvature. It is the cause of centripetal acceleration. The magnitude of centripetal force is directly proportional to the product of the mass and the centripetal acceleration,

$$F_c = ma_c$$

This equation should be recognized as the relationship described by Newton's second law of motion. Substituting (v^2/r) for (a) provides a useful equation with respect to centripetal Force (F_c).

$$F_c = \frac{mv^2}{r}$$

Sample Problem

A 2 kg mass is traveling in a horizontal circle of 0.5 m radius at a speed of 2m/sec. Calculate the centripetal force.

Given $m = 2\text{ kg}$
$v = 2\text{ m/sec}$
$r = 0.5\text{ m}$

Find Fc

Equation $F_c = \dfrac{mv^2}{r}$

Solution $F_c = \dfrac{(2\text{ kg})(2\text{ m/s})^2}{0.5\text{ m}}$

$= \underline{16\text{ kg m/s}^2 = 16\text{ N}}$

QUESTIONS

Base your answers to questions 1 through 5 on the diagram below which represents a ball of mass M attached to a string. The ball moves at a constant speed around a flat horizontal circle of radius R.

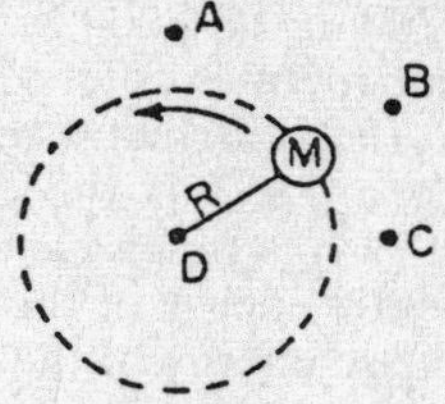

1. When the ball is in the position shown, the direction of the centripetal force is toward point (1) A (2) B (3) C (4) D

2. The centripetal acceleration of the ball is (1) zero (2) constant in direction, but changing in magnitude (3) constant in magnitude, but changing in direction (4) changing in both magnitude and direction

3. If the velocity of the ball is doubled, the centripetal acceleration (1) is halved (2) is doubled (3) remains the same (4) is quadrupled

4. If the string is shortened while the speed of the ball remains the same, the centripetal acceleration will (1) decrease (2) increase (3) remain the same

5. If the mass of the ball is decreased, the centripetal force required to keep it moving in the same circle at the same speed (1) decreases (2) increases (3) remains the same

6. As the time taken for a car to make one lap around a circular track decreases, the centripetal acceleration of the car (1) decreases (2) increases (3) remains the same

Base your answers to questions 7 through 11 on the diagram below which represents a flat racetrack as viewed from above, with the radii of its two curves indicated. A car with a mass of 1,000 kilograms moves counterclockwise around the track at a constant speed of 20 meters per second.

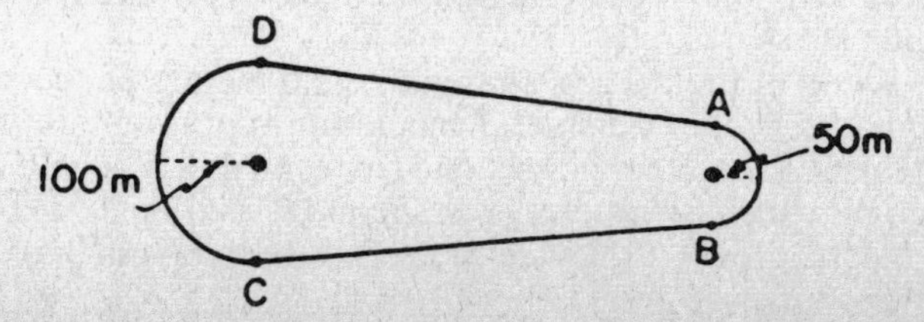

7. The net force acting on the car while it is moving from A to D is (1) 0 N (2) 400 N (3) 8,000 N (4) 20,000 N

8. The net force acting on the car while it is moving from D to C is (1) 0 N (2) 200 N (3) 4,000 N (4) 20,000 N

9. If the car moved from C to B in 20 seconds, the distance CB is (1) 100 m (2) 200 m (3) 300 M (4) 400 m

10. Compared to the centripetal acceleration of the car while moving from B to A, the centripetal acceleration of the car while moving from D to C is (1) the same (2) twice as great (3) one-half as great (4) 4 times greater

11. Compared to the kinetic energy of the car while moving from A to D, the kinetic energy of the car while moving from D to C is (1) less (2) greater (3) the same

Base your answers to questions 12 through 15 on the diagram below which represents a 2.0-kilogram mass moving in a circular path on the end of a string 0.50 meter long. The mass moves in a horizontal plane at a constant speed of 4.0 meters per second.

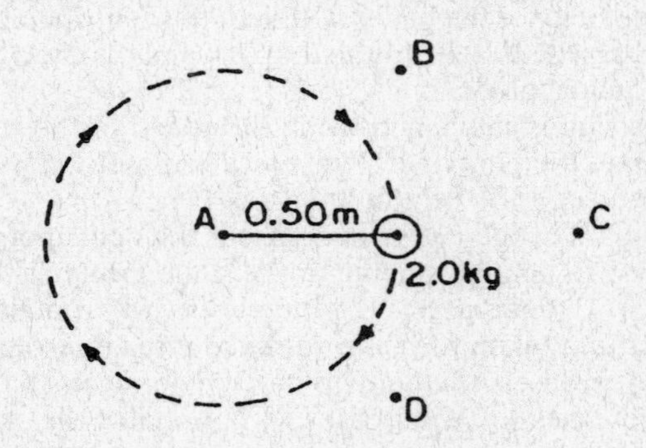

12. The force exerted on the mass by the string is (1) 8 N (2) 16 N (3) 32 N (4) 64 N

13. In the position shown in the diagram, the momentum of the mass is directed toward point (1) A (2) B (3) C (4) D

14. The centripetal force acting on the mass is directed toward point (1) A (2) B (3) C (4) D

15. The speed of the mass is changed to 2.0 meters per second. Compared to the centripetal acceleration of the mass when moving at 4.0 meters per second, its centripetal acceleration when moving at 2.0 meters per second would be (1) half as great (2) twice as great (3) one-fourth as great (4) four times as great

III. Kepler's Laws

Johannes Kepler formulated three laws that provide simple explanations for planetary motions. Previously, astronomers had described each planet's orbit as complicated combinations of circles.

A. Kepler's First Law.

The shape of each planet's orbit is an ellipse. The sun is located at a focus of this ellipse.

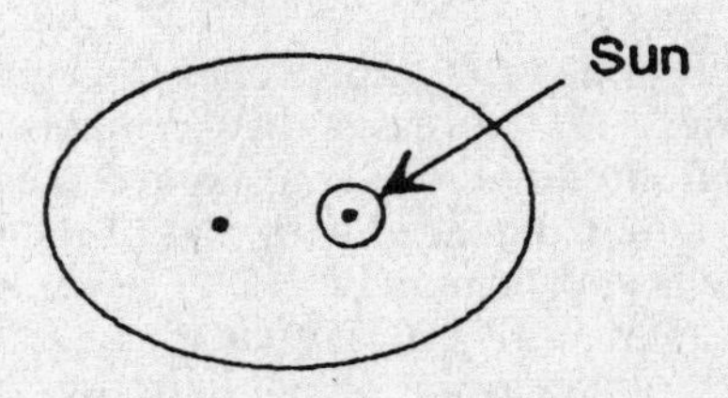

B. Kepler's Second Law

If a line is drawn between the centers of the sun and a planet, the line will sweep the same area in space for any given period of time. For example, assume that AB and CD are distances covered by a planet in a month's time. Sector areas AOB and COB will be equal. Planets always move faster when closer to the sun.

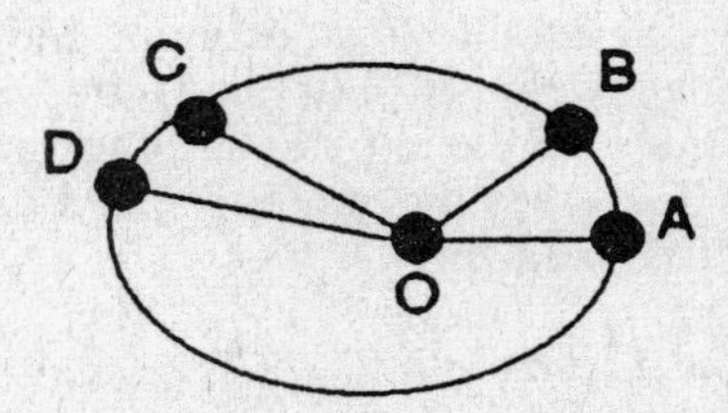

C. Kepler's Third Law

The time need for a planet to orbit the sun (its period) depends on the mean distance between the planet and the sun. This is summarized by the relationship:

$$\frac{R^3}{T^2} = K$$

Where: R is the mean distance between the planet and the sun
T is the period of the planet
K is a constant for any planet that orbits the sun

The value for constant K depends on the mass of the body at the focus of the elliptical orbit. The less the mass, the smaller will be the constant.

For objects orbiting the sun, $K = 3.35 \times 10^{18}\ m^3/s^2$
For objects orbiting the earth, $K = 1.63 \times 10^{12}\ m^3/s^2$

QUESTIONS

1. As the earth in its orbit approaches the sun, its speed (1) decreases (2) increases (3) remains the same

2. An artificial satellite orbits the sun at a mean distance equal to two earth radii. Compared with that of the earth, the period of the satellite is approximately (1) ¼ as great (2) 2 X as great (3) 3 X as great (4) 4 X as great

3. As the mass of a celestial object increases, the Kepler constant for any planet which revolves about the object (1) decreases (2) increases (3) remains the same

IV. Satellite Motion

The moon and the earth revolve around a common center of mass located about one third of the way below the earth's surface toward its center. The moon, being the smaller of the two, is called the earth's satellite.

To send an artificial satellite in an orbit around the earth it must first achieve an escape velocity of about 11 km/s. Once in an orbit it would remain there forever if not for the drag of the earth's atmosphere. This decreases the satellite speed causing it to eventually spiral back to the earth.

A. Geosynchronous Orbits

A satellite at a distance of approximately six radii from the earth's center will orbit the earth every 24 hours. Because the satellite revolves in step with the earth's rotation, it is said to be in a **geosynchronous orbit**. Both move together so the satellite always remains directly above the same spot on the earth's equator.

B. Artificial Satellites

All satellites orbiting the earth have the same Kepler constant as that of our moon. Thus, if either the period or mean orbital radius of an artificial satellite is known, the other parameter can be calculated using the relationship, $R^3/T^2 = K$.

Sample Problem

The moon's orbit has a radius of 3.84×10^8m and a period of 2.35×10^6s. What is the radius of a geosynchronous satellite orbit?

Given $R_m = 3.84 \times 10^8$m
$T_m = 2.35 \times 10^6$s
$T_s = 24$ h $= 8.64 \times 10^4$s

Find R_s

Equation $R_m^3/T_m^2 = R_s^3/T_s^2$

Solution $R_s^3 =$
$(3.84 \times 10^8\text{m})^3 \times (8.64 \times 10^4\text{s})^2/(2.35 \times 10^6\text{s})^2$

Answer $R_s = 4.23 \times 10^7$m

UNIT VII. INTERNAL ENERGY

I. Temperature

The temperature of a substance depends on the **average** kinetic engergy of its particles. It is the property of matter responsible for the direction of the exchange of internal energy between two objects. The object at lower temperature will gain internal energy. The total internal energy of an object includes, but is not limited to, its temperature. It also includes the object's mass, phase and other properties as well.

Absolute temperature is directly proportional to the average kinetic energy of the particles in the substance being studied. Its major application is in the study of the behavior of gases. The definition, strictly interpreted, could only apply to an "ideal gas"; one consisting of particles of negligible size whose only forces on each other result from perfectly elastic collisions.

Absolute zero would be the temperature of the "ideal gas" when all motion is at a minimum. Internal energy can not be transferred from an object at absolute zero. Absolute zero can be approached closely, but can never be reached.

MKS Temperature Scales

Temperature measurements are commonly based on easily reproduced reference points such as the freezing point and the boiling point of water at a pressure of one atmosphere.

1. **Celsius.** On the Celsius scale the freezing point of water is a value of 0 degrees and the boiling point is at 100 degrees.

2. **Kelvin.** The Kelvin scale is based on absolute temperature. Extrapolation of Celsius temperatures of gases as their volume is reduced predicts $-273°C$ as the temperature at which volume is a minimum for an ideal gas. The freezing point of water, 0°C is equivalent to 273K. The Kelvin temperature, therefore, is equal to the Celsius temperature plus 273. The Kelvin degree is the same size as the Celsius, so the boiling point of water at one atmosphere is 100°C or 373K.

II. Internal Energy and Heat

The internal energy of an object is equal to the sum of the kinetic energy and the potential energy of the object's particles. These energies are considered independently of the kinetic and potential energy of the object as a whole.

Kinetic energy is translated as temperature, while potential energy is represented by the phase and the energy of the atomic particles. Therefore, an increase in the internal energy usually increases the average kinetic energy of its molecules, resulting in a rise in temperature. It can also increase the potential energy of its molecules, resulting in a phase change or higher energy levels of its atoms. Internal energy is transferred from one object to another in the form of heat.

According to the law of conservation of energy, whenever substances exchange internal energy, the total energy of the system remains constant. An application of this principle occurs when we chill a liquid by putting ice into it. The internal energy gained by ice is equal to the internal energy lost by the liquid.

The joule, J, is the SI unit of heat. Since mechanical energy and heat energy are both the results of work, they are now measured in the same units.

Until recently, the calorie and the kilocalorie were used as units of heat energy. A kilocalorie is equal to 4185 joules.

A. Specific heat

The specific heat of a substance is the quantity of heat required to raise the temperature of a unit mass of the substance 1 Celsius degree. However, most reference books list specific heats in terms of calories per gram celsius degree (cal/gC°) or kilocalories per kilogram celsius degree (kcal/kgC°): If necessary, convert this to kJ/kg°C using the conversion factor given above. Using this information, the definition of a kcal, and Q as the quantity of heat, provides bases for the equation

$$Q = mc\Delta t$$

where: Q is heat in kilojoules
m is mass in kilograms
c is specific heat in kJ/kg°C
t is temperature in C°

Sample Problem

Given the specific heat of alcohol is 2.43kJ/kg° C, how many joules are needed to raise the temperature of 3.0kg, 10C°?

Given $m = 3.0kg$
$c = 2.43kJ/kg°C$
$\Delta t = 10°C$

Find Q

Equation $Q = mc\Delta t$

Solution $Q = (3.0kg)\ (2.43kJ/kg°C)\ (10°C)$

Answer = 73kJ

Problems involving heat gained and heat lost when two substances of differing initial temperatures and different specific heats come to a final uniform temperature rely on the Conservation of Energy Law for their solution (heat gained = heat lost).

Sample Problem

A heated aluminum block heaving a mass of 0.08 kg is added to 0.50 kg water at 20° C. The temperature of the system stops changing at 24° C. What was the initial temperature of the aluminum? The specific heat of water is 4.19kJ/kg° C and that of aluminum is 0.90kJ/kg° C

Heat gained (water)

Given $m = 0.50kg$
$c = 4.10\ kJ/kg°C$
$\Delta t = 4.0°C$

Find Q

Equation $Q = mc\Delta t$

Solution $Q = (0.50kg)\ (4.19kJ/kg°C)\ (4°C)$

Answer = 8.4kJ

Heat lost (aluminum)

Given $Q = 8.4kJ$
$m = .08kg$
$c = 0.90\ kJ/kg°C$

Find Δ

Equation $Q = mc\Delta t$

Solution $8.4kJ = (0.8kg)\ (0.90kJ/kg°C)\ (\Delta t)$

$$\Delta t = \frac{8.4\ kJ}{(.08kg)\ (0.90\ kJ/kg°C)}$$

$$\underline{Answer = \Delta t = 117°C}$$

Initial temperature $= 20°C + 117°C = 137°C$

B. Exchange of Internal Energy

1. Change of Phase

Whenever a substance undergoes a change in phase internal energy is changed. However, while the change is taking place, the temperature is found to remain constant. All of the transferred energy is involved in the change in internal potential energy required for the phase change. None of it is used to change the average kinetic energy of the particles.

a. **Heat of fusion** is the number of kilocalories required to change one kilogram of a substance from liquid to solid at its freezing point; or solid to liqued at its melting point. The same amount of energy is involved for either freezing or melting. The melting point and the freezing point occur at the same temperature for a given material. Heat of fusion is represented as H_f and it may be expressed in units of kJ/kg. The heat of fusion of ice is 334 kJ/kg. The equation used in problem-solving is $Q = mH_f$.

Sample Problem

How much heat is required to melt .24 kg ice at 0° C?

Given $m = .24kg$
$H_f = 334kJ/kg$

Find Q

Equation $Q = mH_f$

Solution $Q = (.24kg)\ (334\ kJ/kg)$

$$\underline{Answer = 80\ kJ}$$

b. **Heat of vaporization** is the amount of heat **required** to change one gram of a substance from its liquid phase at its boiling point to its gas phase (vapor) with no change in its temperature. The same amount of heat is **liberated** when the change of phase is reversed (condensation). Heat of vaporization is represented as H_v and may be expressed in units of kJ/kg. The heat of vaporization of water is taken to be 2260 kJ/kg. The equation used in problem solving is $Q = mH_v$.

Sample Problem

How much heat must be supplied to vaporize .240kg of water at 100° C at 1 atmosphere?

Given $m = .240\text{kg}$
$H_v = 2260\text{kJ/kg}$

Find Q

Equation $Q = mH_v$

Solution $Q = (.240\text{kg})\ (2260\ \text{kJ/kg})$

Answer = $Q = 542$ kJ

2. Factors Affecting the Boiling and Freezing Points of Water

A dissolved salt lowers the freezing point of water. Spreading salt on an icy road can lower the freezing point below that of the surrounding air. As a result, the ice will melt without becoming any warmer.

A dissolved salt raises the boiling point of water. Therefore, salt water may take longer to boil but once it starts it cooks faster.

Increasing the pressure lowers the melting point of ice. The pressure of an ice skate blade melts the ice below and the skater glides over a thin film of water. Once the pressure has been removed, regelation occurs as the water refreezes.

Increased pressure raises the boiling point of water. Food inside a pressure cooker is subjected to a high temperature that cooks food quickly. If the pressure inside a flask of lukewarm water is reduced, the water can boil at ordinary room temperatures.

III. Kinetic Theory of Gases

The kinetic theory of gases begins with three assumptions:

1. Gases consist of large numbers of molecules in random motion.

2. The molecules are too far apart to attract one another.
3. Collisions of the molecules with each other and with the walls of the container are elastic (kinetic energy is conserved).

A. Pressure

According to the kinetic theory, the **pressure** on the walls of the container is produced by the molecules of gas as they collide and rebound. Therefore, pressure is proportional to the average momentum of the gas molecules: $P \propto \overline{mv}$ (reminder—the bar over the mv means "average value"). Since molecules at higher speeds collide more frequently, the pressure is also proportional to average molecular speed: $P \propto v$

Combining the two proportions: $P \propto \overline{mv^2}$

Since kinetic energy is defined as $\frac{1}{2}mv^2$, it can be written that pressure is proportional to the average kinetic energy of the molecules (the Kelvin temperature).

The pascal is the SI unit of pressure.

$$
\begin{aligned}
1\ \text{Pa} &= 1\ \text{N/m}^2 \\
101\ \text{kPa} &= 1\ \text{atmosphere} \\
&= 760\ \text{torr} \\
&= 760\ \text{mm of mercury}
\end{aligned}
$$

B. Gas Laws

The kinetic theory has provided explanation for the observed regular behavior of gases which have been expressed as the gas laws.

1. **Boyle's Law** states that at constant temperature the product of pressure and volume is constant. $PV = k$ means that volume is inversely proportional to pressure as long as the temperature remains the same, or:

$$\Delta V = \frac{1}{\Delta P}$$

2. **Charles' Law** states that at constant pressure, the volume of a gas is directly proportional to its absolute (Kelvin) temperature.

$$\Delta V \propto \Delta T_k$$

IV. Laws of Thermodynamics

Thermodynamics is the area of physics that relates temperature, heat energy, and mechanical energy. There are three basic laws of thermodynamics:

A. First Law

The first law of thermodynamics restates the law of conservation of energy. No energy is lost when heat is added to a system. All of the heat energy reappears as an increase in the internal energy of the system and external work done by the system.

Conversely, when a system such as a confined hot gas is allowed to expand, it loses internal energy as it performs mechanical work. The operation of gasoline and steam engines are applications of this principle.

The temperature of a gas can be raised by doing mechanical work that will compress it. Any process that can change temperature without the transfer of heat is called an *adiabatic* change.

B. Second Law

There can be no spontaneous flow of heat energy between two systems at the same temperature. If there is a temperature difference, heat always flows from a hot object to one that is colder. To reverse the direction of heat flow, external work must be done.

There is a tendency in nature for all systems to become disordered. Entropy is a quantitative measure of a system's disorder. For example, ice is a solid that has a relatively low entropy. Its molecules are arranged in a highly structured pattern of crystals. Water and steam contain the same molecules but have a much greater entropy.

C. Third Law

It is impossible by any set of finite operations to reduce the temperature of a system to absolute zero. Ordinary helium liquifies at 4.2 K. Temperatures below 1K have been achieved.

QUESTIONS

1. The internal energy of a substance is at a minimum when its temperature is (1) 0°Celsius (2) 0 Kelvin (3) 273°Celsius (4) 273 Kelvin
2. The minimum amount of mechanical energy necessary to raise the temperature of a kilogram of water 1 degree Celsius is (1) 2.10 kJ (2) 4.19 kJ (3) 6.29 kJ (4) 8.38 kJ

3. Which graph best indicates the relationship between the average kinetic energy (E_K) of the molecules of a gas and the absolute temperature (T_{ABS}) of the gas?

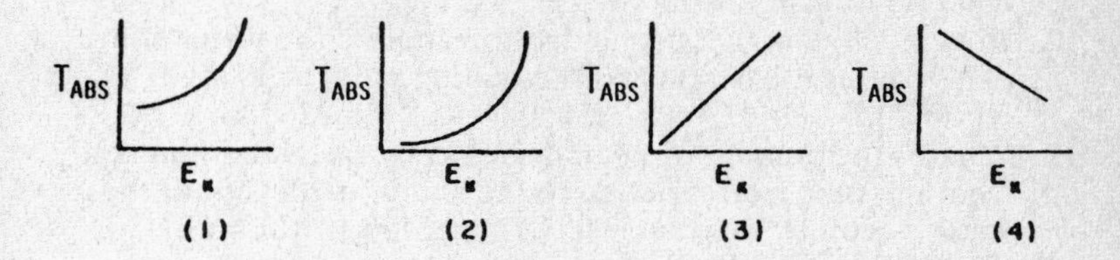

4. A change in temperature of 100 Celsius degrees is equal to a change in Kelvin temperature of (1) 373 K (2) 200 K (3) 100 K (4) 50 K

5. As the temperature of a substance increases, the average kinetic energy of its molecules (1) decreases (2) increases (3) remains the same

6. As the temperature of a constant volume of an ideal gas is increased, its pressure will (1) decrease (2) increase (3) remain the same

7. When a block of ice at zero degrees Celsius melts, the ice absorbs energy from its environment. As the ice is melting, the temperature of the remaining portion of the block (1) decreases (2) increases (3) remains the same

8. Which graph best represents the relationship between the Kelvin temperature scale and the Celsius temperature scale?

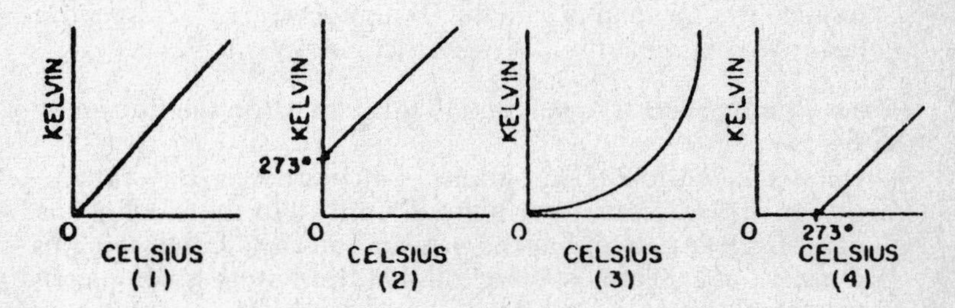

9. The internal energy of water depends on its (1) temperature, only (2) phase, only (3) temperature and mass, only (4) temperature, mass, and phase

Base your answers to questions 10 through 14 on the information below.

A 0.80-kilogram aluminum kettle contains 0.50 kilogram of water at 20°C. The kettle is heated until all of the water changes to vapor at 100°C.

10. How much heat will be needed to vaporize the water completely after it has reached the boiling point at 100°C? (1) 16.7 kJ (2) 1130 kJ (3) 1808 kJ (4) 2260 kJ

11. What is the amount of heat absorbed by the aluminum kettle from the time heat is added to the water at 20°C until the water becomes vapor at 100°C? (1) 40 kJ (2) 58 kJ (3) 150 kJ (4) 310 kJ

12. How much heat is necessary to raise the temperature of the water to its boiling point at a pressure of 1.0 atmosphere? (1) 28 kJ (2) 44 kJ (3) 84 kJ (4) 168 kJ

13. If the kettle were tightly covered during heating, the pressure of the water vapor would (1) decrease (2) increase (3) remain the same

14. Assume a 0.80-kilogram copper kettle were used instead of the aluminum kettle. Compared to the total heat needed to raise the temperature of the water and the aluminum kettle from 20°C to 100°C, the amount of heat needed to raise the temperature of the water and the copper kettle from 20°C to 100°C would be (1) less (2) greater (3) the same

15. Doubling the absolute temperature of an ideal gas will affect the molecules of the gas by doubling their average (1) kinetic energy (2) potential energy (3) momentum (4) velocity

16. The sum of the kinetic energy and potential energy of the molecules of a solid is called (1) temperature (2) specific heat (3) heat energy (4) internal energy

Base your answers to questions 17 through 21 on the information below.

Heat is added to 2.00 kilograms of alcohol at −120°C at a rate of 4.19 kilojoules per minute, until all of the alcohol has vaporized. [Assume that the pressure on the alcohol remains constant at 1.00 atmosphere and that the system is completely insulated.]

17. At its melting point, how long does it take to change all of the alcohol from a solid to a liquid? (1) 52 min (2) 156 min (3) 230 min (4) 408 min

18. Through how many Celsius degrees will the alcohol remain in the liquid phase? (1) 37 (2) 78 (3) 178 (4) 196

19. How long does it take to change the alcohol from −22°C to 28°C? (1) 58 min (2) 71 min (3) 100 min (4) 120 min

20. At which temperature can the alcohol exist as both a liquid and a gas? (1) −115°C (2) 0°C (3) 79°C (4) 100°C

21. At its boiling point, how much heat does it take to evaporate all of the alcohol? (1) 980kJ (2) 1240kJ (3) 1710kJ (4) 2360kJ

22. What temperature reading on the Kelvin scale is equivalent to a reading of zero degrees Celsius? (1) −273°K (2) −100°K (3) 100°K (4) 273°K

Base your answers to questions 23 through 27 on the diagram below which represents a cooling curve for 10. kilograms of a substance as it cools from a vapor at 160°C to a solid at 20°C. Energy is removed from the sample at a rate of 8.0kJ per minute.

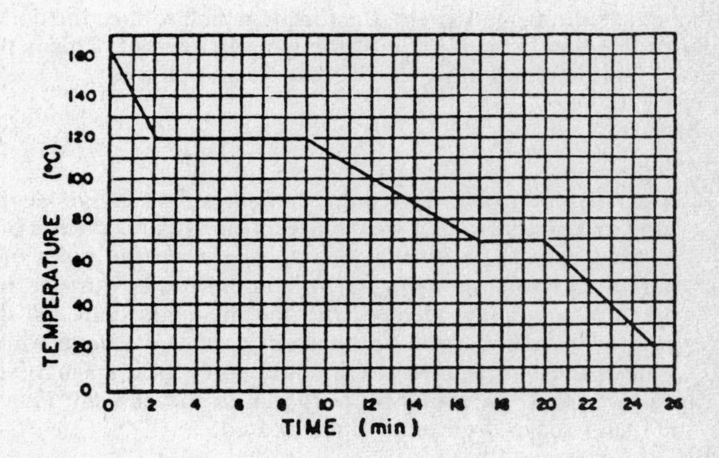

23. The boiling point of the substance is (1) 0°C (2) 70°C (3) 100°C (4) 120°C

24. The heat of vaporization of the substance is (1) 2.3 kJ/kg (2) 5.6 kJ/kg (3) 7.8 kJ/kg (4) 9.4 kJ/kg

25. During the liquid phase, the specific heat of the substance, in kJ/kg°C, is (1) 0.06 (2) 0.09 (3) 0.13 (4) 1.12

26. While the substance is cooling during the liquid phase, the average kinetic energy of the molecules of the substance (1) decreases (2) increases (3) remains the same

27. Compared to the specific heat of the vapor phase of the substance, the specific heat of the liquid phase is (1) less (2) greater (3) the same

UNIT VIII. ELECTROMAGNETIC APPLICATIONS

I. Torque on a Current Carrying Loop

A current carrying loop in a magnetic field will experience a force that will tend to twist the coil. If the coil is free to pivot, it will rotate until the plane of the loop and the magnetic field direction are perpendicular.

In a constant magnetic field, the torque which rotates the coil is proportional to the magnitude of the electric current. This is the operating principle of most electric meters and motors.

A. Meters

1. Galvanometer

A *galvanometer* is a device that indicates the relative magnitude and direction of a weak current. It usually consists of a pivoted coil that is spring mounted between the poles of a permanent horseshoe magnet. When there is no current, the plane of the coil is aligned with the magnetic field and the pointer indicates zero. Any current creates a torque which rotates the coil against the counter force of a spring. The deflection of the coil is proportional to the current. Galvanometer scales are usually uncalibrated.

2. Ammeter

An *ammeter* is a galvanometer with a scale calibrated to read current in amperes. It may have one or more low-resistance conductors, called *shunts,* connected in parallel across its terminals. Shunts make it possible for an ammeter to withstand relatively high currents. They bypass (or shunt) most of the current that would otherwise be in the coil. The small amount of current that remains in the coil provides a deflection which is calibrated to indicate the total current.

Ammeters are always connected *in series* with the circuit being measured. In a series circuit, the current in the ammeter is the same as the current anywhere else in the circuit.

3. Voltmeter

A *voltmeter* is a galvanometer with a scale calibrated to read potential difference in volts. It may have one or more high-resistance conductors, called *multipliers,* connected in series with its coil. The voltage divides between the multiplier and the coil with the greater part of the potential difference across the multiplier. The small potential difference across the coil generates a small current which deflects the coil. The deflection is calibrated to indicate the total potential difference at the voltmeter terminals.

Voltmeters are always connected *in parallel* with the circuit being measured. In a parallel circuit, the potential difference across the voltmeter is the same as the potential difference anywhere else in the circuit.

B. Motors

An electric motor converts electrical energy into rotational mechanical energy. Most motors contain either a permanent magnet or a field coil which can provide a constant magnetic field. Within this field are one or more coils wound on an armature that is free to rotate.

Current in the armature coils generates a magnetic field which interacts with the surrounding field. This produces a torque that forces the armature to rotate. To make this torque greater, an armature has an iron core. The iron has a high permeability which concentrates the magnetic field in the rotating coils.

1. **A-c and D-c Motors.** Most motors today are designed to operate on alternating current. Because the current reverses direction periodically, a new torque is created each time the armature approaches a stable position.

   To provide the necessary current reversal in a d-c motor, a split ring commutator is attached to the armature shaft. This switches the direction of the current in the armature as it rotates.

2. **Back EMF.** As the armature of a motor rotates in a magnetic field, the coils generate a reverse voltage called a *back EMF.* The applied voltage is diminished by the back EMF. Thus, there is a relatively small current in the armature whenever it is rotating rapidly. If the motor should stall, there will be no back EMF. The large current in the armature may cause the motor to overheat and possible catch fire. This is an example of Lenz's Law.

QUESTIONS

1. A current carrying coil is suspended on a uniform magnetic field. The torque on the coil will be greatest when the angle between the plane of the coil and the magnetic field is
(1) 0° (2) 45° (3) 90° (4) 180°

2. An electric meter which is usually not calibrated is the
(1) galvanometer (2) ammeter (3) voltmeter (4) wattmeter

3. No field magnets are found in a(n)(1) ammeter (2) voltmeter
(3) electroscope (4) galvanometer

4. An ammeter may contain a moving coil connected to a
(1) multiplier in series (2) multiplier in parallel
(3) shunt in series (4) shunt in parallel

5. A voltmeter may contain a moving coil connected to a
(1) multiplier in series (2) multiplier in parallel
(3) shunt in series (4) shunt in parallel

6. A feature that is found in d-c motors but not a-c motors is a(n)
(1) armature (2) rotating coil (3) field magnet
(4) commutator

7. As the speed of a motor increases, the back EMF
(1) decreases (2) increases (3) remains the same

8. A 6V power supply operates a motor which produces a 5V back EMF. If the armature coil has a resistance of 2.0 ohms the current in the coil is (1) 0.5 ampere (2) 2.5 amperes
(3) 3.0 amperes (4) 12 amperes

II. Electron Beams

As a metal is heated, electrons escape from its surface by the process of *thermionic emission.* The rate at which they escape depends on the temperature of the surface and the type of metal.

In a television picture tube, electrons are emitted by electrically heating a metal filament at the back of the tube. To create a picture, these electrons are accelerated and strike phosphors on the screen. Light is emitted at the location of each collision while the rest of the screen remains dark.

A. Beam Control by an Electric Field

As an electron beam passes through an electric field, the electrons are accelerated by a force toward the more positive potential of the field. Electrons travelling parallel to the field change speed. Those that enter an electric field at any other angle will be deflected in a curved path toward the more positive potential of the field.

A typical cathode ray tube, used in oscilloscopes, has two pairs of parallel plates mounted in its neck. When a potential difference is applied to one set, it creates a uniform electric field in the vertical direction. This deflects an electron beam either up or down. The other set of plates creates a horizontal electric field that deflects the beam right or left. By adjusting the magnitude and polarity of the voltages on these plates, the electron beam is aimed at any desired spot on the screen.

The potential energy of an electron is the greatest when it is at the most negative portion of the electric field. If the electron is free to move, the system loses electrical potential energy as the electron gain kinetic energy. The maximum amount of kinetic energy each electron can gain is given by the relationship:

$$KE = qV$$

where: KE is the kinetic energy in joules
q is the charge on an electron (1.60×10^{-19}C)
V is the potential difference of the field in volts

Sample Problem

Electrons emitted from the cathode of a television picture tube are accelerated toward the screen by an electric field having a potential difference of 10,000 volts.

a. How much kinetic energy does each electron gain?

b. If the velocity of an electron is zero at the cathode, what is its velocity just before it strikes the screen of the picture tube?

Given $q = 1.60 \times 10^{-19}$
$V = 10{,}000 \text{ volts} = 10{,}000 \text{ J/C}$
$m_c = 9.11 \times 10^{-31}\text{kg}$

Find a) KE b) v

Equation a) $KE = qV$ b) $KE = mv^2/2$

Solution a) $KE = qV$
$= (1.60 \times 10^{-19}\text{C})\ (10{,}000 \text{ J/C})$

Answer $KE = 1.60 \times 10^{-15}\text{J}$

b) $KE = mv^2/2$
solving for v,
$v = 2KE/m$
$= 2(1.60 \times 10^{-15}\text{J})/9.11 \times 10^{-31}\text{kg}$

Answer $v = 5.9 \times 10^{7}$ m/s

B. Beam Control by a Magnetic Field

An electron beam experiences no force when travelling parallel to a magnetic field. At other angles to the field, it encounters a force which is perpendicular to both the beam and the field. The force is maximum when the angle between the beam and the magnetic field is 90 degrees.

The electron beam of large cathode ray tubes is controlled by a yoke which contains two sets of magnetic field coils. One pair of coils, on the left and right sides of the neck, creates a horizontal magnetic field. This deflects the beam in the up-down direction. A second pair of coils above and below the neck creates a vertical magnetic field which deflects the beam in the left-right direction. The combined action of these two magnetic fields aims the beam. When electrons strike a portion of the phosphor-covered screen fluorescense occurs and light is given off at that position.

In addition to electron beams, a beam may consist of other charged particles such as protons or alpha particles. In any case, however, the force exerted by a magnetic field on a perpendicular beam of charged particles is given by the equation:

$$F = qvB$$

where: F is the force in newtons applied to the charge particle
q is the charge of a particle in coulombs
v is the speed of a particle in m/s
B is the magnetic field strength in teslas

Sample Problem

A beam of electrons traveling from north to south at a speed of 6.0×10^4 m/s enters a magnetic field. The field strength is 5.0 teslas and the direction of the field is toward the east.

a. How much force does the magnetic field exert on the beam?

b. In what direction is the beam deflected?

Given $q = 1.60 \times 10^{-19}C$
$v = 6.0 \times 10^4$ m/s
$B = 5.0$ T $= 5.0$ N/Am $= 5.0$ N/(C/s)m
$= 5.0$ Ns/Cm

Find F

Equation $F = qvB$

Solution a) $F = qvB$
$= (1.60 \times 10^{-19}C)\,(6.0 \times 10^{4}\ m/s)\,(5.0\ Ns/Am)$

Answer $F = 4.8 \times 10^{-14}$ newtons

b) Using the left hand rule:
Middle finger points to field direction (East)
Index finger to beam direction (South)
Thumb will be down down

Answer Beam is deflected down

C. Beams of Other Charged Particles

1. Mass Spectrometer

Positively charged particles (ions) of different elements are accelerated by an electric field until they attain a desired velocity. Then they enter a chamber which contains a magnetic field which deflects the particles. Those having a greatest mass and smallest charge are deflected the least and form into a semicircle with a large radius. Other particles having smaller masses and greater charges are deflected into smaller semicircles. Thus, the particles are separated and finish at different positions. Their presence is recorded on photograph plates.

Since the magnetic force that deflects the ions into a semicircle is a centripetal force,

$$qvB = mv^2/r$$

$$\text{and } q/m = v/Br$$

If the charge on the ions are known, the mass can be calculated.

2. Mass of the Electron

The charge of an electron was determined by the Millikan oil drop experiment. Charged oil drops of known weight were sprayed into a chamber containing a vertical electric field. The drops were held motionless by adjusting the electric field until the electric force balanced their weight.

It was found that all the oil drops had charges that were multiples of 1.6×10^{-19}. This is now called the elementary charge. It is the charge on an electron or a proton. Knowing the value of q/m and the value of q, the mass of the electron has been determined to be 9.11×10^{-31} kilogram.

3. Particle Accelerators

Accelerators provide charged particles with sufficient kinetic energy to penetrate the nucleus.

a. A linear accelerator

A linear accelerator consists of a series of straight evacuated tubes placed end to end with gaps between them. Differences in potential in the gaps cause particles to accelerate each time they proceed from one tube to the next.

b. The synchroton

The synchroton provides a means of making accelerators smaller by using magnetic fields to bend the path of the particles being accelerated into circles. The predecessor to the synchroton was the *cyclotron* which operated on a similar principle.

c. The Van de Graaff generator

The Van de Graaff generator is a large electrostatic machine that can produce a large potential difference between two terminals. Acceleration of charged particles is accomplished by allowing the particles to fall through the very large potential difference that is created by the generator.

D. Induced Voltage

When a conductor is moved through a magnetic field, an electrical potential difference may be generated at the ends of the conductor. The magnitude of this potential difference depends on three factors:

a) the strength of the magnetic field. The stronger the magnetic field, the greater the potential difference.

b) the speed of the conductor. The faster the conductor moves, the greater the potential difference.

c) the angle between the conductor path and the magnetic field lines. When the motion is in the same direction as the magnetic flux, (angle of zero degrees) the potential difference is zero. When the angle is 90°, the potential difference is maximum. Between 0 and 90 degrees, the potential varies as the sine of the angle.

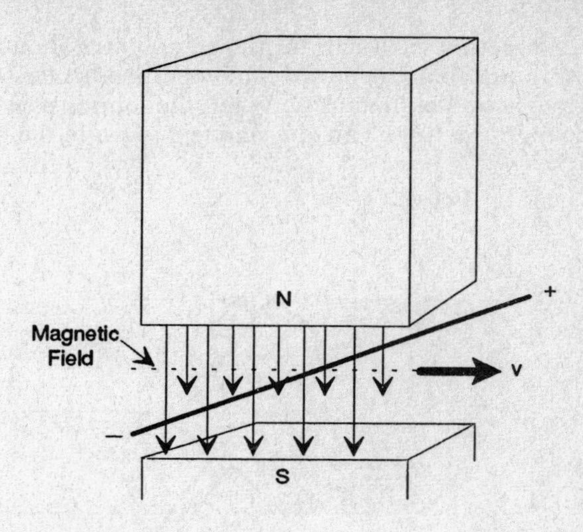

Sample Problem

A straight conductor, 2.0 meters long, is moved at a constant velocity of 4.0 m/s through a uniform 5.0 T magnetic field. Under these conditions, what is the maximum potential difference that can be generated between the ends of the conductor?

Given: $B = 5.0\ T$
$l = 2.0\ m$
$v = 4.0\ m/s$

Find V

$V = Blv$
$= (5.0\ T)(2.0\ m)(4.0\ m/s)$

Answer = 40 volts

E. Generator Principle

A conducting loop rotated in a uniform magnetic field experiences a continual change in the total number of magnetic flux lines linking the loop. A potential is induced at the ends of the loop and alternates direction each half turn. The potential induced is proportioned to the component of velocity perpendicular to the field and the magnetic field intensity. When the loop is part of a complete circuit, there will be an *alternating current* in the loop.

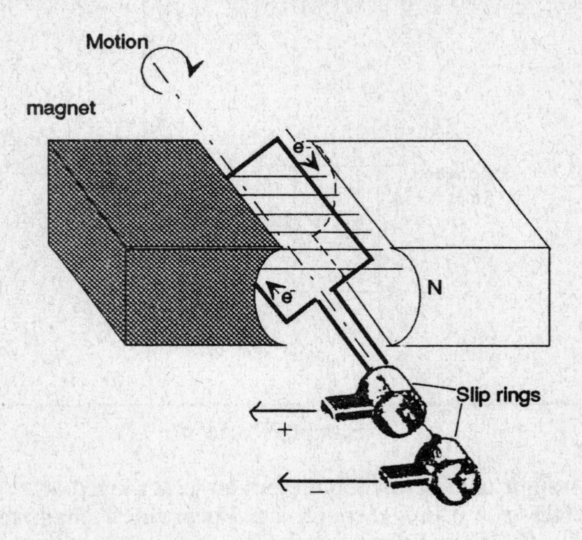

If the loop is rotated at a constant speed in a uniform magnetic field, the magnitude of the alternating current will vary sinusoidally.

F. Transformers

A transformer is a device that can increase or decrease the voltage of an a-c circuit. It consists of two insulated coils that are close together. A changing voltage in the primary coil creates a changing magnetic field which induces an alternating voltage in the secondary coil.

The voltage change of a transformer depends on the number of turns in its secondary coil compared with that of the primary. This is stated in the equation:

$$N_p/N_s = V_p/V_s$$

where: N_p is the number of primary turns
N_s is the number of secondary turns
V_p is the voltage in the primary
V_s is the voltage in the secondary

If the secondary coil is connected in a complete circuit, there will be current in this circuit. The amount of current depends on the voltage induced in the secondary coil and the resistance of the secondary circuit.

$$I_s = V_s/R_s$$

where: I_s is the secondary current in amperes
V_s is the secondary voltage in volts
R_s is the secondary resistance in ohms

The electrical power of the secondary circuit is supplied by the primary coil. In a transformer that is 100% efficient, the power in the primary and secondary are identical. Practical transformers have efficiencies of 99% or less. Part of the transferred energy is wasted by eddy currents that heat the transformer core. To calculate the power in the primary, use the equation:

$$P_p = P_s/\%\ \text{efficiency}$$

where: P_s is the power in the secondary circuit.
This equals the product of V_sI_s.

Sample Problem

120 volts ac is supplied to a step up transformer that has 30 turns on the primary winding and 120 turns on the secondary.

a. Calculate the voltage induced in the secondary coil.
b. Calculate the current in the secondary if it is a closed circuit with a resistance of 240 ohms.
c. Calculate the power in the primary circuit. The efficiency of the transformer is 98%.

Given $N_p = 30$ turns
$N_s = 120$ turns
$V_s = 120V$
$R_s = 240$ ohms; % efficiency = 0.98

Find a) V_s b) I_s c) P_p

Equation $V_s = V_pN_s/N_p$ $I_s = V_s/R_s$ $P_p = V_sI_s/\%eff$

Solution a) $V_s = (120V)(120)/30$

Answer $V_s = 480$ volts

b) $I_s = 480V/240$ ohms

Answer $I_s = 2.0$ amperes

c) $P_p = (480V)(2.0A)/0.98$

Answer $P_p = 1{,}000$ watts

D. Induction Coils

An induction coil is a closely wound coil of many turns. When a direct current is suddenly applied by a switch or a solid state device, each turn generates a magnetic field that induces a counter voltage in every other turn. This delays the establishment of the maximum current in the circuit.

Once the maximum current is established the coil acts like an ordinary electromagnet and provides negligible resistance to the current. However, if the current is suddenly turned off, the magnetic field inside the coil collapses. The effect is similar to that of a permanent magnet being suddenly withdrawn from the coil. The coil generates a very large voltage; larger than that provided by the original dc source.

Induction coils are incorporated in the ignition systems of automobiles. Although the battery of the car only supplies 12 volts dc, the ignition coil can help generate pulses of 20,000 volts that are needed to operate the spark plugs.

IV. The Laser

A laser provides a high intensity of nearly coherent light.

A. Characteristics of the Laser Beam

Light from a laser differs from ordinary light in four important respects that make the laser valuable:

1. **Small divergence.** The beam from a laser does not spread out much (diverge) after it leaves the laser. Thus instead of being dissipated rapidly, the energy is concentrated in a narrow beam.
2. **Monochromatic.** The laser light is said to be monochromatic because it is mostly of one color, or one wavelength.
3. **Coherent.** Ordinary light is incoherent with crests and troughs being emitted at random from different parts of the light source. Laser light, however, is coherent with almost all crests and troughs in phase regardless of where they are generated in the laser tube.
4. **High Intensity.** Laser light is very intense because all its energy is concentrated.

B. Theory of Laser Operation

Fundamentally a helium-neon laser is very much like the familiar neon electric sign.

To make a helium-neon laser that emphasizes only the red light at a wavelength of 633 nanometers, it is only necessary to take a straight section of a neon sign, add some helium gas to the tube, and place a partially reflecting mirror at each end.

1. **Function of Helium.** Helium is included in the laser tube because it enhances the neon's output of red light by several orders of magnitude in a highly efficient energy exchange process. Although neon gas alone can provide some laser action, its output is many times as great when it is mixed with a larger amount of helium in proportions of about 1 to 6.

2. **Energy Absorption and Collision Transfer.** To produce a laser beam, there must be a continous supply of energy which can be converted into electromagnetic radiations. To supply this energy efficiently, a high voltage creates a strong electric field along the length of the laser tube. In this field, free electrons of the laser medium gain kinetic energy as they are accelerated toward the positive terminal.

   Before reaching the positive terminal, there is a high probability that an energetic electron will collide with one of the many helium atoms and give it additional energy. This puts the helium in an excited state as its electrons "jump" to higher energy levels. Think of these levels as irregularly spaced landings on a staircase where the electrons can rest momentarily before falling back toward the ground level.

   The thermal motions of excited helium atoms result in collisions with neon atoms in the laser tube. During such a collision, a second energy exchange occurs. The helium atom reverts to its ground state and the neon atom absorbs just enough energy to raise one of its electrons to its 3S2 level. At this particular landing, called a metastable level, a neon electron can rest for a comparatively long time before "jumping" back down toward ground level.

   Of course, there is always the possibility that an energized electron will supply the correct amount of energy directly to the neon without requiring the helium as an intermediary. However, this does not happen very often because neon has 10 electrons and many different excited levels for each. Thus, there is a rather low probability that an energetic electron can excite a neon atom directly to the desired 3S2 level where laser action can occur.

3. **Spontaneous Emission.** A neon atom can remain in the excited state at the 3S2 level up to several microseconds. Then, for no apparent reason and without any external stim-

ulation, the atom de-excites itself by spontaneously falling toward the ground energy level either in one large step or in a series of several smaller steps.

Each time the neon electron undergoes a transition and falls a step to a lower energy level, it releases energy in the form of an electromagnetic wave burst, or a photon of light. By itself, this photon does not create much light, but together with the combined energy of many others of the same wave length, an intense laser beam can be produced.

4. **Stimulated Emission.** In 1917, Einstein postulated that a photon released spontaneously from an excited atom could interact with another excited atom, and stimulate it to de-excite and produce another photon. The new photon would have the same wavelength, energy and phase as the original photon and the two would join and continue traveling in the same direction. These two photons could then proceed to stimulate other excited atoms to produce additional photons in a cascade effect. The light that they produce is called *coherent* light and it consists of many photons of the same frequency and phase that combine their energies by constructive interference.

5. **Population Inversion.** For stimulated emission to be effective, it is essential that a large number of excited neon atoms is present in the laser tube at any given time. In the normal state of matter, most of the electrons are in the ground state or lowest energy levels. Without a substantial number of excited neon atoms present, photons will pass through the laser medium without encountering a sufficient number of excited atoms to have a noticeable effect. However, when the majority of neon atoms are excited by helium collisions and remain excited for a comparatively long time at a metastable level, "population inversion" is said to exist. In this condition, any photon traversing the laser tube has a high probability of producing many stimulations along the way and an amplified beam of light is produced.

6. **Multiple Reflections.** To further strengthen the output of coherent light in a laser, a partially reflecting mirror is placed at each end of the tube. Any photons that are traveling along the tube axis will be reflected back into the laser medium and receive further reinforcement as they stimulate additional excited atoms to produce 633 nm light. These mirrors reflect most, but not all.

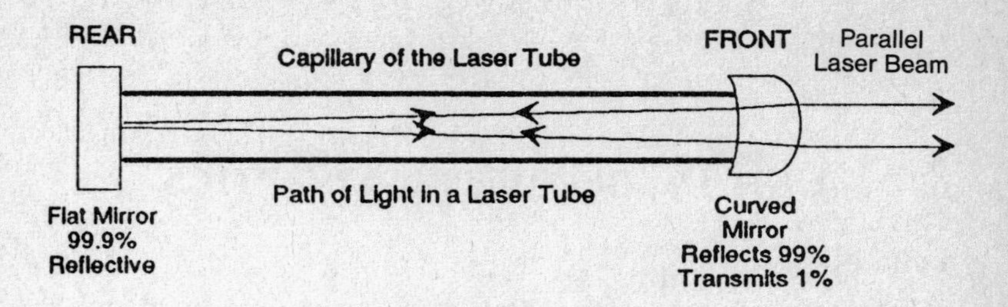

QUESTIONS

1. A horizontal beam of electrons is aimed through an electric field between two horizontal metal plates. The upper plate has a negative charge and the lower plate is positive. The beam will be deflected (1) up (2) down (3) left (4) right
2. Electrons at rest near the cathode of a vacuum tube are accelerated through a potential difference of 100 volts by an electric field. At what velocity will the electrons emerge from the field?
3. A beam of protons travels through a 10 tesla magnetic field in a direction perpendicular to the field. How much force does the field exert on the beam?
4. The composition of a gas can be determined by separating it with a (1) lnear accelerator (2) synchroton (3) mass spectrometer (4) cyclotron
5. A straight conductor 0.50 m long is moved at a constant velocity of 3 m/s across a 4.0 T magnetic field. What is the maximum voltage that can be generated across the ends of the conductor?
6. A step down transformer has 100 volts in the primary and 5 volts in the seondary winding. The current in the secondary is 2.0 amperes. What is the minimum current in the primary?
7. The light beam emitted by a laser is always (1) red (2) perfectly parallel (3) coherent (4) all of the above

UNIT IX. GEOMETRICAL OPTICS

I. Images

Light rays coming from a point on an object tend to diverge as they move outwards. Using an optical system consisting of lenses or mirrors these rays can be made to change direction and converge. At the places where they intersect there will be a likeness of the object, called a *real image.*

Other optical systems can be devised to change the direction of the light rays without converging the light. This change in direction makes is difficult to determine the actual location of the source. At the place where the rays seem to originate, an observer sees a likeness of the object called a *virtual image*.

Real images can be observed by placing a screen at the place where the rays converge. Virtual images will never form on a screen. However, both real and virtual images can be seen and photographed.

II. Images Formed by Reflection

Because reflection is a process which changes the direction of light rays, mirrors can be used to create real or virtual images.

A. Images Formed by a Plane Mirror

A plane mirror can form a virtual image by changing the direction of light rays coming from an object. Because it cannot converge the rays it never forms a real image.

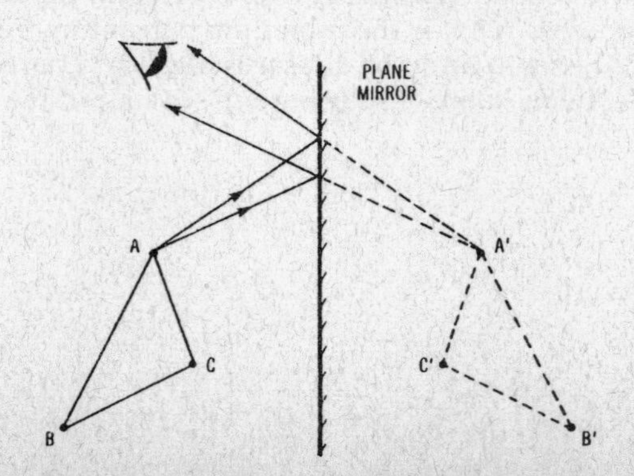

The diagram shows how diverging rays coming from point A of an object are reflected by mirror into the eye of an observer. They appear to originate at point A′ behind the mirror. Rays coming from other points on the object are reflected similarly. The image that is formed by the plane mirror has the following characteristics:

1. It is virtual.
2. It seems to be behind the mirror the same distance as the object is in front.
3. It is erect.
4. It is reversed from right to left.
5. It is the same size as the object.

B. Images Formed by a Spherical Mirror

A spherical mirror is a reflector that is curved like a portion of a spheres' surface. Its center of curvature is located where the center of the complete sphere would be. Its principal axis is a line drawn through the center of curvature and the optical center of the mirror.

Because a spherical mirror converges or diverges incident light coming from an object it can produce real or virtual images of the object.

1. Concave Mirrors

A concave mirror can form either a real or a virtual image of an object. The type, size, and location of the image depends on the curvature of the mirror and the distance between the mirror and the object. These features can be determined by drawing a diagram of the system and two of the light rays coming from the top of the object.

Ray 1 is drawn parallel to the principal axis. It is reflected by the mirror and intersects the principal axis at F, the *principal focus.* This is located about halfway between C and the mirror.

Ray 2 through the center of curvature. The top of the image appears where ray 1 and ray 2 intersect.

Case 1. Object at Infinity

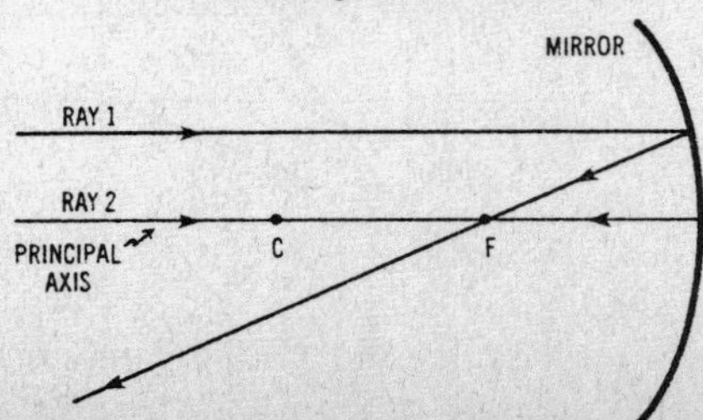

Light rays coming from an object at an infinite distance from the mirror are essentially parallel to each other. They converge to form a point image at F. An example is the image of a star made by a reflecting telescope.

Case 2. Object at a Finite Distance Beyond the Center of Curvature

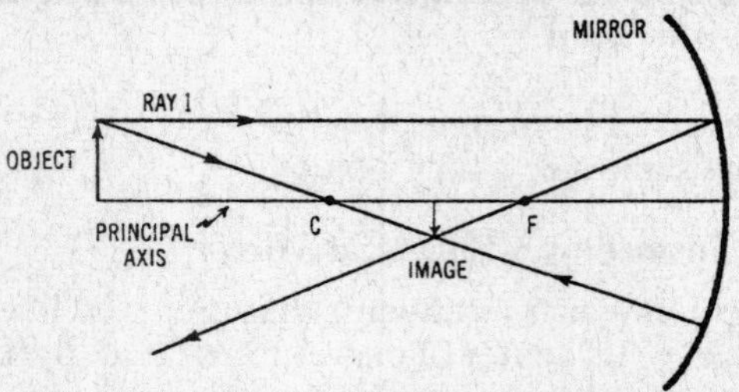

Rays intersect somewhere between C and F. The image is smaller than the object, inverted, and real.

Case 3. Object at the Center of Curvature

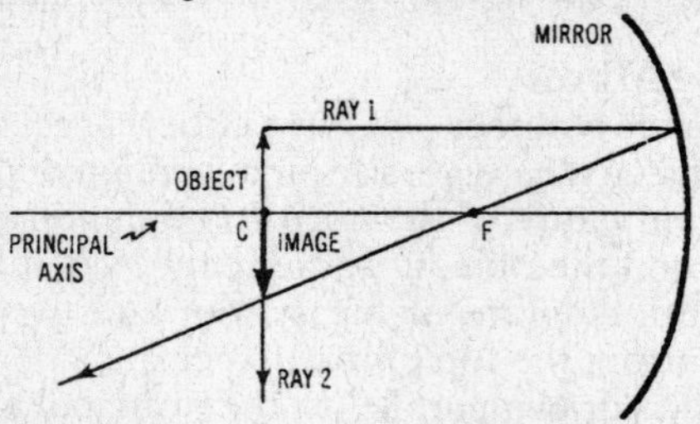

Rays intersect directly below the object. The image is the same size as the object, inverted, and real.

Case 4. Object Between C and F

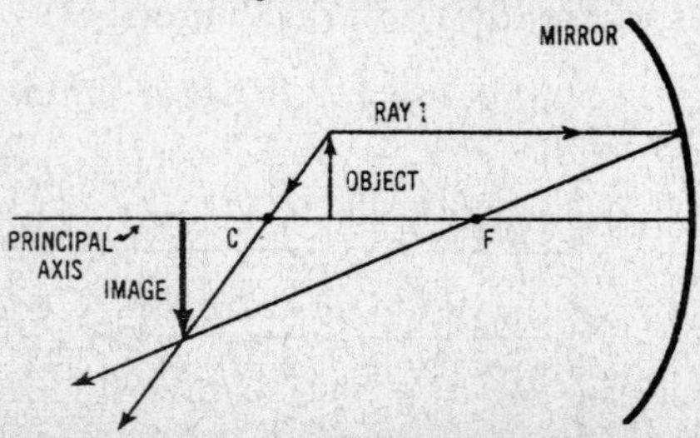

Rays intersect beyond C. The image is larger than the object, inverted and real.

Case 5. Object at F

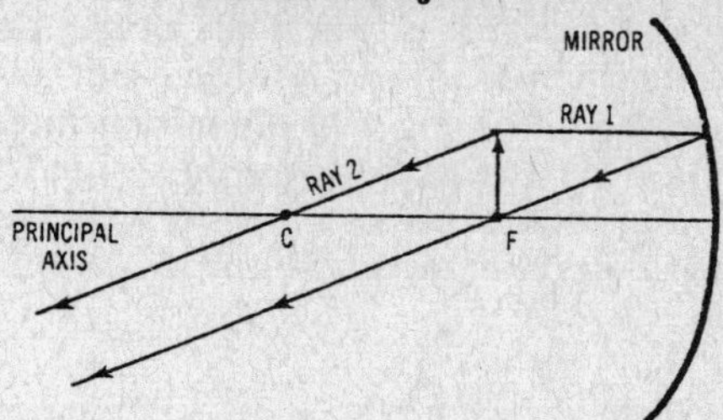

Rays become parallel and never intersect. No image is formed.

Case 6. Object Between F and the Mirror

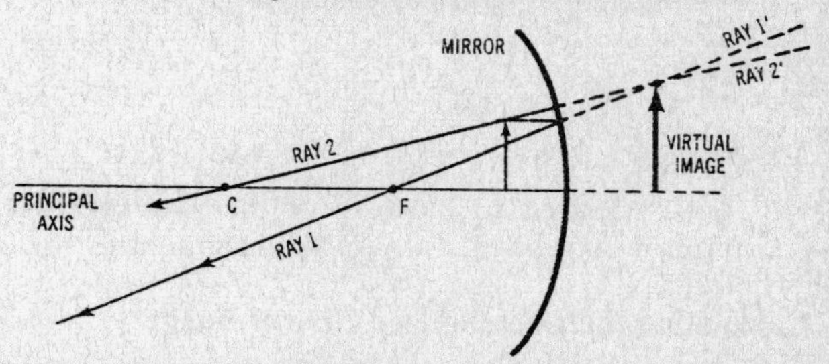

Rays diverge and never intersect. However, they appear to have originated from a point behind the mirror. The image is larger than the object, erect, and virtual. An example is a magnifying mirror used for makeup or shaving.

2. Convex Mirrors

Regardless of the distance between the object and the mirror, rays always diverge and never intersect. However, they appear to originate from a point behind the mirror. The image is always smaller than the object, erect, and virtual. An example is the small image formed by the curved side-view mirror of an automobile.

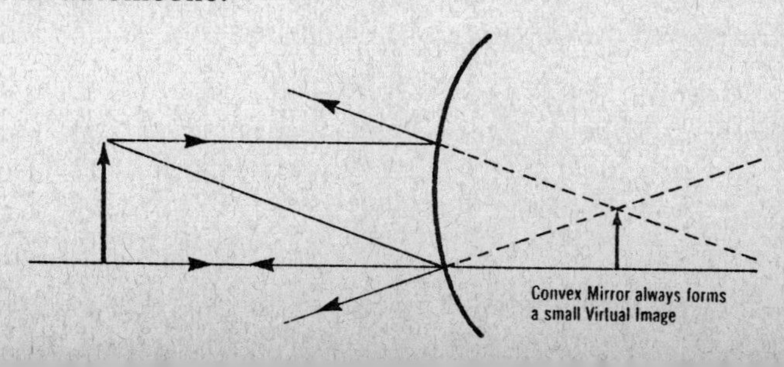

3. Spherical Aberration

If a curved mirror is spherical, incident rays that are parallel to the principal axis do not focus at the same point. Those closer to the outer edges of the mirror intersect the principal axis at a different place than those near the middle of the mirror. This defect is called *spherical aberration.* As a result, mirrors of large diameter will produce blurred images.

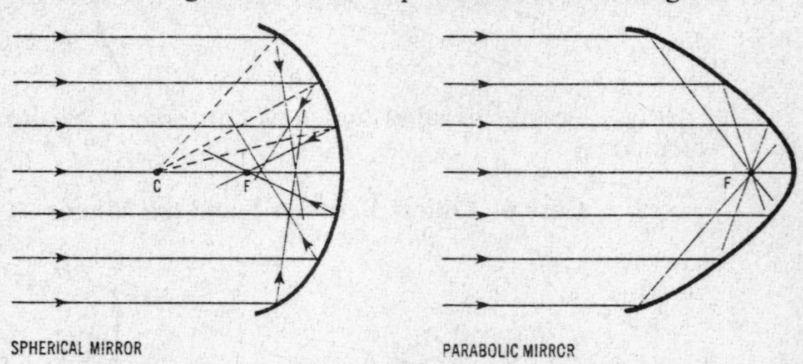

If it is desired to have all of the rays focus at the same point, a mirror having a parabolic shape must be used.

C. Calculating the Location and Size of Image

Where the diameter of a curved mirror is small compared with its radius of curvature, the location of the image can be calculated using the equation:

$$1/d_o + 1/d_i = 1/F$$

where: d_o is the distance between the object and the mirror,
d_i is the distance between the image and the mirror, and
F is the focal length of the mirror, measured between the principal focus and the mirror.

Note: Distances located on the same side of the mirror as the object are positive. Those on the opposite side are negative.

The size of the image can be calculated using the equation:

$$s_o/s_i = d_o/d_i$$

where: d_o is the distance between the object and the mirror,
d_i is the distance between the image and the mirror,
s_o is the size of the object, and
s_i is the size of the image

Note: A positive size indicates that an image is real and inverted. A negative size indicates that it is virtual and erect.

Sample Problem

An object 2.0 cm tall is 25 cm in front of a concave mirror that has a focal length of 5.0 cm.

a. How far from the mirror will the image form?
b. What will be the size of the image?

Given $d_o = 25$ cm
$F = 5.0$ cm
$s_o = 2.0$ cm

Find a) d_i b) s_i

Equation a) $1/d_o + 1/d_i = 1/F$ b) $s_o/s_i = d_o/d_i$

Solution a) $1/d_o + 1/d_i = 1/F$
$1/25 \text{ cm} + 1/d_i = 1/5.0 \text{ cm}$

Answer $d_i = 6.3$ cm

b) $s_o/s_i = d_o/d_i$
$2.0 \text{ cm}/s_i = 25 \text{ cm}/6.3 \text{ cm}$

Answer $s_i = 0.50$ cm

QUESTIONS

1. Which type mirror can produce a real image? (1) convex (2) concave (3) plane (4) all of the above
2. Which type of mirror can produce a virtual image? (1) convex (2) concave (3) plane (4) all of the above
3. As an object at infinity approaches the center of curvature of a concave mirror, the size of the image produced (1) decreases (2) increases (3) decreases, then increases (4) remains the same
4. As an object moves away from the surface of a concave mirror, the focal length of the mirror (1) increases (2) decreases (3) decreases, then increases (4) remains the same
5. A concave mirror produces an image that is the same size as the object. The object must have been located at (1) infinity (2) the center of curvature (3) the principal focus (4) the mirror surface
6. A convex mirror cannot produce an image that is (1) real (2) erect (3) the same size as the object (4) larger than the object
7. As an object approaches a plane mirror the size of its image (1) increases (2) decreases (3) decreases, then increases (4) remains the same

Base your answers to questions 8 through 10 on the information below:

An object that is 3.0 cm tall is 4.0 cm in front of a concave mirror whose focal length is 5.0 cm.

8. The image is (1) real and erect (2) real and inverted (3) virtual and erect (4) virtual and inverted

9. How far from the mirror is the image located?

10. What is the size of the image?

D. Lenses

A lens can be made of any curved transparent material.

1. Converging lenses made of glass are thicker in the middle than at the edges. They will concentrate or **converge** parallel rays of light to a point.

a. **Images.** Converging lenses can form both **real** and **virtual** images.

(1) Real images are those which can be projected on a screen. A real image is formed when diverging rays leaving an object are converged by a lens to meet at corresponding points in space.

(2) Virtual images cannot be projected on a screen but they can be seen and photographed. Virtual images can appear where a real image could not possibly exist because rays do not intersect at the image point.

Virtual images are always erect, while real images are always inverted.

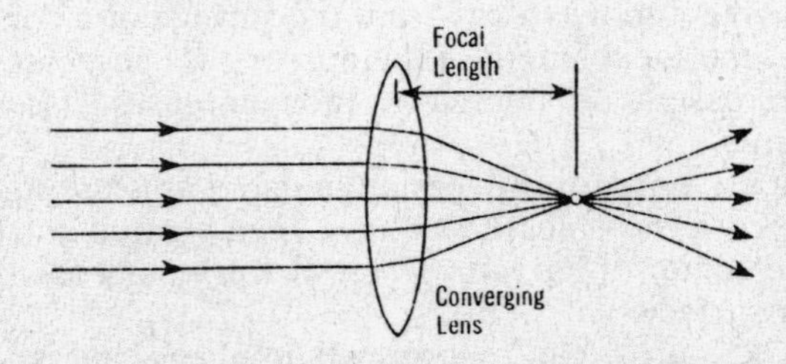

2. Diverging lenses made of glass are thinner at the middle than at the edges. They will spread, or **diverge** parallel rays of light. A diverging lens can produce only small, virtual images. A beam containing parallel rays of light passing through such a lens is changed to a divergent beam. However, if lines are drawn extending the diverging rays back through the lens, they seem to be coming from a point called the virtual focus.

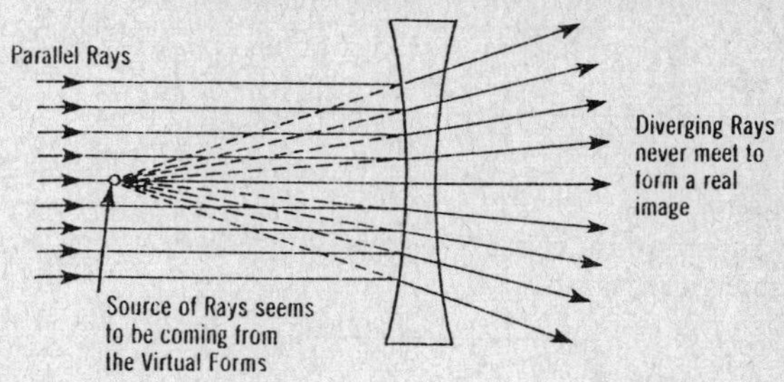

3. **Focal length.** The distance from the focal point to the center of the lens is called the **focal length** of the lens. This applies to converging lenses, too. The focal length is expressed as a positive distance for converging lenses, and a negative distance for diverging lenses. The principal focus, or focal point for a convergent lens is the point at which entering parallel rays intersect after leaving the lens.

4. **Image Formation**

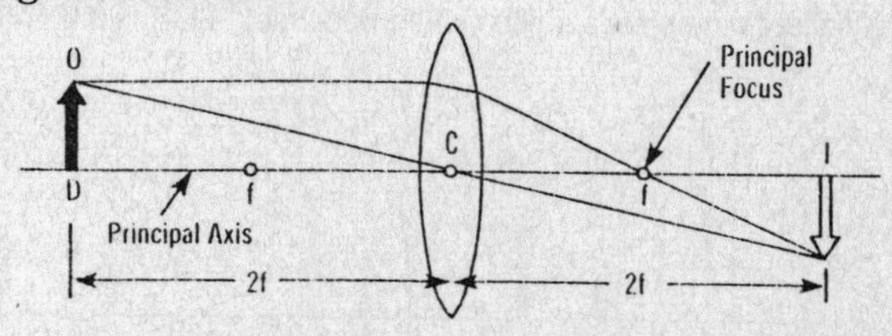

The location of an image can be predicted using a **ray diagram**, if the focal length is known and the distances are drawn to scale. The size of the image can also be predicted from the diagram if the object size is drawn to scale. Two of the rays from an object are shown in the diagram. **Any** two rays will locate the image, but there are two rays that are especially convenient. One is drawn parallel to the principal axis of the lens. This ray is refracted by the lens and crosses the principal axis at the focal point of the lens. The second ray that is drawn through the center of the lens has a negligible amount of refraction. It can be drawn as a straight line. The point of intersection of these two rays helps us locate the position of the top of the object.

5. Size and Distances of Images

a. Ray diagrams:

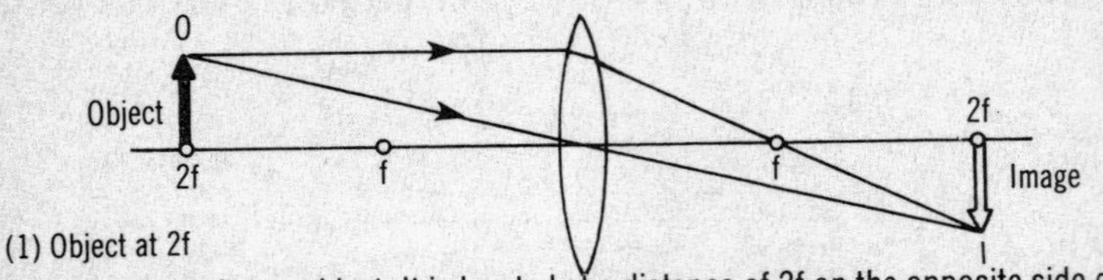

(1) Object at 2f

Image is same size as object. It is located at a distance of 2f on the opposite side of the lens and is inverted.

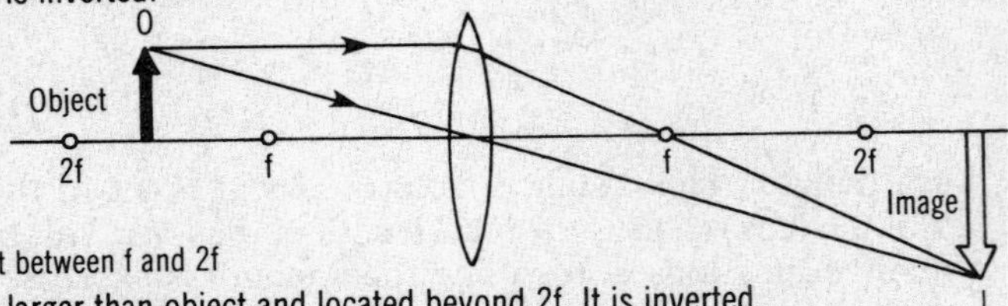

(2) Object between f and 2f

Image is larger than object and located beyond 2f. It is inverted.

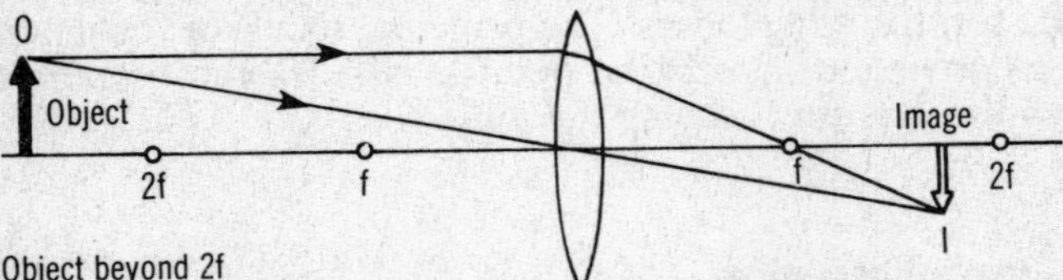

(3) Object beyond 2f

Image is smaller than object and located between f and 2f. It is inverted.

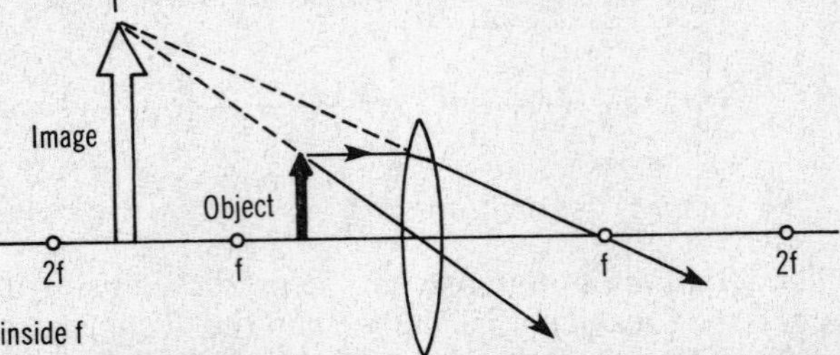

(4) Object inside f

Rays don't converge. Lines drawn back through lens locate a virtual image, enlarged and erect. (A magnifying glass.)

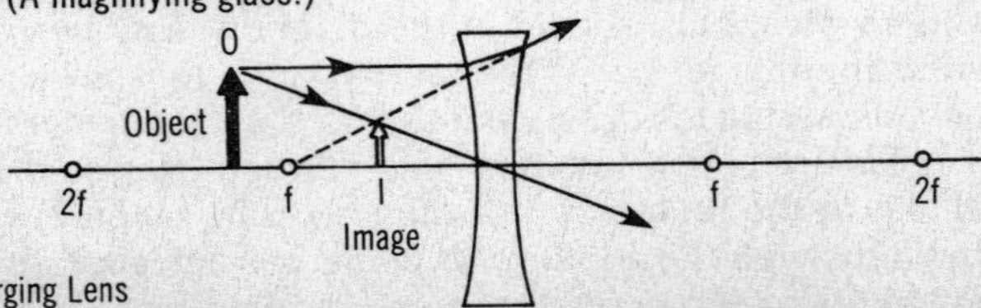

(5) A Diverging Lens

All images are virtual, smaller than the object, and are located on the same side of the lens on the object. They are always erect.

b. **Calculations**

The geometry of image formation provides useful equations:

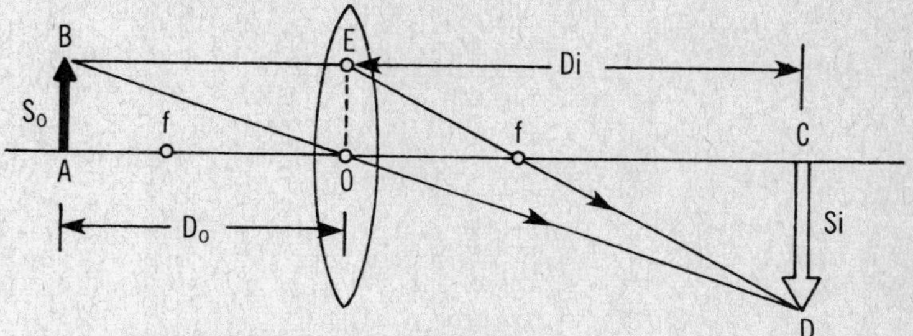

Note that $\triangle ABO \sim \triangle CDO$

$$\text{Then } \frac{\text{size of object, } S_o}{\text{size of image, } S_i} = \frac{\text{object distance, } d_o}{\text{image distance, } d_i}$$

This equation holds for any image formed by a lens or a mirror.

Also $\triangle DOF \sim \triangle CDF$

$$\text{Then } \frac{DO}{S_i} = \frac{FO}{FC} \text{ (focal distance, f)}$$

and, since $DO = BA = S_o$, $EO = S_o$
$CD = S_i$
$FO = f$
$FC = d_i - f$

$$\text{then } \frac{S_o}{S_i} = \frac{f}{d_i - f} = \frac{d_o}{d_i}$$

This can be rearranged to:

$$\frac{1}{f} = \frac{1}{d_o} + \frac{1}{d_i}$$

This relationship among focal length (f), object distance (d_o) and image distance (d_i) applies for all thin lenses and all curved mirrors.

Sample Problem

Find the size and location of the image formed by a +10cm lens of an object 3.0cm high placed 16cm in front of the lens.

Given $f = 10\text{cm}$
$S_o = 3.0\text{cm}$
$d_o = 16\text{cm}$

a) **Find** D_i

Equation $\frac{1}{f} = \frac{1}{D_o} + \frac{1}{D_i}$

Solution $\frac{1}{10\text{ cm}} = \frac{1}{16\text{ cm}} + \frac{1}{D_i}$

$\frac{1}{D_i} = \frac{1}{10\text{ cm}} - \frac{1}{16\text{ cm}}$

$\frac{1}{D_i} = \frac{16 - 10}{160\text{ cm}}$

$6\,D_i = 160\text{ cm}$

Answer $= D_i = 27\text{ cm}$

b) **Find** S_i

Equation $\frac{S_i}{S_o} = \frac{D_i}{D_o}$

Solution $\frac{S_i}{3.0\text{ cm}} = \frac{27\text{ cm}}{16\text{ cm}}$

$S_i = \frac{81\text{ cm}^2}{16\text{ cm}}$

Answer $= S_i = 5.1\text{ cm}$

6. **Lens defects**

   a. **Spherical Aberration.** If the surfaces of a lens are spherical, the rays that arrive near the edges of the lens do not focus at the same place as rays that are closer to the center. If the lens aperture is large, this causes a blurry image. To minimize this defect inexpensively, the aperture is made small to exclude peripheral rays. If a large aperture is needed, the lens must be flattened by additional stages of grinding.

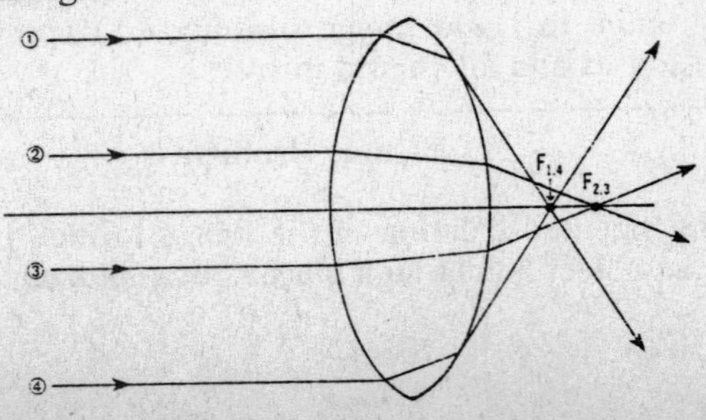

b. **Chromatic Aberration.** Each color focuses at a different distance from a lens. This is due to the fact that shorter wavelengths, such as violet, are refracted more than the longer red wavelengths. This defect, called *chromatic aberration*, is cured by cementing a concave lens to a convex lens that has a different index of refraction. Lenses that are made in this manner are called compound lenses or *achromatic lenses*.

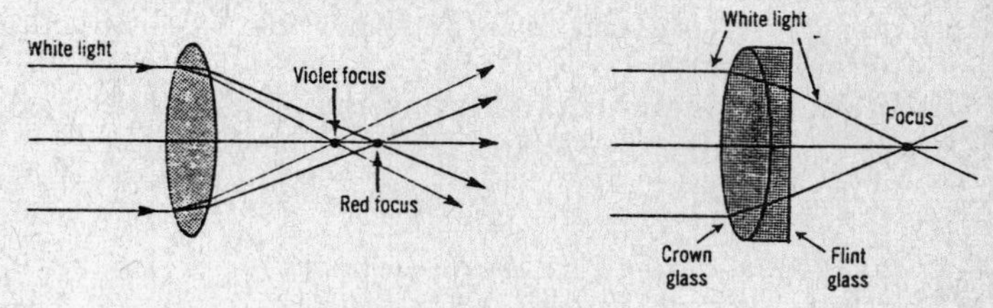

QUESTIONS

1. The diagram represents an object located on the reflecting side of a plane mirror.

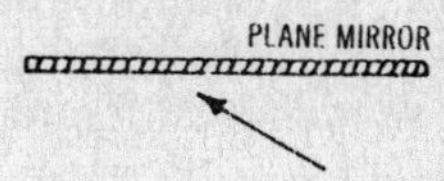

Which arrow below best represents the image produced by the mirror?

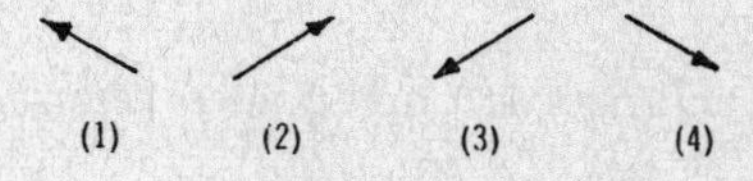

Base your answers to questions 2 through 5 on the following diagram which represents a converging crown glass lens with a focal length of 8.0 centimeters. Ray A represents one of the light rays incident upon the lens.

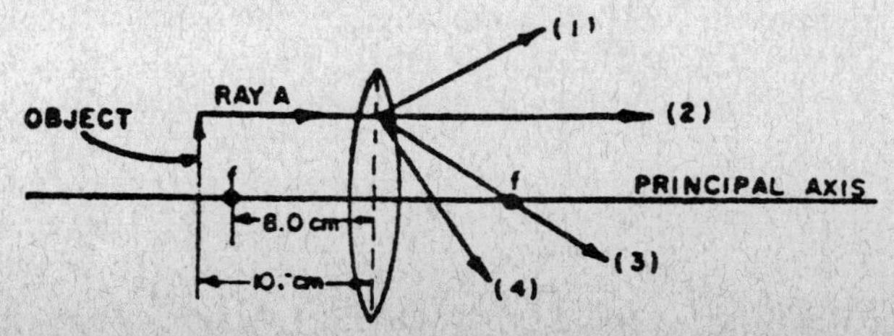

2. Which ray best represents the path of ray A after it emerges from the lens? (1) 1 (2) 2 (3) 3 (4) 4

3. The image of the object formed by the lens is (1) virtual and erect (2) virtual and inverted (3) real and erect (4) real and inverted

4. How far from the lens is the image formed? (1) 18 cm (2) 2 cm (3) 40 (4) 80 cm

5. Which phenomenon best explains why the lens produces an image? (1) diffraction (2) dispersion (3) reflection (4) refraction

6. In which direction does most of the light in ray R pass? (1) A (2) B (3) C (4) D

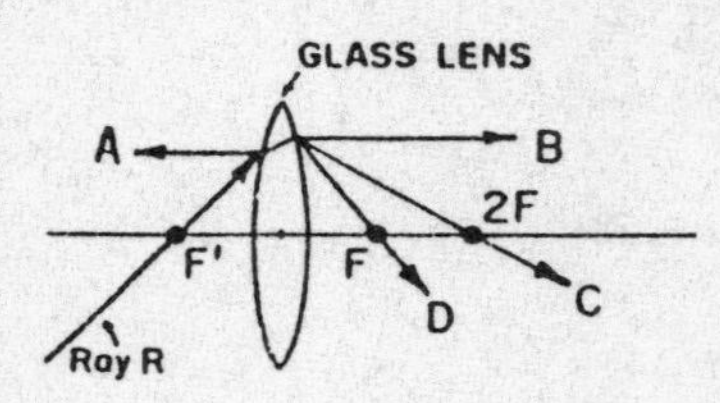

Base your answers to questions 7 through 11 on the information and diagram below. The diagram represents a converging lens made of Lucite, which is used to focus the parallel monochromatic yellow light rays shown. F and F_1 are the principal foci.

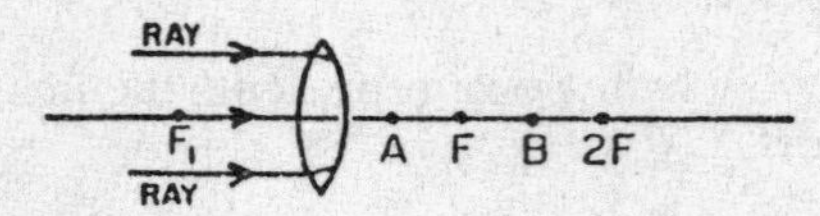

7. The rays will pass through point (1) A (2) B (3) F (4) 2F

8. If an object is placed between F_1 and the lens, the image formed would be (1) real and smaller (2) real and larger (3) virtual and smaller (4) virtual and larger

9. If an object which is placed 0.04 meter to the left of the lens will produce a real image at a distance of 0.08 meter to the right of the lens, the focal length of the lens is approximately (1) 0.015 m (2) 0.027 m (3) 0.040 m (4) 0.080 m

10. As the light emerges from the lens, its speed will (1) decrease (2) increase (3) remain the same

11. The Lucite lens is replaced by a flint glass lens of identical shape. Compared to the focal length of the Lucite lens, the focal length of the flint glass lens will be (1) smaller (2) larger (3) the same

Base your answers to questions 12 through 16 on the diagram below which represent an object placed 0.20 meter from a converging lens with a focal length of 0.15 meter.

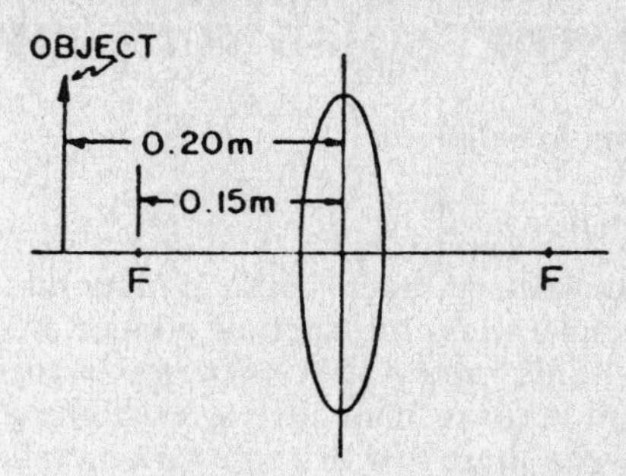

12. Which phenomenon best describes the image formation by the lens? (1) diffraction (2) dispersion (3) polarization (4) refraction

13. The image produced by the lens is (1) enlarged and real (2) enlarged and erect (3) diminished and virtual (4) diminished and inverted

14. If the object distance were increased, the image would become (1) larger and erect (2) smaller and virtual (3) smaller, only (4) larger, only

15. If the object were placed 0.10 meter from the lens, the image would be (1) enlarged and inverted (2) real and inverted (3) reduced and real (4) virtual and erect

16. Which monochromatic light, when used to illuminate the object, would produce the *smallest* image distance? (1) red (2) yellow (3) green (4) blue

UNIT X. SOLID STATE PHYSICS

I. Conduction in Solids

A. Conduction

Some materials are much better conductors of electricity than others. A metal *conductor*, such as copper, may have a resistivity of only 10^{-8} ohm-meter at room temperature. By contrast, an *insulator* such as glass, may have a resistivity as high as 10^{14} ohm-meters. Between these two extremes are metalloids such as germanium and silicon, called *semi-conductors*. The ability of materials to conduct electricity can be explained by theories based on their micro-structures.

B. Theories of Solid Conduction

1. Electron-sea Model

The electron-sea model is simple to understand. According to this model, the atoms of a solid are in fixed positions and form a geometrical pattern called a crystal lattice. If any of the outer, or valence electrons, are loosely bound to their atoms, they are easily dislodged and can drift aimlessly in the spaces between the atoms as electron clouds or an electron sea.

Connecting a battery across the solid produces an electric field which accelerates the drifting electrons toward the positive terminal. This electron current differs somewhat from our familiar ocean currents. After moving a relatively short distance, an electron will collide with another electron or with an ion which can retard its progress.

This model helps to explain why the resistivity of a conductor increases with temperature. At higher temperatures particles of the solid have more kinetic energy and vibrate faster. This increases the number of collisions and makes it more difficult for the electrons to move through the solid.

However, the electron sea model does not give an adequate explanation for the fact that an insulator conducts a *little better* and a semi-conductor conducts *very much better* at higher temperatures.

2. Band model

The band model of conduction is based on the idea that the electrons in any material are confined within discrete regions called energy bands. The three most important of these regions are the *conduction band*, the *valence band* and a forbidden band, sometimes called an *energy gap*.

Electrons can move freely within a solid if they can enter the *conduction band*. Those that are not in the conduction band are held in relatively fixed positions within the solid by atomic bonds.

The outermost electrons of each atom, called *valence electrons*, are found in the *valence band* of the solid. In some cases there is an *energy gap* separating the valence band from the conduction band.

If any electrons in the valence band have sufficient energy to break their bonds and cross the energy gap into the conduction band, they can move freely and be formed into an electric current. Electrons from other bands can also get into the conduction band if they have sufficient energy to jump energy gaps. However, under ordinary circumstances this is comparatively rare.

C. Conductor Characteristics

1. As shown on the diagram at the right, a conductor has a valence band that overlaps the conduction band. This provides the conduction band with a relatively large number of electrons to carry current.
2. All conductors have a great deal of extra space in the conduction band so the electrons can move freely when an electric field is applied.
3. As a general rule, conductors have a low resistivity which increases with temperature.

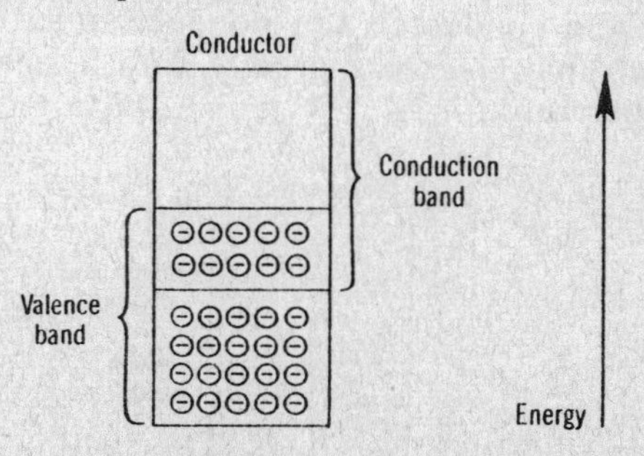

D. Insulator Characteristics

1. As shown on the diagram below, an insulator has a valence band that is almost completely filled with valence electrons.

Little or no conduction can occur in this band because the electrons have little mobility.

2. All insulators have a wide energy gap between the valence band and the conduction band. Very few electrons can obtain sufficient energy to bridge this gap.

3. Insulators have a high resistivity which decreases slightly as temperature increases.

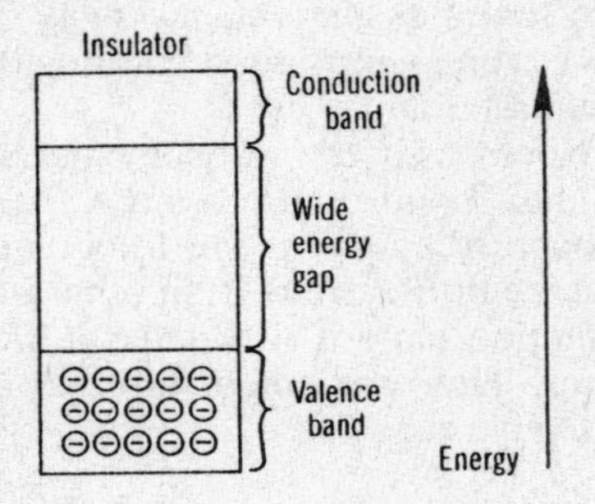

E. Semiconductor Characteristics

1. A semiconductor has a valence band that is *completely filled* with valence electrons.

2. The conduction band of a semiconductor is completely vacant at a temperature of 0 K.

3. There is a very narrow energy gap between the valence and conduction bands. As the temperature increases many electrons can gain sufficient energy to break their bonds and jump across the narrow gap into the conduction band.

4. Unlike a conductor, semiconductors have a high resistivity which may *decrease* greatly with a small increase in the temperature.

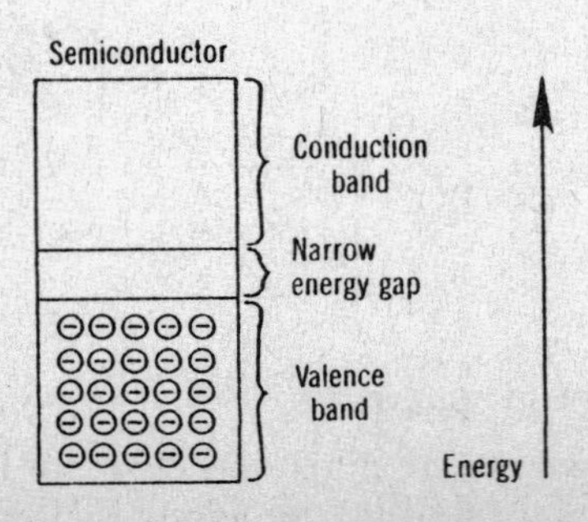

F. Intrinsic Conductivity

A highly purified semiconductor is an insulator. But at temperatures above absolute zero, it will transmit a small current due to the property of *intrinsic conductivity*.

The intrinsic conductivity becomes greater as the temperature increases. It is caused by valence electrons that become thermally excited. They break their bonds, and cross the narrow energy gap into the conduction band. Here, they have a great deal of mobility and can be channeled into an electron current.

Whenever an electron exits the valence band, it leaves behind a vacancy, or *hole*, in its former position. If an electric field is applied across the semiconductor, nearby electrons in the valence band enter these holes, leaving new holes behind them. As the process continues, it appears as if there is movement of holes within the valence band. This is called *hole current*. We may think of hole current as a group of positively charged carriers that move in the opposite direction of electron current.

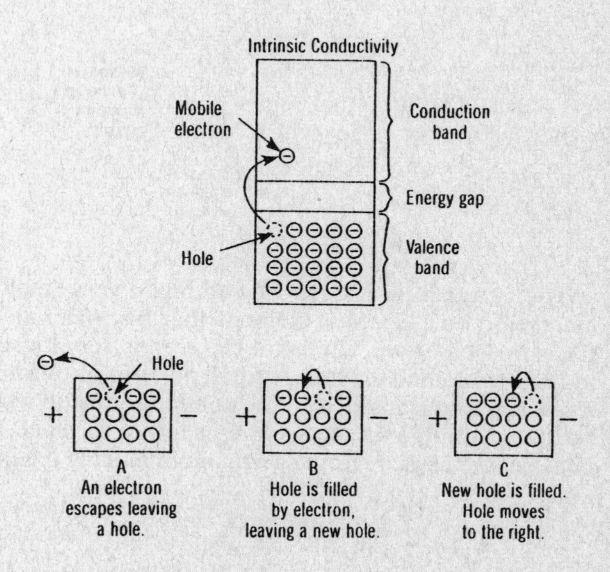

G. Extrinsic Semiconductors

1. Doping

The conduction of a semiconductor can be drastically improved by the process of *doping*. This is done by adding a very small amount of an impurity to the semiconductor.

For example, the outer shell of a geranium atom contains *four* valence electrons. A total of eight electrons is needed to complete this shell and make germanium an insulator.

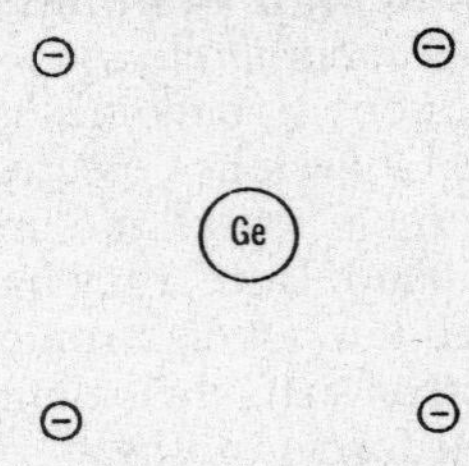

The geranium atoms group into tetrahedral, diamond-shape cells to form a solid germanium crystal. Here the adjacent atoms share their outer electrons by a process called *covalent bonding*. With eight electrons surrounding each atom, the valence band is completely filled making the germanium an insulator.

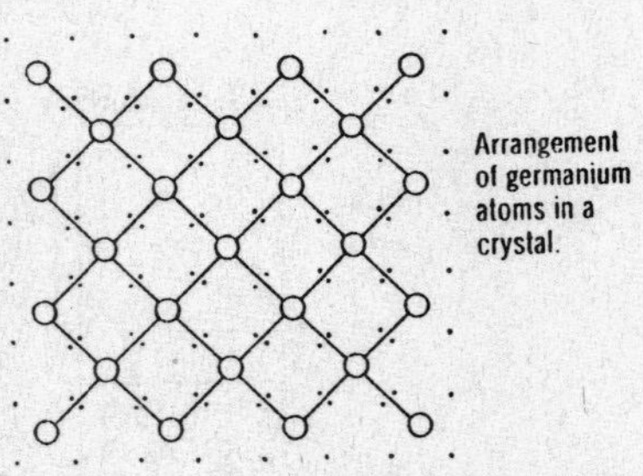

Arrangement of germanium atoms in a crystal.

The germanium is doped by adding a very small quantity of arsenic or any other element that has *five* valence electrons in each atom. The extra electron is very loosely bound in the germanium crystal. A small amount of thermal energy is sufficient to transfer it to the conduction band where it can become a charge carrier. A hole is left in its place. This supplies charge carriers to the germanium making it a conductor.

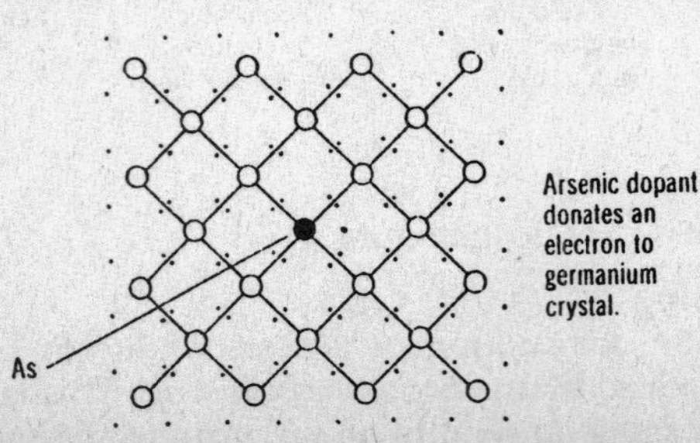

Arsenic dopant donates an electron to germanium crystal.

An alternate doping method calls for adding boron or any other element which has only *three* valence electrons in its outer shell. Three valence electrons is insufficient to form covalent bonds with all the surrounding germanium atoms. In place of one of the bonds, there is a hole in the crystal. This hole can serve as a carrier of positive charge and make the germanium a conductor.

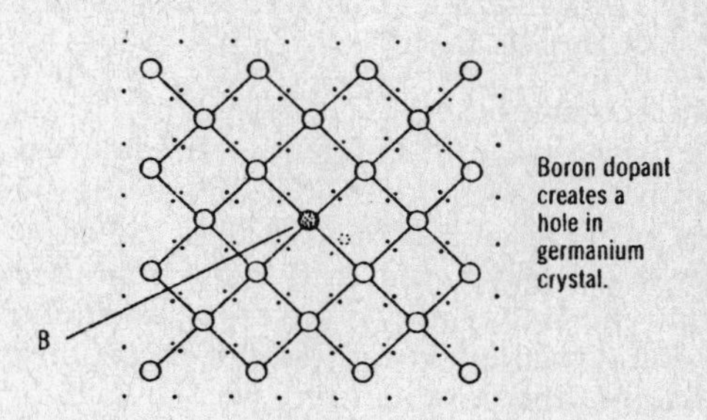

2. N-Type Semiconductors

A dopant that helps to free electrons is called a donor material. When added to an insulator, in minute quantities, it produces a negative, or an N-type, semi-conductor. The holes in an N-type semiconductor are minority charge carriers and the electrons are the majority charge carriers.

Connecting a potential difference across an N-type semiconductor causes the conduction electrons to move toward the higher (more positive) potential.

3. P-Type Semiconductors

A dopant that breaks bonds in an insulator to generate holes is called an *acceptor* material. It produces a positive, or P-type, semiconductor.

Connecting a potential difference across a P-type semiconductor causes the holes to move toward the lower, (more negative) potential.

II. Semiconductor devices

A. The (junction) Diode

A semiconductor diode is a device that allows current to flow easily in one direction but offers a high resistance in the opposite direction. It consists of a p-type semiconductor and an n-type semiconductor that are joined together.

The end of the diode that consists of the P-type material is called the *anode*. The opposite end, which consists of n-type material, is called the *cathode*. Somewhere between the anode and cathode is a very thin region where the p and n-type semiconductors meet. This region is called the *p-n junction.*

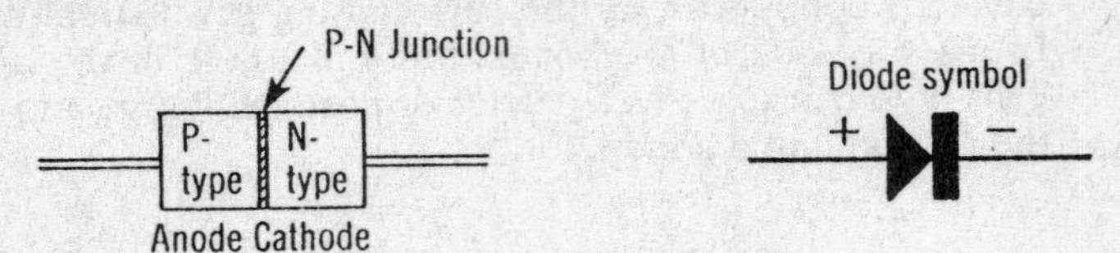

1. Diode conduction

In the absence of an electric field across the diode, electrons from the cathode drift across the p-n junction to fill holes in the anode. This creates an excessive amount of electrons in the p-type material and a deficiency of electrons in the n-type material. As a result, an electric field barrier, (potential barrier) is formed which prohibits further migration across the p-n junction.

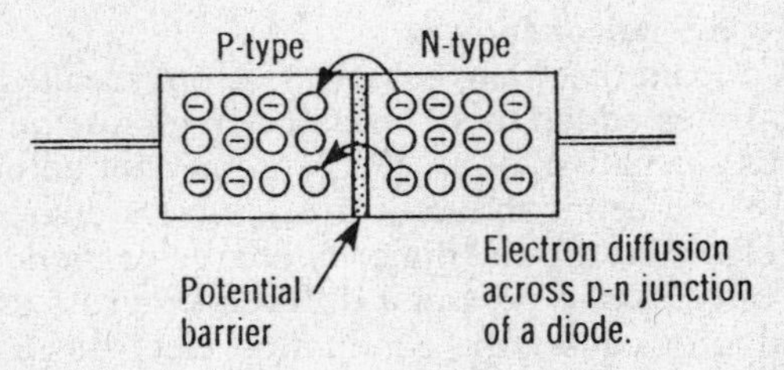

Electron diffusion across p-n junction of a diode.

A diode becomes a good conductor when an electric field is applied by connecting a battery as shown in the diagram below.

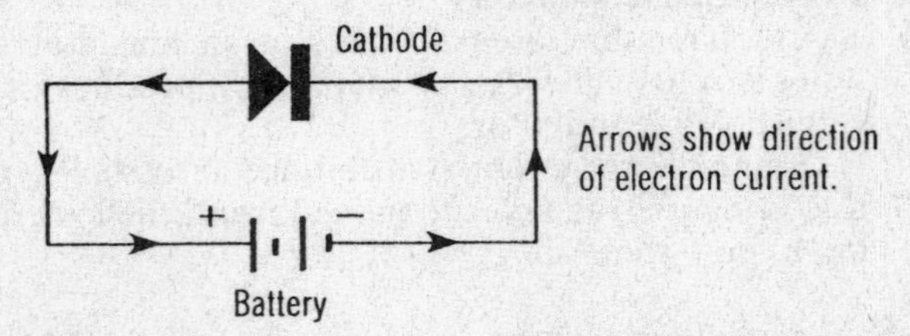

Arrows show direction of electron current.

Electrons from the negative terminal of the battery start to fill the extra holes in the cathode. Meanwhile surplus electrons at the anode are drawn into the positive terminal of the battery. As a result, the potential barrier at the p-n junction is lowered and there is an electron current in the circuit.

It is important to notice that the arrow in the diode symbol *does not* point in the direction of the electron current. By convention, it points in the opposite direction.

Increasing the battery voltage results in a sharply increased current until a maximum number of electrons and holes are involved at saturation. Because this increase of current with voltage is not linear and does not follow ohm's law, the semiconductor is said to have a *non-ohmic resistance*.

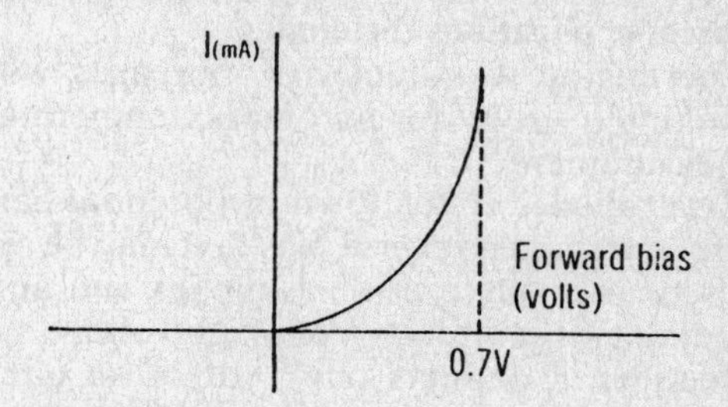

2. Diode non-conduction

Reversing the terminals of the battery (negative terminal to the anode and positive to the cathode) makes the diode a non-conductor. As in most non-conductors, there is an almost imperceptable current in the reversed direction which increases only slightly as higher voltages are applied.

However, when the voltage is somewhere between 200 and 300 volts, the high potential barrier can be broken down by an avalanche of electrons resulting in a suddenly increased current.

The actual voltge at which breakdown occurs is reduced as more doping impurities are added. If the doping is especially heavy, avalanches can occur at 10 volts or less.

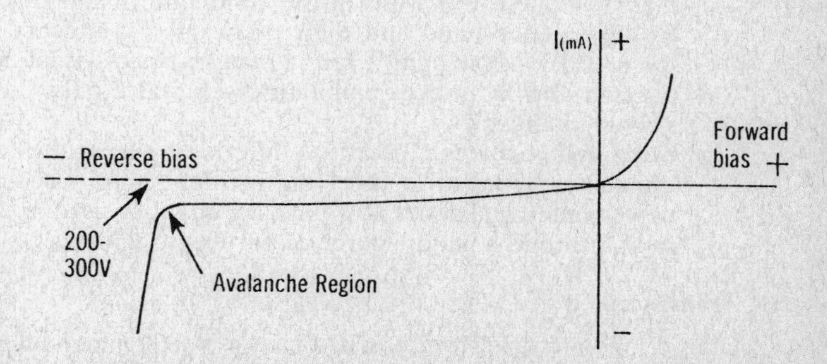

3. Diode bias

When a potential difference is connected across a diode the potential barrier may be either reinforced or diminished depending on the polarity of the potential difference. This phenomenon is known as "biasing".

A diode is *forward biased* when the polarity will support current. It is *reversed biased* when the polarity will not support current.

4. Practical applications for diodes

a. Thermistor. Conductivity increases with temperature making it useful for automatic temperature controls and measurement.

b. Zener diode. Breakdown of potential barrier occurs at a pre-determined reverse bias that may be as low as 3.4 volts. Used to regulate power supplies and automatically protect sensitive circuits from high voltage surges.

c. Rectifier. Conducts only half of an alternating current wave. This converts an alternating current into a pulsating direct current. Used in electronic devices such as radios and television sets to furnish the direct current needed to operate transistors and other circuit elements.

d. Detector. Conducts only half of each radio wave and offers a high resistance that suppresses the reversed portion. The portion that is transmitted through the diode contains all of the information that is needed to operate the speaker of an AM radio or the picture tube of a TV receiver.

e. Liquid crystal diode (LCD). Changes its ability to reflect incident light depending on the direction of an applied electric field. Used to display numbers in pocket calculators and digital clocks.

f. Light emitting diode (LED). Emits light when a reverse bias is applied. Electrons in the conduction band revert to the valence band and their extra energy appears in the form of visible light. Used in many types of visual displays found in pocket calculators, digital clocks, and warning indicators.

g. Solar cell. Creates a potential difference across the terminals in response to receiving incident light. Used to power small appliances now but they might be used in the near future as a major source of our electric power.

B. Transistors

A basic bipolar transistor consists of an n-type semiconductor sandwiched between two p-type semiconductors (PNP); or a p-type semiconductor sandwiched between two n-type semiconductors (NPN).

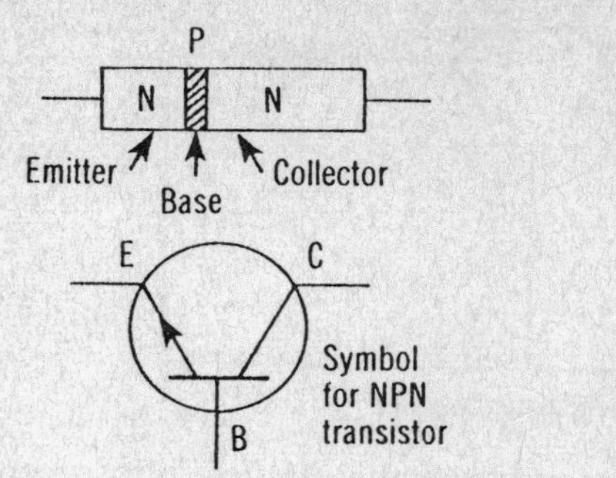

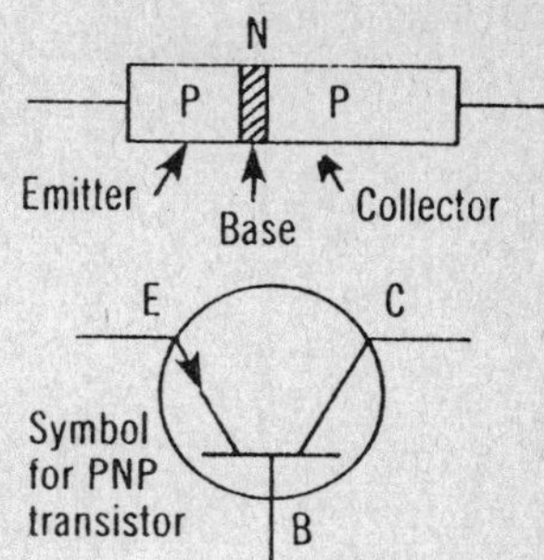

1. **Functional parts.**
   According to function, the three main parts of a transistor are called the *emitter, base, and collector.*
   a. **Emitter.** The emitter is *heavily doped* to serve as a source of charge carriers (electrons or holes).
   b. **Base.** A very thin layer that transmits most of the charge carriers from the emitter to the collector.
   c. **Collector.** Collects charge carriers from the base and makes them available to other circuit elements in an electronic device. It is large and *lightly doped* to supply many minority carriers.

2. **NPN Transistor as an Amplifier**
   The NPN transistor contains an n-p junction followed by a p-n junction.
   a. **Biasing.** A source of low-voltage dc is connected across the n-p junction to provide a forward bias. A high-voltage dc source is connected across the p-n junction to provide a reverse bias.

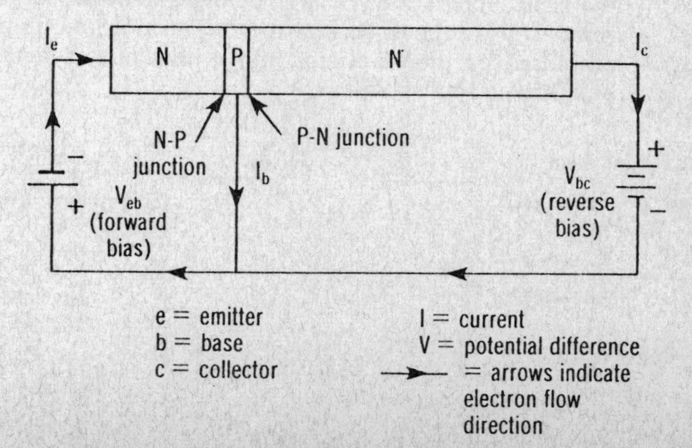

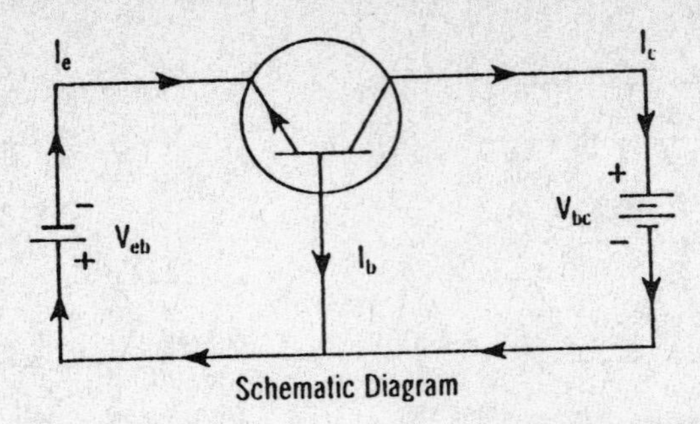

Schematic Diagram

b. **Base current.** Electrons from the emitter diffuse across the n-p junction to fill the holes in the very thin p-type semiconductor base. Some of these electrons are returned to the emitter to replenish its supply. This forms a small current in the emitter-base circuit.

c. **Collector current.** Because the base is very thin, most of the electrons that cross the n-p junction and fill holes in the base are readily available for withdrawal across the p-n junction. Encouraged by the high reverse bias, they cross the junction and a comparatively large collector current is generated.

d. **Amplification.** Removing the forward bias from the emitter-base circuit would leave unfilled holes in the base. The collector current would drop to zero. On the other hand, if the forward bias is increased, the holes in the base will fill faster. Quickly replenished electrons in the base enable the collector current to become much stronger.

A small ac signal can be fed through an input transformer into the emitter of a transistor. Small changes in the base current control large changes in the collector current. Thus, a small ac signal from a microphone can be amplified by the transistor into a high power signal that can drive a loudspeaker.

Microphone
E
C
Speaker
Input Signal
B
Amplified output

Schematic diagram of NPN transistor amplifier

3. PNP transistor operation.
Operation of a PNP transistor is essentially the same as that of an NPN transistor with polarity reversed and holes replacing the electrons.

C. Integrated Circuits

An integrated circuit (IC) is a paper thin silicon chip. Measuring only a few millimeters on a side, it may contain thousands of electronic parts. IC's are made by evaporation and electro-deposition of conductor and semiconductor materials on the silicon.

A pocket calculator or radio usually has only one IC inside. Giant digital computers which used to fill a large size room have been miniaturized with IC's to the size of an attaché case. In many cases these portables can perform more functions than the older giants.

QUESTIONS

1. A doping material containing five valence electrons is classified as (1) an acceptor (2) a thermistor (3) an intensifier (4) a donor
2. What does the circuit diagram below represent? (1) forward-biased diode (2) reverse-biased diode (3) forward-biased transistor (4) reverse-biased transistor

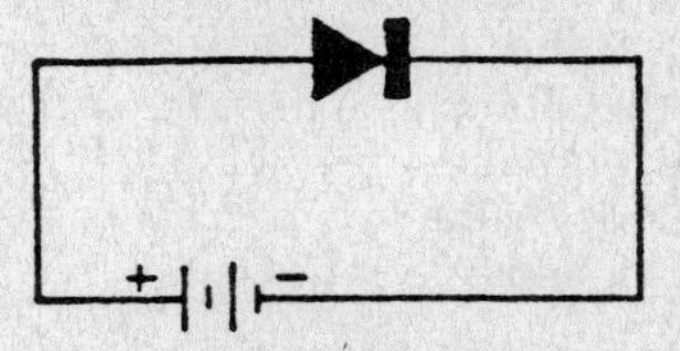

3. As the temperature of a semiconductor increases, its conductivity (1) decreases (2) increases (3) remains the same
4. Which would occur if the connections to the source of potential difference in the diagram below were reversed? (1) The current in the circuit would decrease (2) The current in the circuit would increase (3) The current in the circuit would remain the same

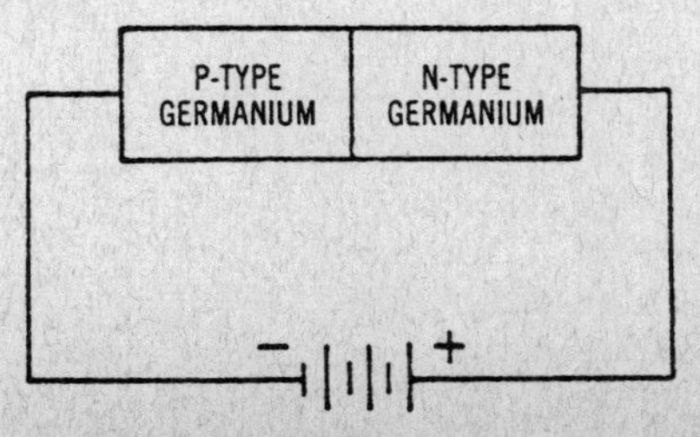

5. A large number of electrons move freely thoughout a solid material. This material is best classified as (1) a conductor (2) an insulator (3) an *N*-type semiconductor (4) a *P*-type semiconductor

6. In an *N*-type semiconductor there is an excess of (1) free electrons (2) protons (3) atoms (4) ions

7. Current carriers in a semiconductor are (1)electrons, only (2) protons, only (3) electrons and positive holes (4) protons and negative holes

8. A potential difference is applied to a *P*-type semiconductor as shown in the diagram below.

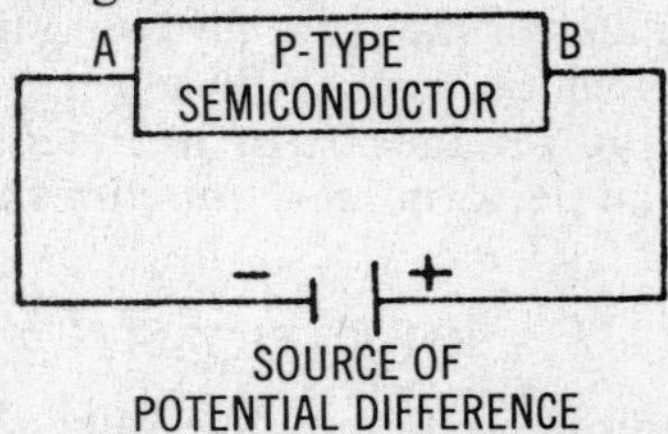

Current is conducted through this semiconductor by (1) positive holes moving from end *A* to end *B* (2) positive holes moving from end *B* to end *A* (3) conducting protons moving from end *A* to end *B* (4) conducting protons moving from end *B* to end *A*

9. The diagram at the right shows the alternating current input signal to a diode. What will the output signal look like?

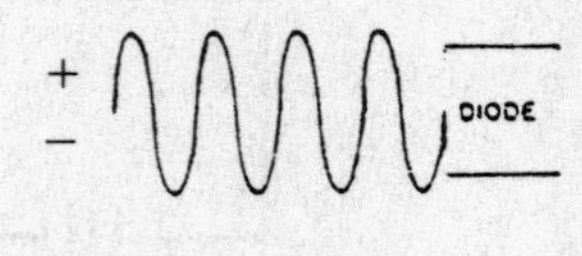

(1) 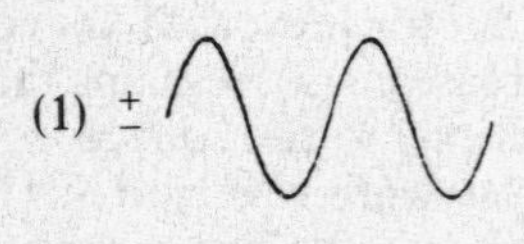

(3) 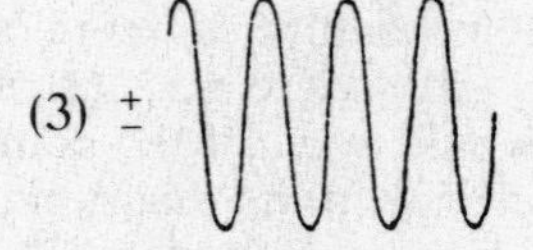

(2) 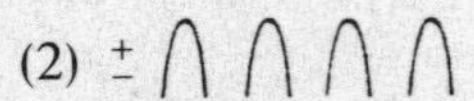

(4) 

10. The diode in the diagram at the right is considered to be (1) open biased (2) closed biased (3) forward biased (4) reverse biased

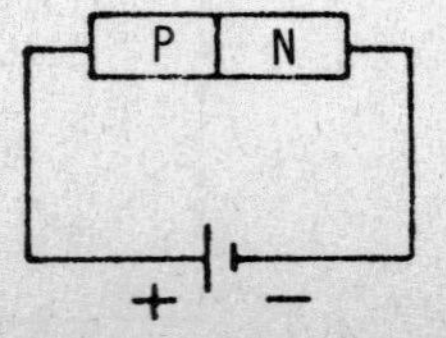

11. An incandescent light bulb is connected in series with a diode and a source of direct current. If the diode is forward biased the light bulb will (1) light and stay lighted (2) flicker with a period dependent on the DC current (3) not light at all (4) light at first, then dim out

12. As the emitter-base bias of an NPN transistor is increased, the base current (1) decreases (2) increases (3) remains the same

13. As the collector-base bias of an NPN transistor is increased, the collector current (1) decreases (2) increases (3) remains the same

14. In a NPN transistor, the greatest amount of semiconductor material is located in its (1) emitter (2) base (3) collector (4) p-n junctions

15. In a PNP transistor, the greatest amount of semiconductor material is located in its (1) emitter (2) base (3) collector (4) p-n junctions

16. Silicon has four valence electrons in its outer shell. Doping silicon with an element having three valence electrons produces a (1) good conductor (2) good insulator (3) p-type semiconductor (4) n-type semiconductor

17. An avalance of electrons in a diode is produced by (1) insufficient forward bias (2) excessive forward bias (3) insufficient reverse bias (4) excessive reverse bias

UNIT XI. NUCLEAR ENERGY

I. The Nucleus

Most of the mass and all of the positive charge of an atom is concentrated in its central core, known as the nucleus.

A. Nucleons

The particles comprising the nucleus are called nucleons. The two major nucleons are:

Protons are those particles considered to have a mass of one atomic mass unit and one unit of positive charge.

Neutrons have a mass about the same as that of protons, but with a charge of zero.

B. Atomic Number

Depending on data collected using X-rays, Henry Mosely showed that each element contains a characteristic number of protons. This number is now considered the *atomic number* of the element. The symbol for atomic number is Z.

C. Mass Number

Mass number is the total number of protons and neutrons in the nucleus. The symbol for mass number is A.

D. Nuclear Force

The *strong nuclear force* as it is now called, holds nucleons together. It acts only when particles are very close together (less than 10^{-15} meter apart). It is always attractive and it's magnitude is the same between protons and protons, protons and neutrons, and neutrons and neutrons. It is the largest force in the nucleus that we know.

E. Nucleus Mass and Binding Energy

The binding energy of a nucleus is defined as the amount of energy required to remove all the nucleons from that nucleus. As a result the sum of the masses of these nucleons is slightly more than the mass of the assembled nucleus. This difference in mass is called the *mass defects* (m). It is the mass equivalent of the binding energy (E) according to the relationship $E = mc^2$. The *binding energy* can also be defined as the energy *released* when the nucleons *form* a nucleus. When binding energies are compared it is frequently in terms of binding energy per nucleon.

F. The Atomic Mass Unit

The atomic mass unit is defined as 1/12th the mass of carbon-12 and its symbol is u

$$1 \text{ u} = 1.66 \times 10^{-27}\text{kg}$$

$$\begin{aligned} \text{Mass of carbon} &= 12.000 \text{ u} \\ \text{proton} &= 1.0073 \text{ u} \\ \text{neutron} &= 1.0087 \text{ u} \\ \text{electron} &= 0.0005 \text{ u} \end{aligned}$$

G. Mass-Energy Relationship

1. **Conservation of mass-energy.** During the process of radioactive disintegration, mass-energy is conserved. Since masses are normally measured in atomic mass units, it will be useful to calculate the energy equivalent of one u (1.66×10^{-27}kg).

2. **Einstein's mass-energy equation** provides the basis for calculating the amount of energy that could be obtained from one atomic mass unit.

$$\begin{aligned} E &= mc^2 \\ E &= \text{energy equivalent (in joules)} \\ m &= \text{mass of one amu (in kilograms)} \\ c &= \text{velocity of light } (3.00 \times 10^8 \text{m/s}) \end{aligned}$$

$$E = (1.66 \times 10^{-27}\text{kg})\,(3.00 \times 10^8\text{m/s})^2$$

$= 1.49 \times 10^{-10}$ joules, or, converting to electron volts:

$$1.49 \times 10^{-10}\text{ joules} \times \frac{1 \text{ eV}}{1.6 \times 10^{-19}\text{ joules}}$$

$$= 9.31 \times 10^8\text{eV (931 MeV)}$$

This value was confirmed experimentally by Cockroft and Walton in 1932.

3. **Pair production.** When a gamma ray is annihilated its energy is transformed into an electron-positron pair. If these particles collide, they annihilate each other and gamma rays are formed. In nuclear physics, mass is usually given in energy units of electron-volts (eV) or million electron volts (MeV).

H. Isotopes

Atoms of a given element that do not have the same number of neutrons in their nuclei are called *isotopes* of the element. For example $^{234}_{92}U$ and $^{238}_{92}U$ are both isotopes of uranium. Each atom of U-234 has 142 neutrons. Those of U-238 have 146 neutrons.

I. Nuclides

There are over a hundred elements including those that are natural and some that are man made. Each of these may have several isotopes. This produces many hundreds of different nucleus varieties, called *nuclides*.

J. Methods of learning about the atom

1. Observational devices

Some of the tools used in the study of nuclear changes (radioactivity) are the electroscope, photographic plates, geiger counters, scintillation counters and closed chambers.

a. The electroscope is used to detect the presence of radiations. A charged electroscope loses its charge when the air around it is ionized. The rate at which the charge is lost is a measure of how many charged particles are nearby.

b. Photographic plates

Photographic plates are used to study the *path* of particles. Ionized regions in the plate become visible when the plate is developed.

c. The geiger counter

The geiger counter detects and measures radio activity. It consists of a metal cylinder filled with a gas with a wire down the center of the cylinder. The wire has a high potential which will attract ions in the cylinder. When charged particles pass into the tube they ionize the gas. These ions move toward the wire causing more ionizations. Electrical pulses created by the ions are directed toward some counting device.

d. Scintillation counters

Scintillation counters record flashes of light produced by the particles. The flashes are counted by a photoelectric cell connected to an electronic counter.

e. Cloud chambers

In cloud chambers vapor trails are produced along the path of ionized particles through air saturated with water vapor. These are photographed. Similar arrangements are used in bubble chambers which trace the path followed by bubbles in a liquid maintained at its boiling point.

2. Particle Accelerators

Accelerators provide charged particles with sufficient kinetic energy to penetrate the nucleus.

a. A linear accelerator

A linear accelerator consists of a series of straight evacuated tubes placed end to end with gaps between them. Differences of potential in the gaps cause particles to accelerate each time they proceed from one tube to the next.

b. The synchroton

The synchroton provides a means of making accelerators smaller by using magnetic fields to bend the path of the particles being accelerated into circles. The predecessor to the synchroton was the cyclotron which operated on a similar principle.

c. The Van de Graaff generator

The Van de Graaff generator is a large electrostatic machine that can produce a large potential difference between two terminals. Acceleration of charged particles is accomplished by allowing the particles to fall through the very large potential difference that is created by the generator.

3. Subatomic Particles

For each subatomic particle, there is an antiparticle that has the same mass but an opposite charge or another opposite property.

Heavier nuclear particles, such as BARYONS (protons, neutrons, and hyperons) are believed to be composed of quarks. Each quark has 1/3 or 2/3 the basic charge of an electron. However, attempts to isolate individual quarks have not been successful so far.

II. Nuclear Reactions

Elements that emit energy naturally without the absorption of energy from an outside source are said to be *naturally radioactive*. Elements with atomic numbers greater than 83 (bismuth) have no known *stable* isotopes. They are all radioactive. Radioactivity is associated with some form of *disintegration*, or *decay* of the nucleus. The disintegration results in the emission of particles and/or radiant energy. In all nuclear reactions, the total charge and mass number on both sides of the equation must balance.

1. **Alpha Decay.** Alpha particles have been identified as helium nuclei that are ejected at high speeds from certain radioactive isotopes. The reaction involved in the nuclear disintegration resulting in the emission of alpha particles is called **alpha decay**. The emitting atoms are called **alpha emitters**. The sym-

bols used to designate isotopes, particles, or radiations in nuclear reactions follow this pattern:

$$^{A}_{Z}X$$

in which X is the symbol of the element or the emission, A is the mass number and Z is the charge.

The alpha particle is designated:

$$^{4}_{2}He$$

An example of an equation for an alpha decay is:

$$^{226}_{88}Ra \rightarrow ^{222}_{86}Rn + ^{4}_{2}He$$

2. **Beta Decay.** Beta particles have the same charge and mass as electrons. They are negatively charged particles ejected at high speeds from certain radioactive isotopes. The reaction is called **beta decay** and the atom is called a *beta emitter*. The equation for the reaction in which beta particles are emitted by Uranium-235 is written:

$$^{235}_{92}U \rightarrow ^{0}_{-1}e + ^{235}_{93}Np$$

Since a beta particle has a charge of minus one, balancing the charges results in an *increase* of one in the atomic number of the product isotope. Beta decay raises two questions:

1. How can the product gain a proton and not gain mass?
2. How can the nucleus emit an electron?

The accepted explanation involves a series of nuclear reactions, called *neutron decay*, which results in a neutron being transformed into a proton and an electron. This electron, called a beta particle is ejected from the nucleus and leaves the extra proton behind.

3. **Gamma Radiation**

Since gamma rays are not affected by an electric field, they are said to possess zero charge. Since they have no mass, they are not considered particles. Because of their penetrating ability, their properties resemble those of X-rays, but they have higher energies than most X-rays.

Characteristics of Alpha, Beta and Gamma Emissions

Name	Symbol	Charge	Mass (amu)	Relative Penetrating Power	Relative Ionizing Power
Alpha	$^{4}_{2}He$	+2	4	low	10,000
Beta	$^{0}_{-1}$	−1	0	moderate	100
Gamma	γ	0	0	high	1

B. Half-life

Each radioactive isotope disintegrates at a characteristic rate. The time required for half the mass of a given sample to disintegrate is called the *half-life* of that isotope. The half-life of iodine-131 ($^{131}_{53}I$) is 8 days. If the initial mass of a sample of iodine-131 was 100 grams, at the end of an eight day interval, only 50 grams of iodine-131 would remain. After *another* eight days had passed, only half of the 50 grams, or 25 grams, would remain. After *another* 8 days, only 12.5 grams would remain, and so on. Other radio isotopes range in value from about 10^{-22} seconds to 10^{17} years. A useful mathematical expression for predicting radioactive mass that will remain after a given time is:

$$m_f = \frac{1}{2^n} m_i$$

m_f = final mass
n = number of half-lives
m_i = initial mass

C. Conservation of Mass-energy

During the process of radioactive decay mass and energy are always conserved.

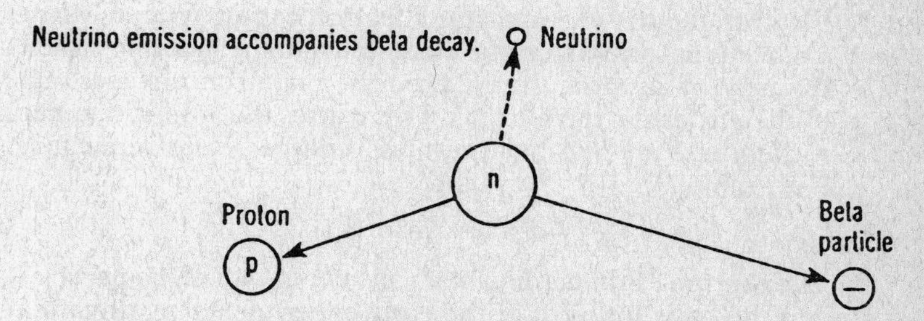

There appears to be a loss of energy whenever a neutron disintegrates into a proton and a beta particle. To maintain the principle of conservation of energy during such beta decay reactions, the emission of *neutrinos* was postulated by Fermi. Neutrinos have no mass or charge, just energy.

D. Artificial Transmutation

When the nuclei of stable atoms are bombarded by accelerated particles, the nuclei become unstable and may be transformed into isotopes or different elements. This process is called *induced*, or *artificial* transmutation. Radio-activity is an example of *natural* transmutation. In the first reported artificial transmutation, nitrogen was bombarded with alpha particles to produce oxygen and hydrogen.

$$^{14}_{7}N + ^{4}_{2}He \rightarrow ^{17}_{8}O + ^{1}_{1}H$$

This was credited to Rutherford and his group in 1919.

In 1934, Irene Curie and Frederick Joliot-Curie produced neutrons by bombarding aluminum with alpha particles to produce a *radioactive* isotope of phosphorus.

$$^{4}_{2}He + ^{27}_{13}Al \rightarrow ^{30}_{15}P + ^{1}_{0}n$$

The $^{30}_{15}P$ was found to emit positrons. Positrons have the same mass as electrons, but they are positively charged. This was the first record of *induced* radioactivity. The reaction:

$$^{30}_{15}P \rightarrow ^{30}_{14}Si + ^{0}_{+1}e$$

1. **Positron emission.**
   Beta decay as it occurs in *induced* radioactivity, is said to include emissions of *positrons* as well as electrons. Two examples of induced beta decay are:

   $$^{64}_{29}Cu \rightarrow ^{64}_{28}Ni + ^{0}_{+1}e \text{ (positron)}$$

   $$\text{and } ^{24}_{11}Na \rightarrow ^{24}_{12}Mg + ^{0}_{-1}e \text{ (electron)}$$

2. **Electron capture (K-capture).** Electron capture occurs when a nucleus absorbs an orbital electron from the innermost shell (K-shell) of an atom. In the process, a neutron replaces one of the protons in the nucleus. This causes the atomic number to decrease by one but the mass number is not changed. For example:

   $$^{40}_{19}K + ^{0}_{-1}e \rightarrow ^{40}_{18}Ar$$

3. **The neutron.** Bombarding many elements with alpha particles was the basis for most of the early experiments investigating radioactivity. When beryllium and boron were bombarded with alpha particles, a radiation more penetrating than any previously encountered, was produced. In 1932, James Chadwick showed that the particles emitted from beryllium had no charge but had about the same mass as protons. This particle was identified as the neutron. Neutrons have several advantages over alpha particles as bombarding agents. Some of these advantages are:

   a. They are not repelled by nuclei, because they are not charged.

   b. It is not necessary to give neutrons very high kinetic energies in order for them to participate in nuclear reactions.

c. When neutrons get close enough to the nucleus, a strong force *attracts* them.
d. Neutrons can be made to stay near the nucleus by slowing them down by collisions with small nuclei with which they do *not* react.

F. Nuclear Fission

Nuclear fission is the splitting of a heavy nucleus into two lighter nuclei that are nearly equal to each other in atomic number. For example, when uranium-235 absorbs neutrons, the nuclei each split usually into two almost equal parts. In the reaction, in addition to the new nuclei produced, two or more neutrons per atom are emitted and energy is released. The amount of energy released per nucleon is the difference between the binding energy per nucleon of the uranium-235 and the average energy per nucleon of the new elements formed.

1. **Nuclear fuels.** Nuclear fuels, such as uranium-235 and polonium-239, are formed into small pellets for use in large-scale nuclear reactors. A three cubic centimeter pellet of uranium has as much energy as a ton of coal. Fuel rods, containing these pellets are inserted into the core of a nuclear reactor.
2. **Thermal neutrons** are required for fission of uranium-235. These are neutrons whose kinetic energies are about equal to the kinetic energies of molecules at ordinary temperatures.
3. **Moderators** are materials that have the ability to slow down fast-moving neutrons so they can be captured by a target nucleus. This slowing down effect can be accomplished by head-on collisions with particles of similar mass. Moderators commonly used include water, heavy water, and graphite.
4. **Chain reaction.** Given a critical mass of nuclear fuel, a self-sustained series of fissions, called a chain reaction may occur. It starts when a slow moving neutron enters a nucleus and causes fission. Among the fission products are several fast neutrons. When they are slowed down by graphite or other moderator they can be captured by other nucleii to continue the process.
   a. Some of the neutrons produced by fission escape and cannot contribute any further to the reaction.
   b. Some of the neutrons are captured by other atoms; i.e., impurities, U^{238} and parts of the reactor.
   c. When a controlled chain reaction is working at a constant level, each fission supplies, on the average, one neutron which initiates another fission.

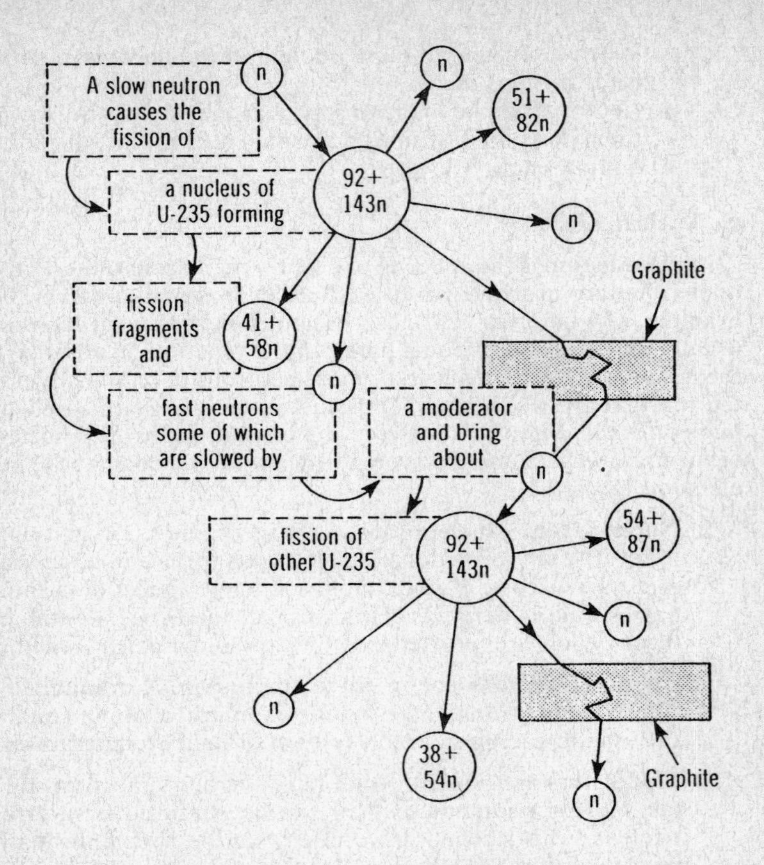

d. Fissions do not consistently form the same fission products.
e. Dozens of different isotopes of intermediate weight have been identified as products of U^{235} fission.
f. Fission fragments as a rule have too many neutrons for stability and are radioactive. They generally emit beta particles until a stable balance of protons and neutrons is achieved. These fission fragments accumulate in a reactor, acting as impurities which reduce the efficiency of the chain reactions, so that they must be periodically removed. Disposal of these highly radioactive wastes has been a most troublesome problem.

5. **Control rods.** Rods, such as those made of boron or cadmium, are partially inserted into the core of a reactor to control the rate of energy release during a chain reaction. They do this by absorbing excess neutrons. Inserting the control

rods completely into the core so many neutrons that the chain reaction stops.

6. **Coolants.** To keep the temperatures generated by fission at reasonable levels, gas or liquid coolants are piped through the reactor. They also carry out energy in the form of heat. This is used to rotate turbines and generate electricity. Examples of coolants are air, water, heavy water, helium, carbon dioxide, molton sodium and moltan lithium. In some reactors, the coolant also serves as the moderator.
7. **Shielding.** To protect the walls of a reactor from radiation damage, they are covered with a steel shield. In addition, one or more thick walls of high density concrete surround the reactor to protect personnel.
8. **Radioactive wastes.** Waste materials from nuclear reactors contain highly radioactive fission products. Special procedures are required for their disposal.
   - **a.** Solids and liquids are encased in special containers that resist corrosion. The containers are kept underground for long term storage in isolated areas.
   - **b.** Wastes are vitrified into solid blocks to make future spills or leakage impossible.
   - **c.** Low-level wastes are diluted until they are harmless. Then, they are released directly into the environment.
   - **d.** Gaseous radioactive wastes may contain dangerous quantities of radon-222, krypton-85, and nitrogen-16. They are allowed to decay in storage until tests show that they can be safely released into the air.
   - **e.** Decomissioned reactors may also contain products that are highly radioactive. If so, the entire structure may have to be isolated and treated as radioactive waste.

F. Fusion

When two or more light nuclei combine to form a single nucleus of greater mass, the reaction is called *nuclear fusion*. The energy released in fusion reactions is much greater than that which is released in fission reactions. The mass of the newly formed nucleus is *less* than the sum of the masses of the light nuclei. The difference in mass represents the mass that was converted to energy during the process. Some of this energy provides for the greater binding energy per nucleon and, therefore, the greater stability of the heavier nucleus formed. An example of a fusion reaction between two isotopes believed to be the source of most stellar energy is:

$$^3_1H + ^1_1H \rightarrow ^4_2He + \text{energy}$$

1. **Fuels.** Since the magnitude of repulsion increases with charge, only nuclei having small positive charges may be used. Ordinary hydrogen has a very low reaction probability. However the hydrogen isotopes deuterium (^2_1H) and tritium (^3_1H) are useful as nuclear fuels.
   a. **Deuterium** is obtained from heavy water (deuterium oxide). All water contains some traces of heavy water. These traces can be concentrated to form fuel for fusion.
   b. **Tritium** (^3_1H) can be produced by the nuclear reaction:

$$^6_3Li + ^1_0n \rightarrow ^3_1H + ^4_2He$$

2. **High energy requirement.** Nuclei repel each other because each has a positive charge. To force them to interact, they must be given enough kinetic energy to overcome their repulsion.

3. **Energy liberated.** Two deuterium nuclei will liberate about 4.0 MeV of energy in the reaction:

$$^2_1H + ^2_1H \rightarrow ^3_1H + ^1_1H + Q$$

A deuterium and a tritium atom will liberate 17.6 MeV in the reaction:

$$^2_1H + ^3_1H \rightarrow ^4_2He + ^1_0n + Q$$

QUESTIONS

Base your answers to questions 1 through 5 on the graph below which represents the disintegration of a sample of a radioactive element. At time $t = 0$ the sample has a mass of 4.0 kilograms.

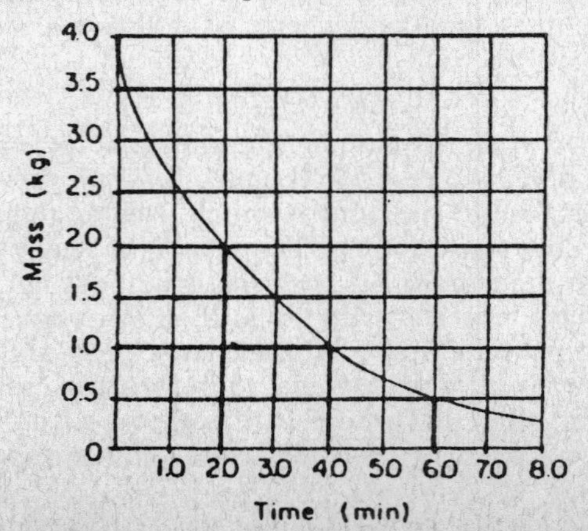

1. What mass of the material remains at 4.0 minutes? (1) 1 kg (2) 2 kg (3) 0 kg (4) 4 kg
2. What is the half-life of the isotope? (1) 1.0 min (2) 2.0 min (3) 3.0 min (4) 4.0 min
3. How many half-lives of the isotope occurred during 8.0 minutes? (1) 1 (2) 2 (3) 8 (4) 4
4. How long did it take for the mass of the sample to reach 0.25 kilogram? (1) 1 min (2) 5 min (3) 3 min (4) 8 min
5. If the mass of this material had been 8.0 kilograms at time $t = 0$, its half-life would have been (1) less (2) greater (3) the same
6. What is the force which holds the nucleons of an atom together? (1) coulomb force (2) magnetic force (3) atomic force (4) nuclear force
7. Which device is used to detect nuclear radiation? (1) synchrotron (2) cloud chamber (3) linear accelerator (4) cyclotron

Base your answers to questions 8 through 12 on the two nuclear equations below.

$$^{27}_{13}\text{Al} + {}^{4}_{2}\text{He} \rightarrow {}^{30}_{15}\text{P} + X + \text{energy}$$

$$^{30}_{15}\text{P} \rightarrow {}^{30}_{14}\text{Si} + Y + \text{energy}$$

8. The first equation indicates that the radioactive phosphorus is produced by bombarding $^{27}_{13}\text{Al}$ with (1) neutrons (2) positrons (3) alpha particles (4) protons
9. In the first equation, particle X is (1) a neutron (2) an electron (3) a positron (4) a neutrino
10. In the second equation, particle Y is (1) an alpha particle (2) a neutron (3) an electron (4) a positron
11. The number of neutrons in the nucleus of $^{27}_{13}\text{Al}$ is (1) 13 (2) 14 (3) 27 (4) 40
12. Compared to the mass of the $^{30}_{15}\text{P}$, the sum of the masses of $^{30}_{14}\text{Si} + Y$ is (1) less (2) more (3) the same
13. Which device could be used to give a positively charged particle sufficient kinetic energy to penetrate the nucleus of an atom? (1) electroscope (2) Geiger counter (3) cloud chamber (4) Van de Graaff generator

Base your answers to questions 14 through 18 on the information below.

A certain radioactive isotope with a half-life of 5.0 minutes decays to a stable (nonradioactive) nucleus by emitting one alpha particle.

14. If the mass of the original sample was 1.00 kilogram, how long would it take to form 0.75 kilogram of the stable material, leaving 0.25 kilogram of the original sample? (1) 2.5 minutes (2) 5.0 minutes (3) 10. minutes (4) 15. minutes

15. The difference between the mass number of the original nucleus and the mass number of the new stable nucleus is (1) 1 (2) 2 (3) 3 (4) 4

16. The difference between the atomic number of the original nucleus and the atomic number of the new stable nucleus is (1) 1 (2) 2 (3) 3 (4) 4

17. Compared to the binding energy per nucleon of the original radioactive nucleus, the binding energy per nucleon of the new stable nucleus is (1) less (2) greater (3) the same

18. As the absolute temperature of the radioactive isotope increases, its half-life (1) decreases (2) increases (3) remains the same

19. The total number of neutrons in the nucleus of any atom is equal to the (1) mass number of the atom (2) atomic number of the atom (3) atomic number minus the mass number (4) mass number minus the atomic number

20. If 100 Mev of energy is released by an atom during fission, the amount of mass that is converted to energy is approximately (1) 1 amu (2) 0.1 amu (3) 9 amu (4) 9×10^4 amu

21. The function of the moderator in a nuclear reactor is to (1) absorb neutrons (2) slow down neutrons (3) speed up neutrons (4) produce extra neutrons

Base your answers to questions 22 through 26 on the information and nuclear equations below.

When nitrogen is bombarded with protons, the first reaction that occurs is ${}^{14}_{7}N + {}^{1}_{1}H \rightarrow {}^{15}_{8}O + X$. The oxygen produced is radioactive, with a half-life of 0.10 second, and decays in the following manner: ${}^{15}_{8}O \rightarrow {}^{15}_{7}N + Y$.

22. The first reaction is an example of (1) alpha decay (2) beta decay (3) induced transmutation (4) natural radioactivity

23. In the first reaction, *X* represents (1) an alpha particle (2) a beta particle (3) a neutron (4) a gamma photon

24. In the second reaction, *Y* represents (1) an electron (2) a neutron (3) a positron (4) a proton

25. If a 4.0-kilogram sample of ${}^{15}_{8}O$ decays for 0.40 second, the mass of ${}^{15}_{8}O$ remaining will be (1) 1.0 kg (2) 2.0 kg (3) 0.50 kg (4) 0.25 kg

26. As the amount of ${}^{15}_{8}O$ decreases, the half-life (1) decreases (2) increases (3) remains the same

27. The nucleus of isotope *A* of an element has a larger mass than isotope *B* of the same element. Compared to the number of protons in the nucleus of isotope *A*, the number of protons in the nucleus of isotope *B* is (1) less (2) greater (3) the same

28. Which device is normally used to accelerate charged particles? (1) cyclotron (2) electroscope (3) cloud chamber (4) Geiger counter

Base your answers to questions 29 through 33 on the equation of a nuclear reaction and information below.

$$^{3}_{1}H + ^{1}_{1}H \rightarrow ^{4}_{2}He + Q$$

The mass of the nuclei is:

$$^{1}_{1}H = 1.00813 \text{ u}$$
$$^{3}_{1}H = 3.01695 \text{ u}$$
$$^{4}_{2}He = 4.00388 \text{ u}$$

29. In the equation, $^{3}_{1}H$ and $^{1}_{1}H$ are (1) nucleons (2) alpha particles (3) isotopes (4) quanta

30. The total number of nucleons in $^{4}_{2}He$ is (1) 1 (2) 2 (3) 3 (4) 4

31. Energy is produced in the reaction. The mass equivalent of this energy is approximately (1) 1 u (2) 2 u (3) .01 u (4) .02 u

32. The equation represents (1) fusion (2) fission (3) alpha decay (4) beta decay

33. In the equation, the alpha particle is represented by the symbol (1) $^{3}_{1}H$ (2) $^{1}_{1}H$ (3) $^{4}_{2}H$ (4) Q

Base your answers to questions 34 and 35 on the information below.

Mass of a proton = 1.007277 u
Mass of a neutron = 1.008665 u
Mass of an electron = 0.0005486 u
Mass of a He nucleus = 4.001509 u

34. How many neutrons are in a $^{4}_{2}He$ nucleus? (1) 1.985567 amu (2) 1.985018 amu (3) 0.030375 amu (4) 0.029278 amu

35. What is the mass defect of a $^{4}_{2}He$ nucleus? (1) 1.985567 amu (2) 1.985018 amu (3) 0.030375 amu (4) 0.029278 amu

Base your answers to questions 36 through 40 on the nuclear equation below.

$$^{30}_{15}P \rightarrow ^{A}_{Z}Si + ^{0}_{+1}X$$

36. In the equation, *X* represents (1) a positron (2) an electron (3) a proton (4) a gamma photon

37. What is the value of *A* in the equation? (1) 28 (2) 29 (3) 30 (4) 31

38. What is the value of *Z* in the equation? (1) 14 (2) 15 (3) 126 (4) 17

39. The nucleus of $^{30}_{15}P$ has a (1) 30 protons (2) 30 neutrons (3) 15 nucleons (4) 15 neutrons

40. If $^{30}_{15}P$ has a half-life of 2.5 minutes, how long will it take for 16 grams of the element to decay to 2.0 grams? (1) 2.5 min (2) 5.0 min (3) 7.5 min (4) 10. min

Base your answers to questions 41 through 43 on the information below.

$^{131}_{53}I$ initially decays by emission of beta particles.

41. Beta particles are (1) protons (2) electrons (3) neutrons (4) electromagnetic waves

42. When $^{131}_{53}I$ decays by beta emission, it becomes (1) $^{130}_{53}I$ (2) $^{129}_{51}Sb$ (3) $^{131}_{54}Xe$ (4) $^{135}_{54}Xe$

43. The half-life of $^{131}_{53}I$ is 8 days. After 24 days, how much of a 100.-gram sample of $^{131}_{53}I$ would remain? (1) 0 g (2) 12.5 g (3) 25.0 g (4) 50.0 g

Base your answers to questions 44 through 48 on the nuclear equation below.

$$^{226}_{88}Ra \rightarrow ^{222}_{86}Rn + ^{4}_{2}He$$

44. What is represented by $^{4}_{2}He$? (1) an alpha particle (2) a beta particle (3) a gamma ray (4) a positron

45. In $^{226}_{88}Ra$, the 88 represents the (1) number of neutrons (2) number of nucleons (3) mass number (4) atomic number

46. In $^{222}_{86}Rn$, the 222 represents the (1) atomic number (2) mass number (3) number of neutrons (4) number of electrons

47. How many nucleons are in an atom of $^{226}_{88}Ra$? (1) 88 (2) 138 (3) 226 (4) 314

48. This equation represents the process of (1) alpha decay (2) beta decay (3) fission (4) fusion

49. A certain radioactive isotope has a half-life of 2 days. If 8 kilograms of the isotope is placed in a sealed container, how much of the isotope will be left after 6 days? (1) 1 kg (2) 2 kg (3) 0.5 kg (4) 4 kg

50. In which reaction does X represent a beta particle?
(1) $^{234}_{92}U \rightarrow {}^{230}_{90}Th + X$ (2) $^{214}_{84}Pa \rightarrow {}^{210}_{82}Pb + X$
(3) $^{226}_{88}Ra \rightarrow {}^{222}_{86}Rn + X$ (4) $^{214}_{82}Pb \rightarrow {}^{214}_{83}Bi + X$

51. How much energy would be produced if 1.0×10^{13} kilogram of matter was entirely converted to energy (1) 9×10^{29} joules (2) 9×10^{21} joules (3) 3×10^{16} joules (4) 1×10^{16} joules

52. An atom of U-235 splits into two nearly equal parts. This is an example of (1) alpha decay (2) beta decay (3) fusion (4) fission

53. Which device makes visual observation of the path of a charged particle possible? (1) Geiger counter (2) Van de Graff generator (3) cyclotron (4) cloud chamber

54. Which statement most accurately describes the interaction which binds a nucleus together? (1) long-range and weak (2) long-range and strong (3) short-range and weak (4) short-range and strong

55. Particle accelerators that depend solely on electric and magnetic fields to add energy to nuclear particles can *not* accelerate (1) electrons (2) protons (3) neutrons (4) alpha particles

56. The synchrotron and cyclotron are examples of (1) high-energy particles (2) radiation detectors (3) beta-emitting nuclei (4) particle accelerators

57. A lithium nucleus contains three protons and four neutrons. What is its atomic mass number? (1) 1 (2) 7 (3) 3 (4) 4

58. Which is an isotope of $^{214}_{83}X$? (1) $^{214}_{82}X$ (2) $^{214}_{84}X$ (3) $^{210}_{83}X$ (4) $^{210}_{81}X$

NOTES

FREE-RESPONSE QUESTIONS

1. You wish to visit planet "X" and measure the acceleration of gravity there using instruments borrowed from your school laboratory.
   a. What instruments would you take from your school laboratory? Justify your choice.
   b. Describe how you would use your instruments to measure the acceleration of gravity on "X".
   c. Local scientists on "X" do not use our standard SI units of measurement. They claim that the acceleration of gravity at the testing site is 0.85 whipples/ploob^2. What questions should you ask to compare this acceleration value with the one you determined with Earth instruments?

2. Sounds generally travel faster through solid doors than through air. However, closing the classroom door prevents most of the noise from the hallway from entering the classroom.
   a. Draw a diagram of the hallway and classroom with the door *open*. On the diagram, draw arrows to show the path of the soundwaves.
   b. Draw a second diagram of the hallway and classroom with the door *closed*. On the diagram, draw arrows to show the path of the soundwaves.
   c. Using your knowledge of wave phenomena, explain why less noise from the hallway reaches the classroom when the door is closed.
   d. Explain how closing the classroom door will affect the level of noise that remains in the hallway.

3. An automobile manufacturer can design one of two cars: one that is rigid or one that crushes easily except for the passenger compartment.
   a. Which of the two designs would provide the greatest safety for the passenger?
   b. Compare the impulses that would be received by each car during a crash.
   c. Compare the changes of momentum that each car would experience during a crash.
   d. Explain why the passengers in the safer car would receive less force during a collision.

4. A positively charged sphere is placed between two parallel charged plates in a vacuum. The electric field is adjusted so the particle remains suspended.
   a. Draw a labeled diagram indicating the forces acting on the sphere.
   b. The positively charged sphere is replaced by a negatively charged sphere with the same magnitude of charge but half the mass of the positively charged sphere. Describe how the forces on the negatively charged sphere will be different from those on the positively charged sphere.

5. A student follows the path described in the table below which shows the direction and the distance that she recorded before each turn.

Number	Direction	Paces
1	N	2
2	W	1
3	S	6
4	E	2
5	N	1

   a. Draw a vector diagram of the student's route.
   b. What was the total distance traveled?
   c. What was the total displacement?

6. A block slides down an inclined plane onto a table. It then slides along the table and strikes a spring, compressing the spring.
   a. Draw a labeled diagram illustrating how the speed of the block changes until it comes to a final stop.
   b. The initial height of the block above the table will affect how much the spring is compressed. List three other variables that might affect the amount by which the spring is compressed when the block strikes it.
   c. Now consider the height variable. Describe a simple experiment that shows how the initial height of the block is related to the compression of the spring.

7. The length of a coiled spring is recorded as different weights are hung from it. The data is shown on the chart below.

Weight (N)	Length (cm)
3.0	5.5
5.0	6.4
6.0	7.1
11.0	9.6

a. Sketch a graph with properly labeled axes.
b. Determine the length of the unstretched spring.
c. On your graph, draw a dotted line to show the graph that you might expect if the experiment were repeated with a stiffer spring.

8. As you exit your automobile, you experience an electric shock when you touch the exterior of the car.
a. Assuming that you have no electrical charge while driving the car, explain how you might have become electrically charged.
b. You wish to determine the sign of the charge on you. You have access to object "A" with a negative charge and object "B" with a positive charge. You may use any other device available in your physics laboratory. Outline a procedure that you will use to discover your charge.
c. Describe one "safeguard" you might use to avoid such shocks when touching your automobile in the future.

9. In a laboratory exercise, a student collected the following data as the unbalanced force applied to a body of mass M was changed.

Data Table

Force (newtons)	Acceleration (meters per second^2)
4.0	2.1
8.0	4.0
12.0	6.0
16.0	7.9
20.0	10.0

a. Label the axes of the graph with the appropriate values for force and acceleration.
b. Plot an acceleration versus force graph for the laboratory data provided.
c. Using the data or your graph, determine the mass M of the body. [Show all calculations.]

10. Base your answers to parts *a* through *c* on the experiment described below and on your knowledge of physics.

A photoemissive metal was illuminated successively by various frequencies of light. The maximum kinetic energies of the emitted photoelectrons were measured and recorded on the table shown below.

Data Table

Frequency ($\times 10^{14}$Hz)	Maximum Kinetic Energy (eV)
8.2	1.5
7.4	1.2
6.9	.93
6.1	.62
5.5	.36
5.2	.24

a. Using the information and the data table, construct a graph *on your answer paper*, following the directions below:
   (1) Mark an appropriate scale on the axis labeled "Maximum Kinetic Energy (eV)."
   (2) Plot the data on the grid and draw the best-fit line.
b. Based on the graph, what is the threshold frequency of the metal used in this experiment?
c. If data from a different photoemissive metal were graphed, which characteristic of the graph would remain the same?

11. An aluminum block weighing 20 newtons, sliding from *left to right* in a straight line on a horizontal steel surface, is acted on by a 2.4-newton friction force. The block will be brought to rest by the friction force in a distance of 10. meters.

a. On the diagram of the block, draw an arrow to identify the direction of *each* force acting on the block while it is still moving, but is being slowed by the friction force. Identify *each* force by appropriately labeling the arrow that represents its line of direction.

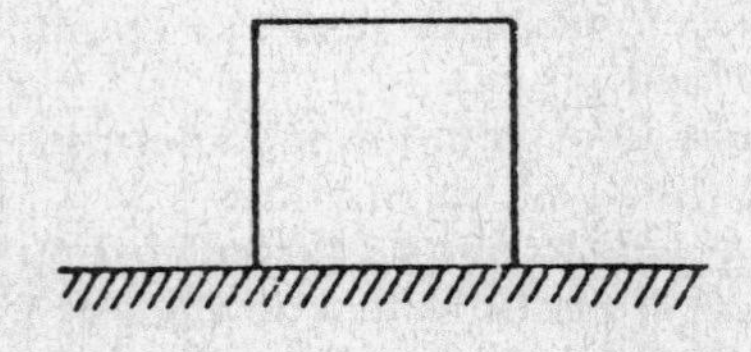

b. Determine the magnitude of the acceleration of the block as it is brought to rest by the friction force. [Show all work.]

12. A student walks from her house toward the bus stop, located 50 meters to the east. After walking 20 meters, she remem-

bers that she has left her lunch at the door. She runs home, picks up her lunch, walks again, and arrives at the bus stop.
a. Sketch a displacement versus time graph for the student's motion
b. Label your graph with appropriate values for time and displacement.

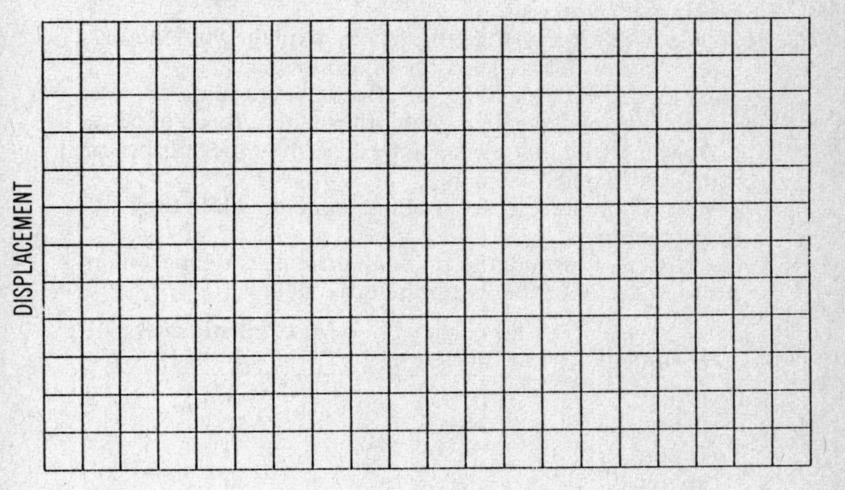

13. a. Draw a circuit diagram showing the following elements connected in parallel:

Elements

One 12.0-volt battery
One 2.0-ohm resistor
One 3.0-ohm resistor

Place an ammeter in the circuit to read the total current. Use the symbols shown below. [Assume availability of any number of wires of negligible resistance.]

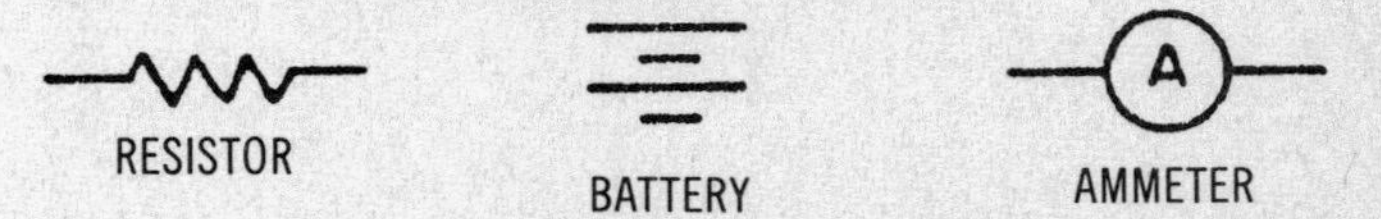

b. Determine the total circuit resistance. [Show all calculations.]

14. Imagine that the wind has died down. You are stranded in the middle of a lake in a sailboat. You have on board an elecric fan, a battery to operate the fan, and several friends with lots of advice. One friend suggests you should connect the fan to the battery and point it at the sail. Another friend suggests that you take down the sail and point the fan to the rear of the boat.

a. Draw a free body diagram showing all the horizontal forces on the boat when the fan is pointing at the sail.

b. Draw a free body diagram showing all the horizontal forces on the boat when the sail is down and the fan is pointed to the rear.

c. Using your knowledge of physics, explain why one of procedures fails to start the boat moving.

15. An 80 kg boy faces a 40 kg girl. Both are standing on roler skates. They push against each other with a force of 60 N.

a. Does it really matter if one pushes, the other pushes or if they both push? Explain.

b. What is the acceleration of each person while they are pushing?

16. A student performing the photoelectric effect experiment recorded the data shown on the table below.

Color of light	Frequency ($\times 10^{14}$ Hz)	Max. kinetic energy ($\times 10^{-19}$ J)
Violet	7.4	1.5
Blue	6.9	1.3
Green	5.5	0.7
Yellow	5.2	0.6

a. Draw a graph of the maximum kinetic energy vs frequency using properly labeled axes.

b. From your graph, determine the threshold frequency and the work function of the photo-emissive surface.

c. From the slope of your graph, calculate Planck's constant.

d. Calculate the percent error between the value of Planck's constant that was determined by this experiment and the standard value (6.6×10^{-34} J.s)

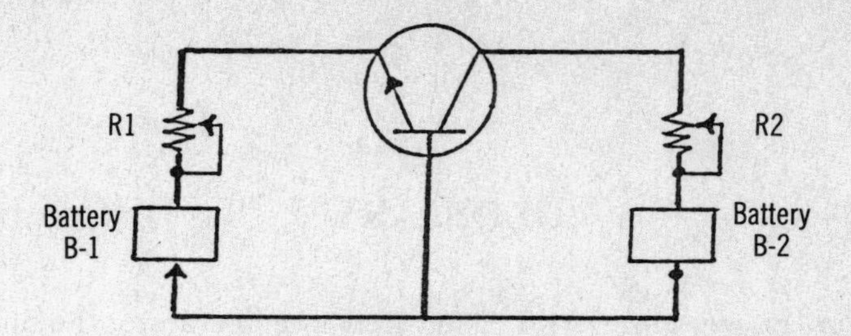

17. The diagram above is a transistor circuit containing an NPN transistor, two batteries, and two variable resistors.

a. Label the base, emitter, and collector of the NPN transistor.

b. On the diagram, add + or − signs at each of the battery terminals to show that the emitter circuit is forward biased and the collector circuit is reversed biased.

c. On the diagram, draw arrows to show the direction of the current through each of the circuit elements.

d. Explain how you could *increase* the collector-base current in this circuit.

18. A fielder throws a baseball to first base, 25 meters away. There is no wind but the ball encounters normal air friction as it travels.

a. Describe in complete sentences any changes that are likely to occur in the flight of the ball after it leaves the fielder's hand regarding its:

1. velocity in the vertical direction
2. acceleration in the vertical direction
3. velocity in the horizontal direction
4. acceleration in the horizontal direction

b. Draw a labeled diagram showing the trajectory of the ball after it is released.

GLOSSARY

Absolute temperature: Also called Kelvin temperature; the number of degrees above absolute zero.

Absolute zero: The lowest possible temperature: −273.16°C or zero kelvin.

Absorption spectrum: A continuous spectrum, like that of white light, interrupted by dark lines or bands. Produced by the absorption of certain wavelengths by a substance through which the light or other radiation passes.

Acceleration: A vector quantity that represents the time-rate of change in velocity.

Acceleration of gravity: Rate of change of velocity due to gravitational attraction of the earth.

Accuracy: Closeness of a measurement to the standard value of a quantity.

Adhesion: Attraction between unlike particles.

Alpha decay: The emission of an alpha partical from a nucleus.

Alpha particles ($^{4}_{2}He$): The nucleus of helium atoms, consisting of 2 protons and 2 neutrons.

Alternating current (AC): A current that reverses its direction at a regular frequency.

Ammeter: A meter for measuring electric current.

Ampere: The SI unit of current. It is a fundamental unit.

Amplifier: An electronic device which changes a weak electrical signal into a strong one.

Amplitude: The maximum displacement of a vibrating particle or wave from its rest position.

Angle of incidence: The angle between the incident ray and the normal.

Angle of reflection: The angle between the reflected ray and the normal, or perpendicular to the surface from which it is reflected.

Angle of refraction: The angle between the refracted ray and the normal, or perpendicular to the surface from which it is refracted.

Angular momentum: For a particle moving in a circular orbit, the product of its momentum and the radius of its orbit.

Anode: The electrode connected to the positive terminal of a battery or other source of EMF.

Antielectron ($_{+1}^{0}e$): A positive electron or positron.

Antimatter: Matter composed of antiparticles
Antineutrino: Antiparticle of the neutrino.
Antineutron($^1_0$n): Antiparticle of the neutron.
Antiparticle: The counterpart of subnuclear particle of matter, whose main property is that it and the particle annihilate each other on coming together, liberating their energy as radiation.
Antiproton($_{-1}^{1}$H): Antiparticle of the proton, having the same mass but a negative charge.
Armature: Usually the rotating coil of an electric generator or motor on an iron core.
Atom: The smallest particle of an element that has all its chemical properties. **The atomic number** is the number of protons in the nucleus. The symbol for atomic number is Z.
Atomic mass unit: On twelveth the mass of a carbon-12 atom. the symbol is u.
Avogadro's number: The number of molecules in one mole of any substance, equal to 6.02×10^{23} molecules.
Avogadro's principle: Equal volumes of all gases all the same temperature and pressure contain the same number of molecules.

Back EMF: The self induced electromotive force in the rotating armature of a motor that opposes the EMF applied to the motor.
Balmer series: A series of related lines in the visible part of the spectrum of hydrogen. It is produced by the electrons in excited hydrogen atoms which pass from higher energy levels to the one whose quantum number n = 2.
Barometer: A device for measuring atmospheric pressure.
Baryons: Nuclear particles with masses equal to or greater than the mass of the proton.
Beats: A series of alternate reinforcements and cancellations produced by two sets of superimposed sound waves of close but different frequencies heard as a throbbing effect.
Beta decay: The emission of an electron from a nucleus.
Beta particles: High-speed electrons emitted by a radioactive nucleus.
Binding energy: Energy that must be supplied to a nucleus to separate it into its nucleons. The binding energy is the energy equivalent of the mass defect.
Black hole: Collapsed astronomical object of sufficient mass to prevent the escape of light.
Boiling Point: Temperature at which a rapid change of phase occurs from a liquid to a gas. At the boiling point, the vapor pressure is equal to the atmospheric pressure.
Boyle's law: Volume of an enclosed mass of gas at a constant temperature varies inversely with the pressure.

Breeder reactor: A nuclear reactor that converts nonfissionable material to fissionable material with the production of energy.
Bubble chamber: A device which makes the paths of ionizing particles visible as trails of tiny bubbles in a superheated liquid.
Candella: The SI unit of luminous intensity. It is 1/60 the amount of light emitted by a square centimeter of a black body radiator at the temperature of freezing platinum (1772°C).
Celsius Temperature scale: Scale with 0° and 100° equal to the freezing and boiling points of pure water.
Center of curvature: Center of the sphere from which a spherical mirror is taken.
Centripetal acceleration: Acceleration always at right angles to the velocity of a particle.
Centripetal force: Force directed toward the center of a circle; keeps particles moving in uniform circular motion.
Charles' law: The volume of a fixed mass of gas at constant pressure varies directly with the absolute temperature.
Chain reaction: A self sustained series of nuclear fissions that can occur when a critical mass of nuclear fuel is present.
Chromatic aberration: Failure of a lens to bring all wavelengths of light to focus at the same point.
Cloud Chamber: An atomic radiation detector that produces visible vapor trails along the paths of charged particles.
Coherent light: Light in which all waves leaving the source are in phase.
Complimentary colors: Primary and secondary color that when added produce white light.
Components of a vector: Two or more vectors (usually perpendicular) that when added together, produce the original vector.
Compression: Portion of a longitudinal wave where particles of the medium are squeezed together.
Compton effect: Interaction of X-rays and electrons as the X-rays traverse matter resulting in a lengthening of the X-ray wavelength.
Concave: Curved surface with the edges closest to the observer.
Concurrent forces: Forces acting on the same point.
Conductance: Ability of a material to conduct electric charge. It is the reciprocal of resistance and is measured in siemens.
Conductor: A material that has a low resistivity for electric current.
Control rods: Devices is a nuclear reactor used to regulate the rate of the nuclear reaction.
Converging lens: A lens that converges parallel rays of light.
Converging mirror: Concave mirror capable of causing parallel rays to converge.
Convex: Curved surface with the edges furthest from the observer.

Coulomb: Unit of quantity of electric charge equal to the charge found on 6.25×10^{18} electrons. It is the charge that is transferred by a current of one ampere in one second.

Coulomb's law: The force between fixed point charges is directly proportional to the product of the charges and inversely proportional to the square of the distance between them.

Critical angle: The angle of incidence for which a ray passing obliquely from an optically more dense to an optically less dense medium has an angle of refraction of 90° and hence does not pass through the interface.

Critical mass: The minimum quantity of radioactive material in a reactor or in a nuclear bomb that will sustain a chain reaction.

Current: A flow of a liquid or a gas or of particles such as electric charges.

Cycle: One complete vibration of any mechanical oscillation or other periodic change.

Cyclotron: A particle accelerator that uses majnetic and electric forces to impart very high velocities to charged sub-atomic particles.

De Broglie principle: Material particles have wavelike characteristics; wavelength varies inversely with momentum.

Deceleration: A negative acceleration where the velocity is reduced in the direction of travel.

Derived unit: Unit of measurement defined in terms of other units.

Destructive interference: The superposition of two waves approximately in opposite phase so that then combined amplitude is the difference between their amplitudes and smaller than either.

Deuterium ($^{2}_{1}H$): The hydrogen isotope whose mass number is 2.

Deuteron ($^{2}_{1}H$): The nucleus of a deuterium atom consisting of one proton and one neutron.

Diamagnetism: A property of a substance that causes it to be repelled by a magnet.

Diffraction: The spreading of waves around an obstacle and into the region behind it.

Diffraction grating (transmission type): An optically transparent surface on which are ruled thousands of equidistant opaque parallel lines; it uses the diffraction effects of the slits between these lines to separate light passing through it into its spectrum.

Diffusion: The continuous random migration of molecular particles of one substance through another resulting from molecular motion.

Diffuse reflection: The scattering of light rays by the reflection of light from a rough surface.

Diode: A two-element device that conducts electric current more easily in one direction than in the opposite direction.

Direct current (DC): The movement in electrons in one direction around a circuit.

Direct proportion: Relationship between two quantities whose ratio is a constant.

Dispersion: Is the separation of polychromatic light into its component wavelengths.

Displacement: Is a vector quantity that represents the length and direction of a straight line path from one point to another between which motion of an object has taken place.

Distance: A scalar quantity that represents the length of a path from one point to another.

Diverging lens: Usually glass that is thinner at the middle than at the edge and diverges parallel rays of light. A diverging lens can produce only a virtual image.

Domain: A tiny section of a ferromagnetic substance, such as iron, in which the atoms are lined up with their north poles facing one direction. The entire section acts like a single tiny magnet with its own north and south poles.

Donor: An element that is added to an insulator to make it an n-type semiconductor.

Dopant: An impurity added to a semiconductor to produce charge carriers.

Doppler effect: The change in frequency or pitch of sound waves, heard when the source of sound and the observer are moving toward or away from each other. A similar change is observed in the frequency or color of light, when the source of light and the observer are moving toward or away from each other.

Dynamics: Deals with the relation between the forces acting on an object and the resulting change in motion.

Echo: Rebound of a pulse or wave from a distant surface.

Effective resistance: Resistance of a single resistor that could replace a combination of resistors.

Efficiency: Ratio of output work to input work expressed as a percent.

Effort: The force applied to a machine.

Elasticity: Ability of object to return to its original form after removal of deforming forces.

Electric circuit: Continuous path that can be followed by charged particles.

Elastic collision: A collision in which both the total momentum and the total kinetic energy of the colliding bodies have the same values before and after the collision.

Elastic limit: The largest stress that can be applied to a body without permanently deforming it.

Electric current: The flow of electric charges, such as electrons in a metallic conductor, or ions in a liquid or gas, through a circuit.

Electric field: The space around an electrically charged body which exerts an electric force on a charge placed within it.

Electric field intensity at point: The force exerted by an electric field on a unit charge at that point.

Electrification: The process of giving a body an electric charge.

Electrode: A positively or negatively charged terminal of a device, such as an electrolytic cell, a gas discharge tube, or a vacuum tube.

Electromagnet: A coil of wire wound around a soft iron core, whose magnetic field is produced by passing an electric current through the coil.

Electromagnetic induction: The process of producing an EMF in a conductor by changing the magnetic flux passing through the circuit.

Electromagnetic interactions: Forces of attraction and repulsion between charged particles.

Electromagnetic spectrum: The entire family of electromagnetic radiations ranging from short wavelength, high-energy gamma rays to long-wavelength, low-energy radio waves.

Electromagnetic waves: Transverse waves moving at the speed of light and consisting of rapidly alternating electric and magnetic fields at right angles to each other and to the dirction in which the waves are traveling.

Electron: Subatomic particle of small mass and negative charge.

Electron cloud: Region of high probability if finding an electron.

Electron collision excitation: Collision between an electron and an atom resulting in an excited atom.

Electron gas: Free electrons present in a metallic conductor.

Electron volt: A unit of energy equal to 1.6×10^{-19} joule.

Electron shell: A region surrounding an atomic nucleus which contains orbiting electrons.

Electroscope: Device used to detect the presence of electric charges.

EMF: The potential difference generated by electromagnetic induction or the voltage produced by a battery or an electric generator.

Emission spectrum: Spectrum produced by the excited atoms of an element.

Enrichment: Process in which the number of fissionable nuclear is increased.

Energy: Capacity to do work.

Energy gap: A band or zone surrounding an atomic nucleus which cannot be occupied by an orbiting electron.

Entropy: Measure of disorder of a system.
Equilibrant: Force equal in magnitude to a resultant, but opposite in direction.
Equilibrium: Condition in which the net force on an object is zero.
Equivalence principle: Gravitational and inertial masses are equal.
Evaporation: Change from liquid to gas.
Excited atom: Atom with one or more electrons in a higher than normal energy level.
Extrapolation: Extending graph beyond measured points.

Farad (F): The SI unit of capacity equal to one coulomb per volt.
Ferromagnetism: The ability of iron, nickel, and cobalt to be strongly attracted by magnets.
Field: A region in space where particles are subject to being influenced by forces such as those of gravity, magnetism or electricity.
Fission: The splitting of the nucleus of a heavy atom, such as Uranium 235, into two main parts, accompanied by the release of much energy.
Fixed points: Temperatures such as those of melting ice and boiling water, used in calibrating thermometers.
Fluid: A substance having no definite shape and being able to flow; a liquid or gas.
Fluorescence: The process whereby a substance emits radiation (usually as visible light) when struck by charged particles, such as electrons or alpha particles, or when radiation of a higher frequency (usually ultraviolet light) falls on it.
Forward bias: An electric potential that facilitates the conduction of a semiconductor.
Freezing point: The temperature at which a liquid changes phase to a solid.
Flux density: The number of magnetic flux lines per unit area. It is the force exerted per unit current per unit length when the current is perpendicular to the field.
Focal length: Distance from the focal point to a mirror or lens.
Focal point: Point of convergence, real or apparent (virtual), of rays reflected by a mirror or refracted by a lens.
Force: A vector quantity that may be defined as a push or pull.
Frame of reference: Coordinate system used to describe motion.
Fraunhofer lines: Absorption lines in the sun's spectrum due to gases in the solar atmosphere.
Frequency: The number of cycles occurring per unit time.
Friction: A force opposing the relative motion of two objects in contact. When an object moves against friction, work is done.
Fusion: The process of combining two light nuclei to form a heavier one.

Galvanometer: An instrument used to detect and measure very small electric currents.

Gamma rays (γ): Highly penetrating electromagnetic radiations of very short wavelengths emitted by the nucleus of radioactive atoms.

Gas: The diffuse physical state of a substance in which it has no definite volume or shape.

Geiger counter: An instrument that detects radiations from radioactive substances by their ability to ionize the matter through which they pass.

General gas law: The relationship PV = RT, where R is the universal gas constant, P is the pressure, V is the volume and T is the Kelvin temperature of one mole of an ideal gas.

Generator: A device that converts mechanical energy into electrical energy.

Gram: A small unit of mass, equal to 1/1000 of the standard kilogram.

Gravitational force: The force of attraction that every mass exerts upon every other mass.

Gravitational mass: The mass determined by measuring an object on an equal arm balance.

Graviton: The particle assumed to be the carrier of the gravitational force.

Ground state: The condition of an atom when its electrons are at the lowest possible energy levels.

Grounding: Connecting a charged object to the earth to remove the object's charge.

Half life: The time required for one-half of the nuclei of a sample of radioactive material to disintegrate.

Heat: Quantity of thermal energy transferred from one object to another because of a difference in temperature.

Heat of fusion: The number of kilojoules required to change one kilogram of a substance from the solid to the liquid phase at its melting point, with no change in temperature.

Heat of vaporization: The number of kilojoules required to change one kilogram of a substance from the liquid to the gaseous phase at the boiling point, with no change in temperature.

Hertz: Unit of frequency equal to one event (cycle) per second.

Hooke's law: The strain produced in an elastic body is directly proportional to stress as long as the elastic limit is not exceeded.

Horsepower: A unit of power equal to 746 watts.

Huygens' principle: Every point on a wave front may be considered a source of wavelets with the same speed.

Hypothesis: A plausable but unproven explanation for a scientific observation or phenomenon.

Ideal gas: An imaginary gas which conforms exactly to the universal gas law.

Ideal machine: A theoretical machine in which there are no frictional or other losses of the work put in so that the output equals the input

Image: The likeness of an object made by a lens or mirror; it may be real or virtual.

Impulse: A vector quantity with a magnitude equal to the product of the unbalanced force and the time the force acts. Its direction is the same as that of the force.

Incident ray: The ray which falls upon a surface between two substances.

Inclined plane: A simple machine consisting essentially of a sloping surface.

Index of refraction: The ratio of the speed of light in a vacuum to its speed in the material medium.

Induced EMF: A potential difference applied to a conductor by a changing magnetic field.

Induced magnetism: Magnetism produced in a magnetic substance when brought into the field of a magnet.

Induced radioactivity: Radioactivity resulting from the bombardment of a nonradioactive element with high-speed protons, neutrons, etc.

Inductance (L): The property of a circuit of part of a circuit whereby it sets up a back EMF opposing any change in the current flowing through it, measured in henrys.

Induction: Is a process by which a charged object causes a redistribution of the charges of another object without contact.

Inertia: The property of matter by which it resists any change in its state of motion or rest.

Inertial mass: The ratio of the force applied to an object to the acceleration it produces.

Infrared light: Electromagnetic radiations whose wavelengths lie between those of visible light and radio waves.

Interface: Common boundary between two materials having different properties.

Interference: Is the effect produced by two or more waves which are passing simultaneously through a region.

Internal energy: Is the total kinetic and potential energy associated with the motions and relative positions of the molecules of an object, apart from any kinetic or potential energy of the object as a whole.

Inverse proportion: The relationship between two variables whose product is constant.
Ion: An atom or group of atoms having an unbalanced electric charge.
Ionic bonding: Attraction between atoms due to electron transfer from one to another.
Ionization energy: The energy required to detach one of its electrons from an atom and thus turn the atom into an ion.
Isotopes: Different forms of an element whose atomic nuclei have the same number of protons but different numbers of neutrons.

Joule: The SI unit of work. It is the work done when a force of one newton acts through a distance of one meter. Energy is also measured in joules.

Kelvin temperature: Temperature of an object based on the Kelvin or absolute, temperature scale.
Kelvin temperature scale: Scale with 0 = absolute zero and 273.15 = freezing point of water.
Kilogram: The SI unit of mass. It is a fundamental unit.
Kilowatt-hour: Amount of energy equal to 3.6×10^6 joule.
Kinematics: The mathematical methods of describing motion without regard to the forces which produce it.
Kinetic energy: The energy an object has because of its motion. Like all energy it is a scalar quantity. Kinetic energy is equal to one-half the product of the mass and the speed squared.
Kinetic theory: Concept that all matter is made of small particles that are in constant motion.
Kirchhoff's laws: Two laws that describe current and voltage relationships in circuits consisting of networks of resistances.

Laser: Device for producing intense coherent light.
Law of conservation of charge: Electric charge can be neither created nor destroyed.
Law of conservation of energy: In nonnuclear changes, energy can be neither created nor destroyed.
Law of conservation of momentum: When no resultant external force acts on a system, the total momentum of a system remains unchanged. When bodies interact their total momentum remains unchanged.
Law of universal gravitation: Every particle of matter in the universe attracts every other particle with a force that is directly proportional to the product of their masses and inversely proportional to the square of the distance between them.
Law of work: In an ideal machine, the work output is equal to the work input.

Lenz's law: An induced current flows in such a direction as to oppose by its magnetic field the motion by which the current was induced.

Leptons: Subatomic particles of little or no mass including electrons, muons, and neutrinos.

Light: An electromagnetic disturbance that can produce the sensation of sight.

Line spectrum: A spectrum emitted by an incandescent gas under low pressure and consisting of a series of colored lines separated by dark spaces.

Lines of force: Lines drawn to map electric or magnetic fields to show the direction and the intensity of the field from point to point.

Longitudinal waves: Wave motion in which the disturbance is parallel to the direction of travel of the wave.

Lumen: Unit of luminous flux.

Luminous body: An object emitting light.

Luminous flux: Flow of light from a source.

Luminous intensity: Measure of brightness of a light emitted by a source.

Magnetic field: Space around a magnet in which magnetic forces can be detected.

Magnetic flux density: Number of magnetic flux lines per unit area.

Magnetic flux lines: Imaginary lines indicating the magnitude and direction of a magnetic field.

Magnetic force: Force between two objects due to the magnetic flux of one or both objects.

Magnetic induction: Strength of a magnetic field.

Magnification: The ratio of image size to object size.

Mass: Quantity of matter in an object measured by its resistance to a change in tis motion (inertia).

Mass defect: Difference in mass between the actual atomic nucleus and the sum of the particles from which the nucleus was made.

Mass number: Is the total number of protons and neutrons in the nucleus. The symbol for mass number is A.

Mass spectrograph: An electromagnetic instrument for measuring the masses of atoms.

Matter: Bodies having mass and volume.

Matter waves: de Broglie waves associated with particles, which are responsible for certain wavelike behaviors of those particles.

Mechanical advantage (ideal): The number of times a machine would multiply a force in the absence of friction.

Mechanical advantage (real): The ratio of the resistance overcome by a machine to the effort applied.

Mechanical equivalent of heat: The quantity of mechanical energy equal to 1 unit of heat; (1 cal = 4.2J or 1J = 0.24 cal).
Melting point: The temperature at which a solid changes to a liquid at normal pressure.
Meson: A particle with a mass between that of the electron and that of a proton. It may be positively charged, negatively charged, or neutral.
Meter: The SI unit of length. It is fundamental unit.
Moderator: Material used to decrease speed of fast neutrons in a nuclear reactor.
Mole: The abbreviation for gram molecular mass. It is the sum of the sum of the atomic masses that make up a molecule of a given material.
Molecule: The smallest piece of an element or compound that can exist independently.
Momentum: Is a vector quantity. Its magnitude is equal to the prod- to show the direction of the intensity of the field from point to point.
Monochromatic light: Light of a single color or narrow range of frequencies.

Negative acceleration: Acceleration that acts to slow down a moving body.
Neutrino: A neutral particle having a mass number of zero.
Neutron (^1_0n): A neutral particle having about the same mass as a proton and present in all atomic nuclei other than ordinary hudrogen.
Newton: The force which will impart to a mass of one kilogram an acceleration of one meter per second per second. It is a derived unit.
Nodal line: Line connecting nodes.
Node: Point in a medium or field that remains unchanged when acted upon by more than one disturbance simultaneously.
Normal: Direction that is perpendicular to a surface.
Nuclear fission: Splitting a large atomic nucleus into two approximately equal parts.
Nuclear force: A strong force which holds the nucleons together, Nuclear forces operate when the distance between nucleons is less than 10^{-15} meters.
Nuclear reactor: Device for obtaining energy from a controlled fission reaction.
Nucleon: Proton or neutron.
Nucleus: Core of an atom containing the protons and neutrons.
Nuclide: The nucleus of an isotope of any element having a given mass.

N-type semiconductor: A semiconductor in which the majority charge carriers are electrons.

Objective lens: Light-gathering and image-forming lens of a microscope or telescope.

Ohm (Ω): The SI unit of electrical resistance through which it takes 1 volt to produce a current of 1 ampere.

Ohm's law: At constant temperature, the ratio of the difference of potential across a metallic resistor to the current is constant.

Order of magnitude: The power of ten that is nearest to a given number.

Oscillating circuit: A circuit consisting essentially of an inductance and a capacitor used to generate high-frequency alternating currents.

Output: The work obtained from a machine equal to the resistance times the distance it moves.

Parallel circuit: An electric circuit in which there is more than one current path.

Paramagnetism: The property of a substance that causes it to be weakly attracted by a magnet.

Particle accelerator: An apparatus used to impart very hugh velocities and energies to charged subatomic particles, such as protons and deuterons, used as projectiles to penetrate atomic nuclei.

Pascal (Pa): The SI unit of pressure equal to a force of 1 newton acting on 1 square meter of area.

Pendulum: A mass suspended from a point so that it swings freely by the influence of gravity.

Period: The time required for the completion of a cycle. It is the reciprocal of the frequency.

Periodic wave: A series of regular disturbances.

Permeability: The ability of a substance to concentrate the lines of force when placed in a magnetic field.

Photoelectric effect: The emission of electrons from an object when certain electromagnetic radiation strikes it.

Photoelectron: An electron emitted when a photosensitive material is illuminated by light.

Photon: Is a quantum of light energy.

Planck's constant: A universal constant (h) relating the energy of a photon to the frequency of the radiation from which it comes $h = 6.63 \times 10^{-34}$ joule-second.

Plasma: An ionized gas containing free electrons and positive ions that conduct current.

P-n diode: A rectifier consisting of a p-type semiconductor and an n-type semiconductor bonded together.

P-n junction: The interface between a p-type and an n-type semiconductor.

Polarization: Transverse wave vibration in a single plane. Longitudinal waves cannot be polarized.

Polychromatic light: A mixture of many colors.

Positron: The antiparticle of the electron. It has the same mass as the electron but a positive instead of a negative charge.

Potential difference: The change in energy per unit charge as a charge is moved from one point to the other in an electric field.

Potential energy: The energy an object has because of its position or condition. Under ideal conditions it is equal to the work required to bring the object to that position or condition.

Power: The time-rate of doing work. It is a scalar quantity.

Pressure: The force per unit of area.

Primary: The coil to a transformer to which the voltage to be stepped up or down is applied.

Principal axis: The line joining the center of curvature and the center of a mirror or lens.

Principal focus: The point to which incident rays parallel to the principal axis of a lens or mirror converge, or from which they diverge.

Paramagnetism: The property of a substance that causes it to be weakly attracted by a magnet.

P-type semiconductor: A semiconductor in which the majority carriers are holes.

A pulse: A single vibratory disturbance which moves from point to point transferring energy.

Quantum: The unit of energy associated with each given frequency of radiation; a photon.

Quantum numbers: Numbers associated with each of the possible energy levels of an atom. They include the principal quantum number, n, the angular momentum quantum number, 1, the magnetic quantum number m, and the spin quantum number, s.

Quantum theory: A theory which assumes that radiant energy is emitted and absorbed by matter in discrete minimum packets equal to Planck's constant times the frequency of the radiation.

Quarks: Theoretical particles that make up all subatomic particles.

Radioactive materials: Materials that exhibit the phenomenon of radioactivity.

Radioactivity: Spontaneous decay of unstable nuclei.

Ray: Line drawn to represent the path traveled by light.
Real images: Likeness of objects that may be projected on a screen.
Receiver: Device used to detect electromagnetic waves.
Rectifier: An electronic component that changes alternating current into pulsating direct current.
Reflection: The process of returning a wave or particle to its original medium from the interface of two media.
Refraction: Change in direction of a wave front as it passes from one medium to another.
Resistance: Opposition to flow of electric current.
Resistivity: The ability of a material to conduct electric charges.
Resolution: Measure of ability of an optical instrument to separate images of objects that are close together. The resolution varies directly as the diameter of the opening.
Resonance: The setting up of strong vibrations in a body at its natural vibration frequency.
Rest mass: The mass that a body has when it is at rest.
Resultant: The single vector whose effect is the same as the combined effects of two or more similar vectors.
Reverse bias: A potential difference that inhibits charge movement through a semiconductor.

Satellite: Object in orbit around a planet.
Scalar quantity: Quantity having magnitude (size) only.
Schematic diagram: Diagram of an electric circuit using symbols.
Scientific notation: Expressing numbers in a form: $M \times 10^n$ where $1 \leq M \leq 10$ and n is an integer.
Scintillation: Flash of light emitted when a substance is struck by radiation.
Second: The SI unit of time. It is afundamental unit.
Secondary winding: An isolated coil in a transformer that usually has a different number of turns than the primary winding.
Series circuit: Is one in which there is only one current path.
Series connection: Connecting electrical devices in a line so that they make a single path for the current to pass through them.
Shadow: The space behind an illuminated object from which it excludes light.
Short circuit: An electric circuit in which the resistance is so low as to permit a dangerous current.
Significant figures: Digits which reflect the accuracy and precision of physical measurements.
Simple harmonic motion: A vibratory or periodic motion, in which the force acting on the vibrating body is proportional to its

displacement from its central equilibrium position and always acts toward that position.

Snell's law: A light ray passing from one medium to another is refracted so that the ratio of the sine of the angle of incidence to the sine of the angle of refraction is equal to a constant for all angles of incidence.

Sound waves: Longitudinal waves in air and other material media set up by vibrating bodies.

Spark chamber: A device that detects charged subatomic particles by a trail of electric sparks they set off in it.

Specific heat: The quantity of heat required to raise the temperature of a unit mass of the substance one Celsius degree.

Speed: A scalar quantity that represents the magnitude of the velocity.

Speed of light: 3.00×10^8 m/s (in vacuum).

Spherical aberration: Failure of a spherical mirror or lens to bring all rays parallel to the principal axis to focus at the same point.

Spherical concave mirror: A converging mirror that is formed of a spherical segment of one base.

Standard pressure: 1.01325×10^5 Pa = 101.325 kPa = 1 atm = 760 mm Hg = 760 Torr.

Standing wave: A wave whose nodes are stationary.

State: Physical condition of a material.

Static electricity: Electrical charges at rest.

Strain: The deformation produced by a stress.

Stress: A deforming force applied to a body.

Strong nuclear interaction: The force of attraction between two nucleons in which the meson serves as the carrier.

Superconductivity: The complete loss of electrical resistance by certain materials when cooled to near absolute zero.

Superposition: THe process whereby two or more waves combine their effects when passing through the same parts of a medium at the same time.

Sympathetic vibration: The vibration of a body at its natural frequency set up by the vibration of another body having the same natural frequency.

Synchrotron: A particle accelerator in which the frequency of the accelerating voltage is kept in step with the frequency of revolution of particles in a cyclotron-like device.

Temperature: That property of matter which determines the direction of the exchange of internal energy between objects. The object at lower temperature will gain internal energy. Absolute temperature is directly proportional to the average kinetic energy of random motion of the molecules of an ideal gas.

Theory: An imagined conceptual scheme, mechanism, or model that provides a plausible explanation of a series of experimental observation.

Thermal energy: Internal energy.

Thermonuclear reaction: A nuclear reaction that can take place only at very high temperatures involving the fusion of light nuclei into a heavier one with the release of energy.

Thought experiment: An imagined experiment assuming ideal conditions that may never be achieved experimentally.

Threshold frequency: The minimum frequency of incident light that will eject a photoelectron from a given metal.

Total internal reflection: The total reflection of a beam of light traveling in an optically dense medium when it falls upon the surface of a less dense medium.

Tracer: A radioactive element used to observe the path and distribution of the atoms of that element in a chemical, biological, or physical process in which it takes part.

Transformer: A device for changing the voltage of an alternating current to a desired value.

Transistor: An electronic component commonly used as an amplifier in a solid state device.

Transmutation: A change from one isotope to another of the same or different atomic number because of a gain or loss of protons and/or neutrons by the nucleus.

Translucent substance: One through which light is transmitted diffusely so that bodies cannot be seen through it.

Transparent substance: One through which light passes without diffusion so that bodies can be seen through it.

Transverse waves: Waves such as light, in which the vibrations are perpendicular to the direction in which the waves are traveling.

Triple point: The temperature and pressure at which the solid, liquid, and gaseous states of a substance can exist at the same time and remain in equilibrium.

Tritium (^3_1H): The isotope of hydrogen having a mass number of 3.

Ultraviolet light: The range of invisible radiations in the electromagnetic spectrum between violet light and X rays.

Uncertainty principle: It is not possible to measure exactly both the position and the momentum of a particle at the same time. The product of the uncertainty in position and the uncertainty in momentum is of the order of h.

Uniform motion: Displacement at constant velocity.
Universal gas constant: The constant R in the general gas law, PV = nRT.

Valence band: The outermost shell of an atom. When this shell is completely filled with electrons, a material is an insulator.
Valence electron: An electron in an outer incomplete shell of an atom that takes part in ionic and covalent bonding.
Van de Graaff generator: A device that builds up a high potential difference by accumulating electric charges on an insulated hollows metal sphere. It is used to accelerate electrically charged particles.
Vaporization: The change of state from a liquid to a gas.
Vector: A quantity having both magnitude and direction.
Velocity: A vector quantity which represents the time-rate of change of displacement.
Virtual image: Likeness of an object which can be seen or photographed but connot be projected on a screen.
Volt: A potential difference that exists between two points if one joule of work is required to transfer one coulomb of charge from one point to the other in an electric field.

Watt: Is the SI unit of power. It is equal to one joule per second.
Wave: A vibratory disturbance that is propagated from a source.
Wave front: The locus of adjacent points of the wave which are in phase.
Wavelength: Is the distance between two consecutive points in phase.
Weak force: Force involved in the decay of atomic nuclei and nuclear particles. A type of electromagnetic force.
Weight: Gravitational attraction of the earth or celestial body for an object.
Work: A scalar quantity that is equal to the product of the component of force acting in the direction of the motion and the distance that the object is moved.
Work function: The minimum energy needed to eject a photoelectron from a given metal.
W-particle: Carrier of the weak nuclear interaction.

X-rays: A range of deeply penetrating electromagnetic radiations whose wavelengths lie between those of ultraviolet light and gamma rays.

The University of the State of New York
THE STATE EDUCATION DEPARTMENT
Albany, New York 12234

REFERENCE TABLES FOR PHYSICS*

LIST OF PHYSICAL CONSTANTS

Name	*Symbol*	*Value(s)*
Gravitational constant	G	6.7×10^{-11} N•m²/kg²
Acceleration due to gravity (up to 16 km altitude)	g	9.8 m/s²
Speed of light in a vacuum	c	3.0×10^{8} m/s
Speed of sound at STP		3.3×10^{2} m/s
Mass-energy relationship		1 u (amu) = 9.3×10^{2} MeV
Mass of the Earth		6.0×10^{24} kg
Mass of the Moon		7.4×10^{22} kg
Mean radius of the Earth		6.4×10^{6} m
Mean radius of the Moon		1.7×10^{6} m
Mean distance from Earth to Moon		3.8×10^{8} m
Electrostatic constant	k	9.0×10^{9} N•m²/C²
Charge of the electron (1 elementary charge)		1.6×10^{-19} C
One coulomb	C	6.3×10^{18} elementary charges
Electronvolt	eV	1.6×10^{-19} J
Planck's constant	h	6.6×10^{-34} J•s
Rest mass of the electron	m_e	9.1×10^{-31} kg
Rest mass of the proton	m_p	1.7×10^{-27} kg
Rest mass of the neutron	m_n	1.7×10^{-27} kg

ABSOLUTE INDICES OF REFRACTION

($\lambda = 5.9 \times 10^{-7}$ m)

Air	1.00
Alcohol	1.36
Canada Balsam	1.53
Corn Oil	1.47
Diamond	2.42
Glass, Crown	1.52
Glass, Flint	1.61
Glycerol	1.47
Lucite	1.50
Quartz, Fused	1.46
Water	1.33

WAVELENGTHS OF LIGHT IN A VACUUM

Violet	$4.0 - 4.2 \times 10^{-7}$ m
Blue	$4.2 - 4.9 \times 10^{-7}$ m
Green	$4.9 - 5.7 \times 10^{-7}$ m
Yellow	$5.7 - 5.9 \times 10^{-7}$ m
Orange	$5.9 - 6.5 \times 10^{-7}$ m
Red	$6.5 - 7.0 \times 10^{-7}$ m

*These revised Reference Tables for Physics should be used in the classroom beginning September 1988. The first examination to be based on these tables is the June 1989 Regents Examination in Physics.

DET 511 (7-88-100,000)
88-7122

HEAT CONSTANTS

	Specific Heat (average) (kJ/kg•C°)	*Melting Point* (°C)	*Boiling Point* (°C)	*Heat of Fusion* (kJ/kg)	*Heat of Vaporization* (kJ/kg)
Alcohol (ethyl)	2.43 (liq.)	−117	79	109	855
Aluminum	0.90 (sol.)	660	2467	396	10500
Ammonia	4.71 (liq.)	−78	−33	332	1370
Copper	0.39 (sol.)	1083	2567	205	4790
Iron	0.45 (sol.)	1535	2750	267	6290
Lead	0.13 (sol.)	328	1740	25	866
Mercury	0.14 (liq.)	−39	357	11	295
Platinum	0.13 (sol.)	1772	3827	101	229
Silver	0.24 (sol.)	962	2212	105	2370
Tungsten	0.13 (sol.)	3410	5660	192	4350
Water: ice	2.05 (sol.)	0	—	334	—
Water: water	4.19 (liq.)	—	100	—	2260
Water: steam	2.01 (gas)	—	—	—	—
Zinc	0.39 (sol.)	420	907	113	1770

ENERGY LEVEL DIAGRAMS FOR MERCURY AND HYDROGEN

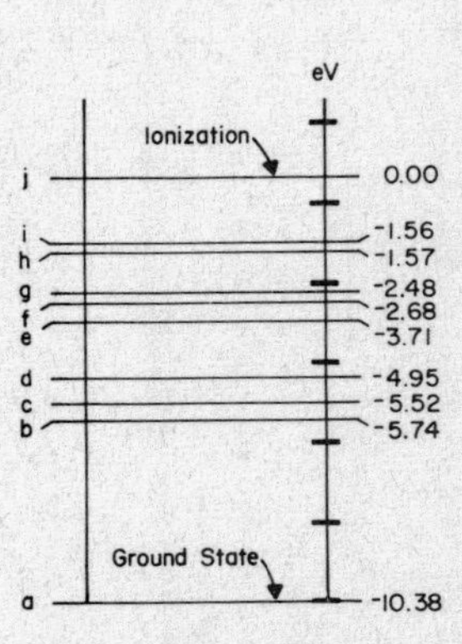

A few energy levels for the mercury atom

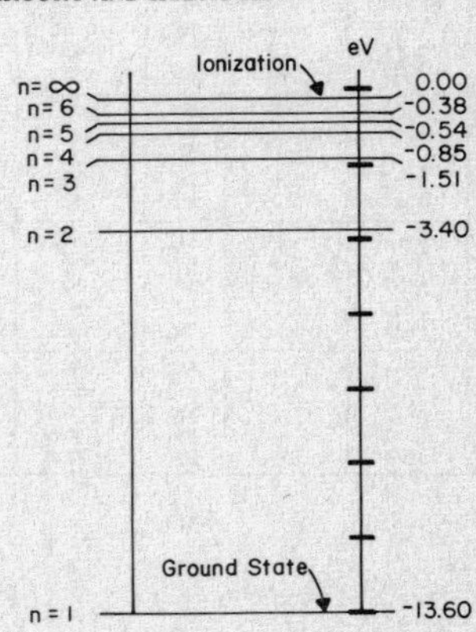

Energy levels for the hydrogen atom

VALUES OF TRIGONOMETRIC FUNCTIONS

Angle	Sine	Cosine	Angle	Sine	Cosine
1°	.0175	.9998	46°	.7193	.6947
2°	.0349	.9994	47°	.7314	.6820
3°	.0523	.9986	48°	.7431	.6691
4°	.0698	.9976	49°	.7547	.6561
5°	.0872	.9962	50°	.7660	.6428
6°	.1045	.9945	51°	.7771	.6293
7°	.1219	.9925	52°	.7880	.6157
8°	.1392	.9903	53°	.7986	.6018
9°	.1564	.9877	54°	.8090	.5878
10°	.1736	.9848	55°	.8192	.5736
11°	.1908	.9816	56°	.8290	.5592
12°	.2079	.9781	57°	.8387	.5446
13°	.2250	.9744	58°	.8480	.5299
14°	.2419	.9703	59°	.8572	.5150
15°	.2588	.9659	60°	.8660	.5000
16°	.2756	.9613	61°	.8746	.4848
17°	.2924	.9563	62°	.8829	.4695
18°	.3090	.9511	63°	.8910	.4540
19°	.3256	.9455	64°	.8988	.4384
20°	.3420	.9397	65°	.9063	.4226
21°	.3584	.9336	66°	.9135	.4067
22°	.3746	.9272	67°	.9205	.3907
23°	.3907	.9205	68°	.9272	.3746
24°	.4067	.9135	69°	.9336	.3584
25°	.4226	.9063	70°	.9397	.3420
26°	.4384	.8988	71°	.9455	.3256
27°	.4540	.8910	72°	.9511	.3090
28°	.4695	.8829	73°	.9563	.2924
29°	.4848	.8746	74°	.9613	.2756
30°	.5000	.8660	75°	.9659	.2588
31°	.5150	.8572	76°	.9703	.2419
32°	.5299	.8480	77°	.9744	.2250
33°	.5446	.8387	78°	.9781	.2079
34°	.5592	.8290	79°	.9816	.1908
35°	.5736	.8192	80°	.9848	.1736
36°	.5878	.8090	81°	.9877	.1564
37°	.6018	.7986	82°	.9903	.1392
38°	.6157	.7880	83°	.9925	.1219
39°	.6293	.7771	84°	.9945	.1045
40°	.6428	.7660	85°	.9962	.0872
41°	.6561	.7547	86°	.9976	.0698
42°	.6691	.7431	87°	.9986	.0523
43°	.6820	.7314	88°	.9994	.0349
44°	.6947	.7193	89°	.9998	.0175
45°	.7071	.7071	90°	1.0000	.0000

URANIUM DISINTEGRATION SERIES

Atomic Number and Chemical Symbol

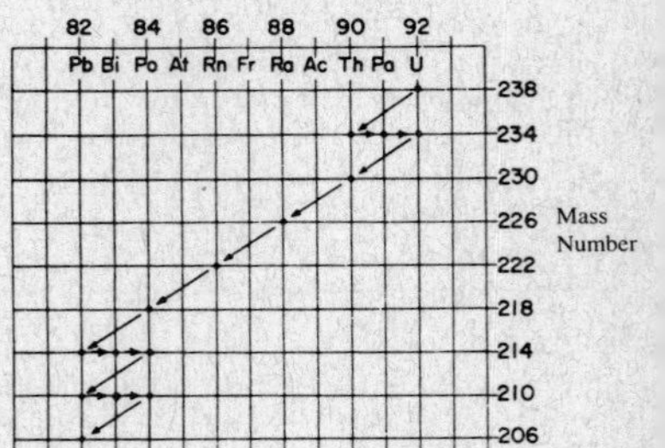

SUMMARY OF EQUATIONS

MECHANICS

$\bar{v} = \frac{\Delta s}{\Delta t}$

$\bar{v} = \frac{v_f + v_i}{2}$

$\bar{a} = \frac{\Delta v}{\Delta t}$

$\Delta s = v_i \Delta t + \frac{1}{2}a(\Delta t)^2$

$v_f^2 = v_i^2 + 2a\Delta s$

$F = ma$

$w = mg$

$F = \frac{Gm_1m_2}{r^2}$

$p = mv$

$J = F\Delta t$

$F\Delta t = m\Delta v$

a = acceleration
r = distance between centers
F = force
g = acceleration due to gravity
G = universal gravitation constant
J = impulse
m = mass
p = momentum
Δs = displacement
t = time
v = velocity
w = weight

ENERGY

$W = F\Delta s$

$P = \frac{W}{\Delta t} = \frac{F\Delta s}{\Delta t} = F\bar{v}$

$\Delta PE = mg\Delta h$

$KE = \frac{1}{2}mv^2$

$F = kx$

$PE_s = \frac{1}{2}kx^2$

F = force
g = acceleration due to gravity
h = height
k = spring constant
KE = kinetic energy
m = mass
P = power
PE = potential energy
PE_s = potential energy stored in a spring
Δs = displacement
t = time
v = velocity
W = work
x = change in spring length from the equilibrium position

ELECTRICITY AND MAGNETISM

$F = \frac{kq_1q_2}{r^2}$

$E = \frac{F}{q}$

$V = \frac{W}{q}$

$E = \frac{V}{d}$

$I = \frac{\Delta q}{\Delta t}$

$R = \frac{V}{I}$

$P = VI = I^2R = \frac{V^2}{R}$

$W = Pt = VIt = I^2Rt$

Series Circuits:

$I_t = I_1 = I_2 = I_3 = \ldots$

$V_t = V_1 + V_2 + V_3 + \ldots$

$R_t = R_1 + R_2 + R_3 + \ldots$

Parallel Circuits:

$I_t = I_1 + I_2 + I_3 + \ldots$

$V_t = V_1 = V_2 = V_3 = \ldots$

$\frac{1}{R_t} = \frac{1}{R_1} + \frac{1}{R_2} + \frac{1}{R_3} + \ldots$

d = separation of parallel plates
r = distance between centers
E = electric field intensity
F = force
I = current
k = electrostatic constant
ℓ = length of conductor
P = power
q = charge
R = resistance
t = time
V = electric potential difference
W = energy

INTERNAL ENERGY

$Q = mc\Delta T_c$

$Q_f = mH_f$

$Q_v = mH_v$

c = specific heat
H_f = heat of fusion
H_v = heat of vaporization
m = mass
Q = amount of heat
T_c = Celsius temperature

WAVE PHENOMENA

$T = \frac{1}{f}$

$v = f\lambda$

$n = \frac{c}{v}$

$\sin \theta_c = \frac{1}{n}$

$n_1 \sin \theta_1 = n_2 \sin \theta_2$

$n_1 v_1 = n_2 v_2$

$\frac{\lambda}{d} = \frac{x}{L}$

- c = speed of light in a vacuum
- d = distance between slits
- f = frequency
- L = distance from slit to screen
- n = index of absolute refraction
- T = period
- v = speed
- x = distance from central maximum to first-order maximum
- λ = wavelength
- θ = angle
- θ_c = critical angle of incidence relative to air

ELECTROMAGNETIC APPLICATIONS

$F = qvB$

$\frac{N_p}{N_s} = \frac{V_p}{V_s}$

$V_p I_p = V_s I_s$ (ideal)

% Efficiency = $\frac{V_s I_s}{V_p I_p} \times 100$

$V = B\ell v$

- B = flux density
- F = force
- I_p = current in primary coil
- I_s = current in secondary coil
- N_p = number of turns of primary coil
- N_s = number of turns of secondary coil
- q = charge
- v = velocity
- V_p = voltage of primary coil
- V_s = voltage of secondary coil
- ℓ = length of conductor
- V = electric potential difference

MODERN PHYSICS

$W_o = hf_o$

$E_{photon} = hf$

$KE_{max} = hf - W_o$

$p = \frac{h}{\lambda}$

$E_{photon} = E_i - E_f$

- c = speed of light in a vacuum
- E = energy
- f = frequency
- f_o = threshold frequency
- h = Planck's constant
- KE = kinetic energy
- p = momentum
- W_o = work function
- λ = wavelength

MOTION IN A PLANE

$v_{iy} = v_i \sin \theta$

$v_{ix} = v_i \cos \theta$

$a_c = \frac{v^2}{r}$

$F_c = \frac{mv^2}{r}$

- a_c = centripetal acceleration
- F_c = centripetal force
- m = mass
- r = radius
- v = velocity
- θ = angle

GEOMETRIC OPTICS

$\frac{1}{d_o} + \frac{1}{d_i} = \frac{1}{f}$

$\frac{S_o}{S_i} = \frac{d_o}{d_i}$

- d_i = image distance
- d_o = object distance
- f = focal length
- S_i = image size
- S_o = object size

NUCLEAR ENERGY

$E = mc^2$

$m_f = \frac{m_i}{2^n}$

- c = speed of light in a vacuum
- E = energy
- m = mass
- n = number of half-lives

INDEX

Regents

Examinations

PHYSICS
Thursday, June 15, 1989

Part I

Answer all 60 questions in this part. [70]

Directions (1-60): For *each* statement or question, select the word or expression that, of those given, best completes the statement or answers the question. Record your answer on the separate answer paper in accordance with the directions on the front page of this booklet.

1. A person travels 6 meters north, 4 meters east, and 6 meters south. What is the total displacement? (1) 16 m east (2) 6 m north (3) 6 m south (4) 4 m east

Base your answers to questions 2 through 4 on the information and diagram below. The diagram represents a block sliding along a frictionless surface between points *A* and *G*.

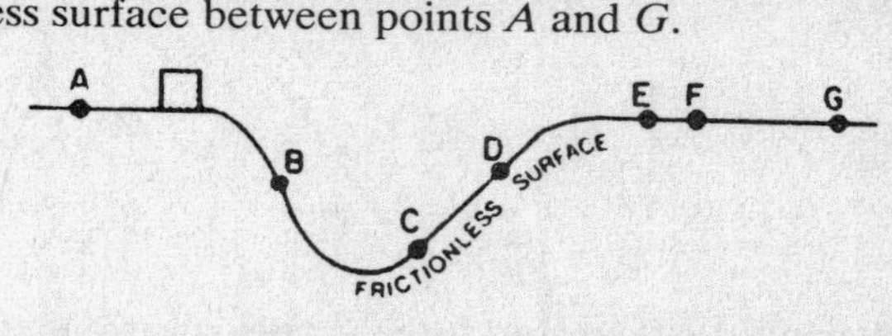

2. As the block moves from point *A* to point *B*, the speed of the block will be (1) decreasing (2) increasing (3) constant, but not zero (4) zero

3. Which expression represents the magnitude of the block's acceleration as it moves from point *C* to point *D*? (1) $\frac{m}{F}$ (2) $\frac{\Delta r}{\Delta t}$ (3) $m\Delta r$ (4) $\frac{2\Delta s}{\Delta t}$

4. Which formula represents the velocity of the block as it moves along the horizontal surface from point *E* to point *F*? (1) $\overline{v} = \frac{\Delta s}{\Delta t}$ (2) $\overline{v} = \frac{\Delta v}{2}$ (3) $v_f^2 = 2a\Delta s$ (4) $\Delta v = \frac{1}{2}a(\Delta t)^2$

5. The diagram below shows a graph of distance as a function of time for an object in straight-line motion. According to the graph, the object most likely has

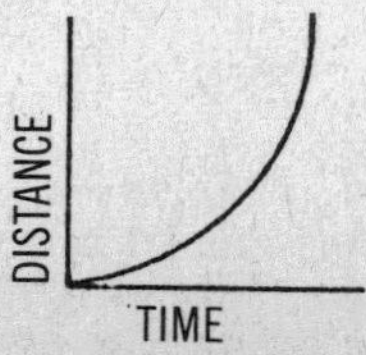

(1) a constant momentum (2) a decreasing acceleration (3) a decreasing mass (4) an increasing speed

6. If an object's velocity changes from 25 meters per second to 15 meters per second in 2.0 seconds, the magnitude of the object's acceleration is (1) 5.0 m/s² (2) 7.5 m/s² (3) 13 m/s² (4) 20. m/s²

7. An object initially traveling in a straight line with a speed of 5.0 meters per second is accelerated at 2.0 meters per second squared for 4.0 seconds. The total distance traveled by the object in the 4.0 seconds is (1) 36 m (2) 24 m (3) 16 m (4) 4.0 m

8. Which vector below represents the resultant of the concurrent vectors *A* and *B* in the diagram below?

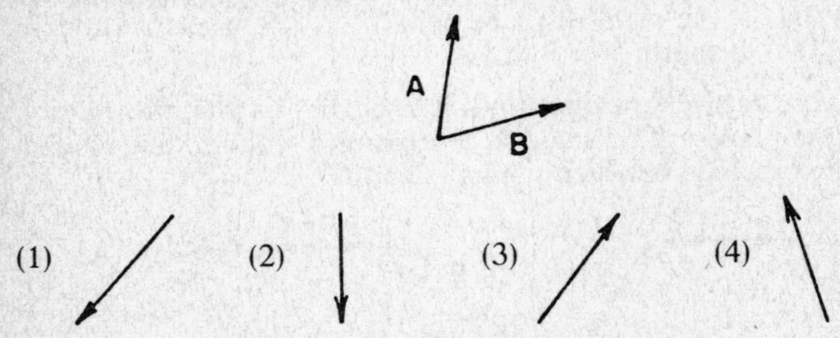

9. An unbalanced 6.0-newton force acts eastward on an object for 3.0 seconds. The impulse produced by the force is (1) 18 N•s east (2) 2.0 N•s east (3) 18 N•s west (4) 2.0 N•s west

10. An object near the surface of planet *X* falls freely from rest and reaches a speed of 12.0 meters per second after it has fallen 14.4 meters. What is the acceleration due to gravity on planet *X*? (1) 2.50 m/s² (2) 5.00 m/s² (3) 9.80 m/s² (4) 10.0 m/s²

11. Four forces are acting on an object as shown in the diagram below. If the object is moving with a constant velocity, the magnitude of force *F* must be

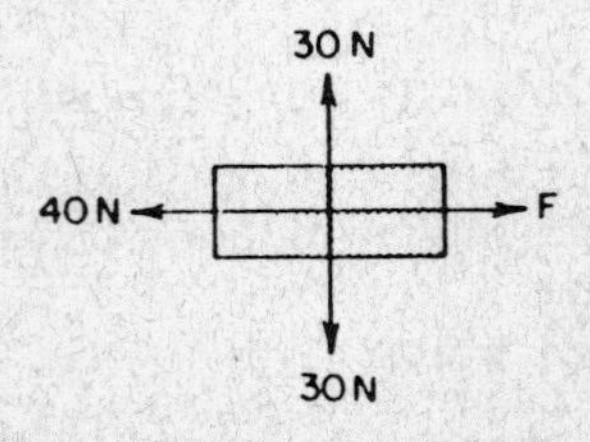

(1) 0 N (2) 20 N (3) 100 N (4) 40 N

12. A force of 50. newtons causes an object to accelerate at 10. meters per second squared. What is the mass of the object? (1) 500 kg (2) 60. kg (3) 5.0 kg (4) 0.20 kg

13. Which graph best represents the relationship between the mass of an object and its distance from the center of the Earth?

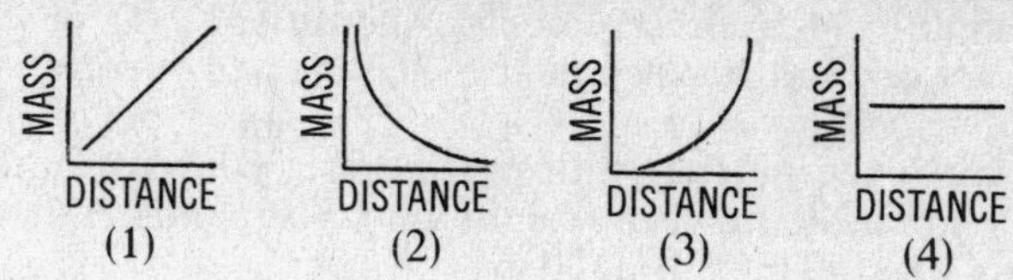

14. Gravitational force of attraction F exists between two point masses A and B when they are separated by a fixed distance. After mass A is tripled and mass B is halved, the gravitational attraction between the two masses is (1) $1/6F$ (2) $2/3F$ (3) $3/2F$ (4) $6F$

15. A 50.-kilogram student stands on the surface of the Earth. What is the magnitude of the gravitational force of the Earth on the student? (1) 490 N (2) 50. N (3) 9.8 N (4) 6.7×10^{-11} N

16. A cart rolls down an inclined plane with constant speed as shown in the diagram below. Which arrow represents the direction of the frictional force?

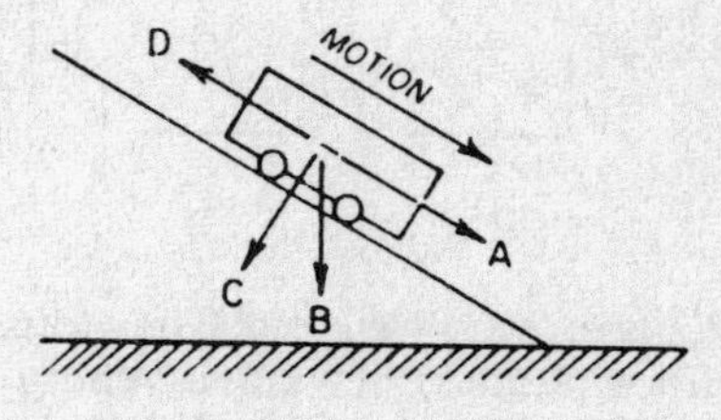

(1) *A* (2) *B* (3) *C* (4) *D*

17. What is the magnitude of the velocity of a 25-kilogram mass that is moving with a momentum of 100. kilogram-meters per second? (1) 0.25 m/s (2) 2500 m/s (3) 40. m/s (4) 4.0 m/s

18. The diagram below shows spheres A and B with masses of M and $3M$, respectively. If the gravitational force of attraction of sphere A on sphere B is 2 newtons, then the gravitational force of attraction of sphere B on sphere A is

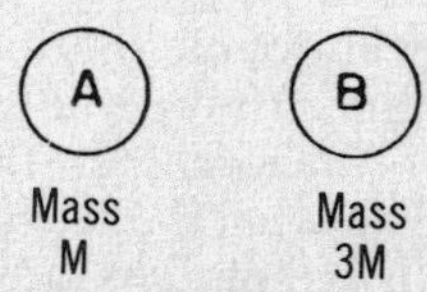

(1) 9 N (2) 2 N (3) 3 N (4) 4 N

19. A 2.0-kilogram rifle initially at rest fires a 0.002-kilogram bullet. As the bullet leaves the rifle with a velocity of 500 meters per second, what is the momentum of the rifle-bullet system? (1) 2.5 kg•m/s (2) 2.0 kg•m/s (3) 0.5 kg•m/s (4) 0 kg•m/s

20. What is the approximate thickness of this piece of paper? (1) 10^1 m (2) 10^0 m (3) 10^{-2} m (4) 10^{-4} m

21. Which terms represent scalar quantities? (1) power and force (2) work and displacement (3) time and energy (4) distance and velocity

22. What is the maximum distance that a 60.-watt motor may vertically lift a 90.-newton weight in 7.5 seconds? (1) 2.3 m (2) 5.0 m (3) 140 m (4) 1100 m

23. An object gains 10. joules of potential energy as it is lifted vertically 2.0 meters. If a second object with one-half the mass is lifted vertically 2.0 meters, the potential energy gained by the second object will be (1) 10. J (2) 20. J (3) 5.0 J (4) 2.5 J

24. A spring of negligible mass with a spring constant of 200 newtons per meter is stretched 0.2 meter. How much potential energy is stored in the spring? (1) 40 J (2) 20 J (3) 8 J (4) 4 J

25. Which graph best represents the relationship between the kinetic energy (KE) of a moving object as a function of its velocity *(v)*?

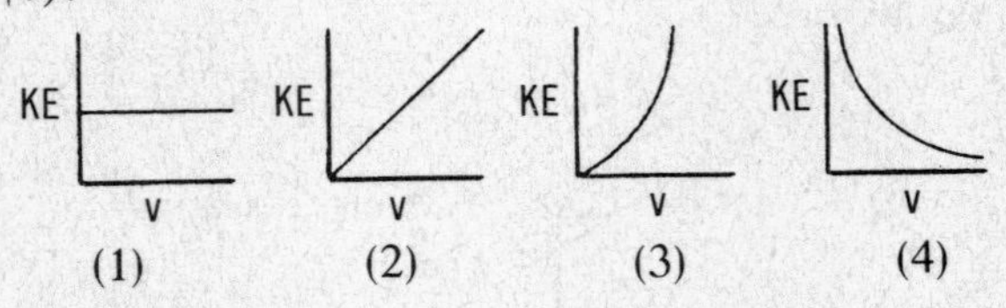

26. A basketball player who weighs 600 newtons jumps 0.5 meter vertically off the floor. What is her kinetic energy just before hitting the floor? (1) 30 J (2) 60 J (3) 300 J (4) 600 J

27. A negatively charged object is brought near the knob of a negatively charged electroscope. The leaves of the electroscope will (1) move closer together (2) move farther apart (3) become positively charged (4) become neutral

28. Sphere *A* carries a charge of +2 coulombs and an identical sphere *B* is neutral. If the spheres touch one another and then are separated, the charge on sphere *B* would be (1) +1 C (2) +2 C (3) 0 C (4) +4 C

29. Which procedure will double the force between two point charges? (1) doubling the distance between the charges (2) doubling the magnitude of one charge (3) halving the distance between the charges (4) halving the magnitude of one charge

30. How much energy is required to move 3.2×10^{-19} coulomb of charge through a potential difference of 5 volts? (1) 5 eV (2) 2 eV (3) 10 eV (4) 20 eV

31. A charge of 5.0 coulombs moves through a circuit in 0.50 second. How much current is flowing through the circuit? (1) 2.5 A (2) 5.0 A (3) 7.0 A (4) 10. A

32. Which graph represents a circuit element at constant temperature that obeys Ohm's law?

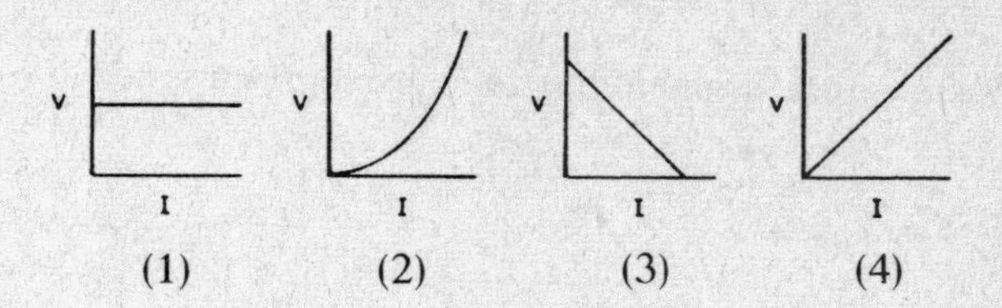

33. A glass rod is given a positive charge by rubbing it with silk. The rod has become positive by (1) gaining electrons (2) gaining protons (3) losing electrons (4) losing protons

34. Which diagram below shows correct current direction in a circuit segment?

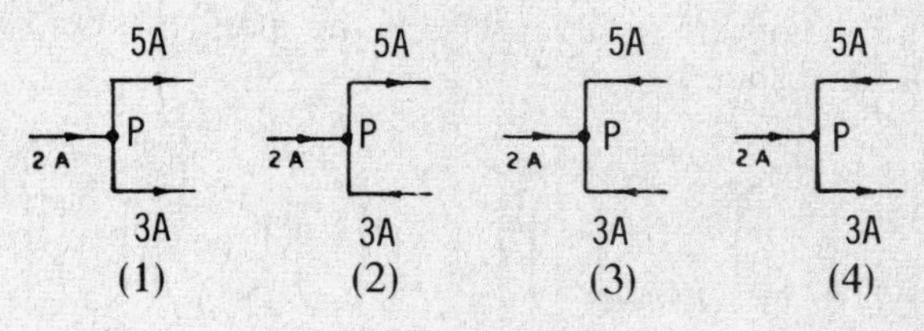

35. Which circuit below would have the *lowest* volt-meter reading?

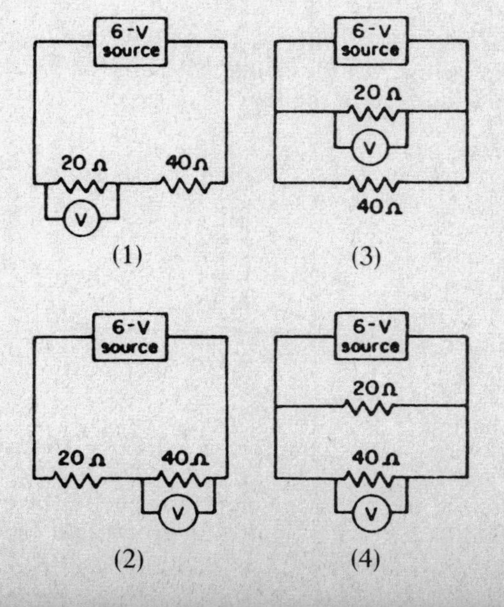

36. A toaster connected to a 120-volt outlet draws a current of 6.0 amperes. How much electrical energy does the toaster use in 5.0 seconds? (1) 1.4×10^2 J (2) 7.2×10^2 J (3) 3.6×10^3 J (4) 2.2×10^4 J

37. Which circuit segment has an equivalent resistance of 6 ohms?

38. Which diagram best represents the magnetic field near the poles of a horseshoe magnet?

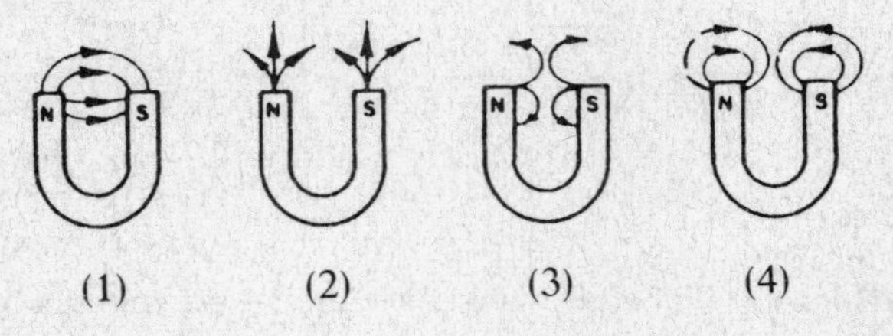

39. A bar magnet is dropped through a wire loop as shown in the diagram below. As the south pole approaches the loop, the electron flow induced in the loop

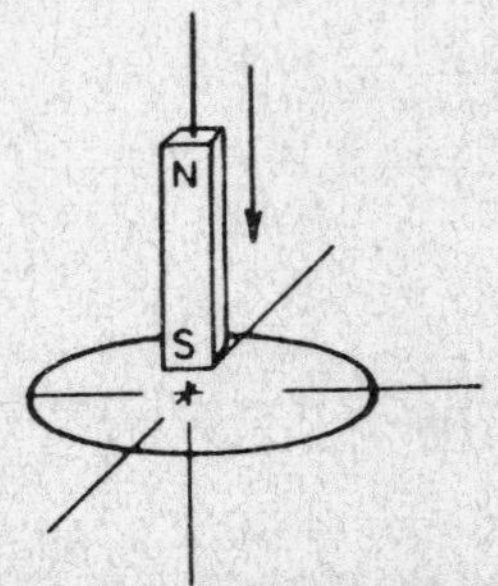

(1) is clockwise (2) is counterclockwise (3) attracts the south pole (4) speeds up the magnet

40. Which of the following electromagnetic waves has the *lowest* frequency? (1) violet light (2) green light (3) yellow light (4) red light

41. Which characteristic of a wave changes as the wave travels across a boundary between two different media? (1) frequency (2) period (3) phase (4) speed

42. What is the period of a wave with a frequency of 2.0×10^2 hertz? (1) 6.0×10^{-10} s (2) 2.0×10^{-3} s (3) 5.0×10^{-3} s (4) 1.5×10^6 s

43. The wavelength of the periodic wave shown in the diagram below is 4.0 meters. What is the distance from point *B* to point *C*?

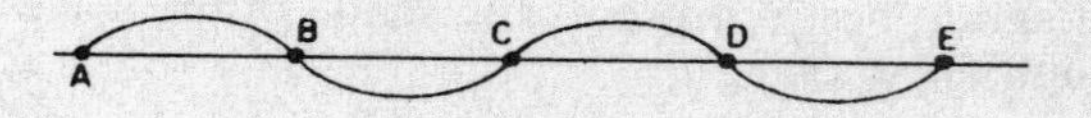

(1) 1.0 m (2) 2.0 m (3) 3.0 m (4) 4.0 m

44. Sound waves with a constant frequency of 250 hertz are traveling through air at STP. What is the wavelength of the sound waves? (1) 0.76 m (2) 1.3 m (3) 250 m (4) 83,000 m

45. An observer detects an apparent change in the frequency of sound waves produced by an airplane passing overhead. This phenomenon illustrates (1) the Doppler effect (2) the refraction of sound waves (3) wave amplitude increase (4) wave intensity increase

46. Which phenomenon *must* occur when two or more waves pass simultaneously through the same region in a medium? (1) refraction (2) interference (3) dispersion (4) reflection

47. The time required for light to travel a distance of 1.5×10^{11} meters is closest to (1) 5.0×10^2 s (2) 2.0×10^{-3} s (3) 5.0×10^{-1} s (4) 4.5×10^{19} s

48. Two waves of the same wavelength (λ) interfere to form a standing wave pattern as shown in the diagram. What is the straight-line distance between consecutive nodes?

(1) 1λ (2) 2λ (3) $\frac{1}{2}\lambda$ (4) $\frac{1}{4}\lambda$

49. When a ray of light strikes a mirror perpendicular to its surface, the angle of reflection is (1) 0° (2) 45° (3) 60° (4) 90°

50. Total internal reflection can occur as light waves pass from (1) water to air (2) Lucite to crown glass (3) alcohol to glycerol (4) air to crown glass

51. The diagram below represents straight wave fronts approaching an opening in a barrier.

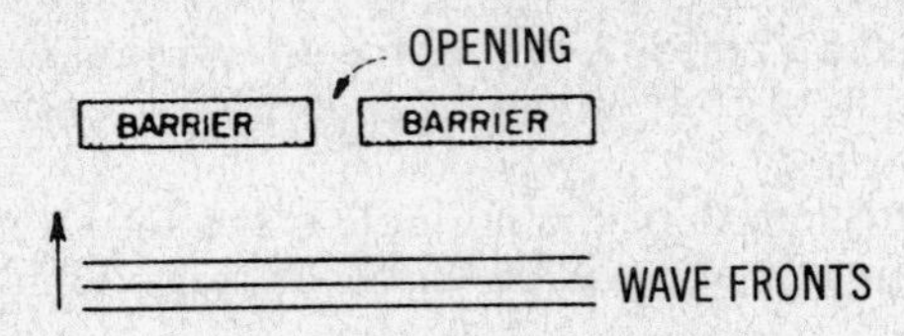

Which diagram best represents the shape of the waves after passing through the opening?

52. Coherent light of wavelength 6.4×10^{-7} meter passes through two narrow slits, producing an interference pattern on a screen with the first-order bright bands 2.0×10^{-2} meter from the central maximum. The screen is 4.0 meters from the slits. What is the distance between the slits? (1) 2.6×10^{-6} m (2) 3.2×10^{-9} m (3) 1.3×10^{-4} m (4) 7.8×10^{3} m

53. Rutherford observed that most of the alpha particles directed at a thin metal foil passed through without deflection. Based on this observation, he concluded that the (1) atom is very dense (2) atom is mostly empty space (3) nucleus is positively charged (4) nucleus is stationary

54. A beam of blue light causes photoelectrons to be emitted from a photoemissive surface. An increase in the intensity of the blue light will cause an increase in the (1) maximum kinetic energy of the emitted photoelectrons (2) number of photoelectrons emitted per unit of time (3) charge carried by each photoelectron (4) work function of the photoemissive surface

55. Photons with a frequency of 1.0×10^{20} hertz strike a metal surface. If electrons with a maximum kinetic energy of 3.0×10^{-14} joule are emitted, the work function of the metal is (1) 1.0×10^{-14} J (2) 2.2×10^{-14} J (3) 3.6×10^{-14} J (4) 6.6×10^{-14} J

56. A mass m moving with a velocity v has a wavelength of (1) $h \times \frac{1}{2}mv^2$ (2) $\frac{h}{\frac{1}{2}mv^2}$ (3) $h \times mv$ (4) $\frac{h}{mv}$

Note that questions 57 through 60 have only three choices.

57. As the resistance of a constant-voltage circuit is increased, the power developed in the circuit (1) decreases (2) increases (3) remains the same

58. As an electron moves between two charged parallel plates from point *B* to point *A*, as shown in the diagram below, the force of the electric field on the electron

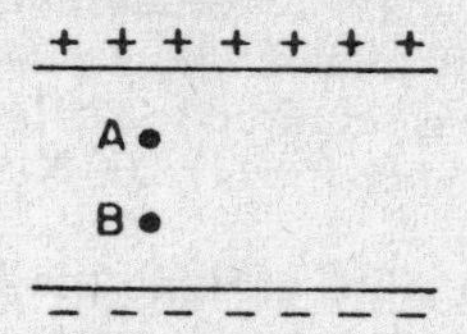

(1) decreases (2) increases (3) remains the same

59. Compared to the wavelength of a wave of green light in air, the wavelength of this same wave of green light in Lucite is (1) less (2) greater (3) the same

60. The wavelength of photon *A* is greater than that of photon *B*. Compared to the energy of photon *A*, the energy of photon *B* is (1) less (2) greater (3) the same

Part II

This part consists of six groups, each containing ten questions. Each group tests an optional area of the course. Choose two of these six groups. Be sure that you answer all ten questions in each group chosen. [20]

Group 1–Motion in a Plane

If you choose this group, be sure to answer questions 61-70.

61. A 1-kilogram object is thrown horizontally and a 2-kilogram object is dropped vertically at the same instant and from the same point above the ground. If friction is neglected, at any given instant both objects will have the same (1) kinetic energy (2) momentum (3) total velocity (4) height

62. Two masses, *A* and *B*, move in circular paths as shown in the diagram below.

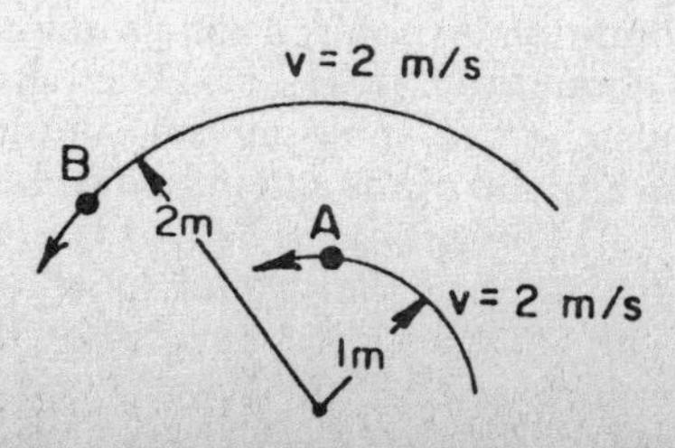

The centripetal acceleration of mass A, compared to that of mass B, is (1) the same (2) twice as great (3) one-half as great (4) four times as great

63. A motorcycle of mass 100 kilograms travels around a flat, circular track of radius 10 meters with a constant speed of 20 meters per second. What force is required to keep the motorcycle moving in a circular path at this speed? (1) 200 N (2) 400 N (3) 2000 N (4) 4000 N

64. The path of a satellite orbiting the Earth is best described as (1) linear (2) hyperbolic (3) parabolic (4) elliptical

65. The diagram below shows an object traveling clockwise in a horizontal, circular path at constant speed. Which arrow best shows the direction of the centripetal acceleration of the object at the instant shown?

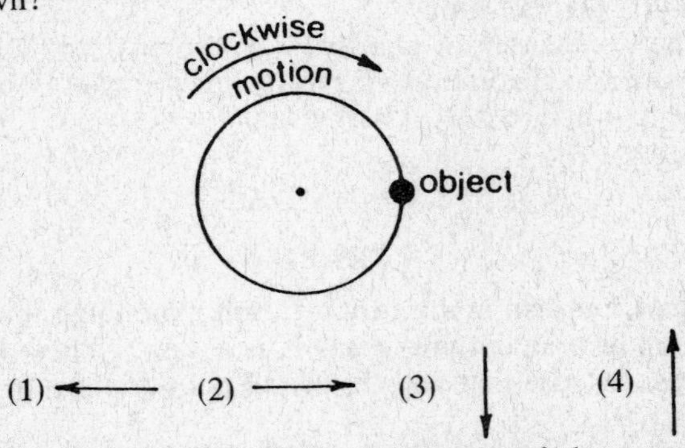

(1) ← (2) → (3) ↓ (4) ↑

66. For planets orbiting the Sun, the ratio of the mean radius of the orbit cubed to the orbital period of motion squared is (1) greatest for the most massive planet (2) greatest for the least massive planet (3) constantly changing as a planet rotates (4) the same for all planets

67. A satellite orbits the Earth in a circular orbit. Which statement best explains why the satellite does not move closer to the center of the Earth? (1) The gravitational field of the Earth does not reach the satellite's orbit. (2) The Earth's gravity keeps the satellite moving with constant velocity. (3) The satellite is always moving perpendicularly to the force due to gravity. (4) The satellite does not have any weight.

68. Which condition is required for a satellite to be in a geosynchronous orbit about the Earth? (1) The period of revolution of the satellite must be the same as the rotational period of the Earth. (2) The altitude of the satellite must be equal to the radius of the Earth. (3) The orbital speed of the satellite around the earth must be the same as the orbital speed of the Earth around the Sun. (4) The daily distance traveled by the satellite must be equal to the circumference of the Earth.

Note that questions 69 and 70 have only three choices.

69. A projectile is launched at an angle of 60° above the horizontal. Compared to the vertical component of the initial velocity of the projectile, the vertical component of the projectile's velocity when it has reached its maximum height is (1) less (2) greater (3) the same

70. The diagram below shows the movement of a planet around the Sun. Area 1 equals area 2.

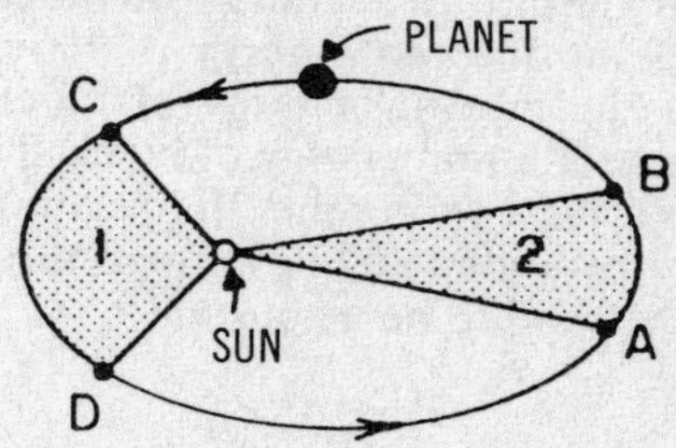

Compared to the time the planet takes to move from C to D, the time it takes to move from A to B is (1) less (2) greater (3) the same

Group 2–Internal Energy

If you choose this group, be sure to answer questions 71-80.

71. A temperature change of 51 Celsius degrees would be equivalent to a temperature change of (1) 51 K (2) 324 K (3) 222 K (4) −222 K

72. For object A to have a higher absolute temperature than object B, object A must have a (1) higher average internal potential energy (2) higher average internal kinetic energy (3) greater mass (4) greater specific heat

73. The graph below represents the relationship between the temperature of a gas and the average kinetic energy (KE) of the molecules of the gas.

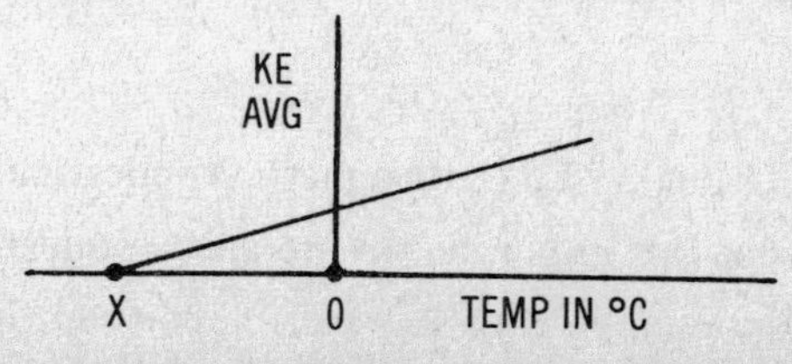

The temperature represented at point X is approximately (1) 273°C (2) 0°C (3) −273°C (4) −373°C

74. If 73 kilojoules of heat energy is added to 1.50 kilograms of ethyl alcohol initially at 20.°C, what will be the final temperature of the liquid? (1) 20.°C (2) 22°C (3) 40.°C (4) 60.°C

75. Equal masses of copper, iron, lead, and silver are heated from 20°C to 100°C. Which substance absorbs the *least* amount of heat? (1) lead (2) iron (3) copper (4) silver

76. Block *A*, at 100°C, and block *B*, at 50°C, are brought together in a well-insulated container. The internal energy of block *A* will
(1) decrease and the internal energy of block *B* will decrease
(2) decrease and the internal energy of block *B* will increase
(3) increase and the internal energy of block *B* will decrease
(4) increase and the internal energy of block *B* will increase

77. Which graph best represents the relationship between the heat absorbed *(Q)* by a solid and its temperature *(T)*? [Assume the specific heat of the solid to be a constant.]

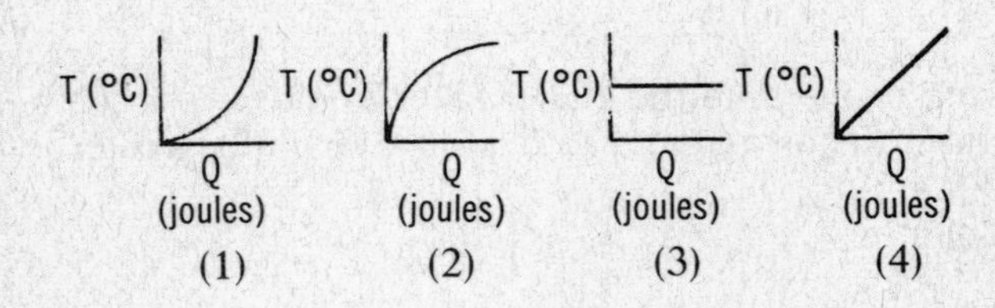

78. A given mass of gas is enclosed in a rigid container. If the velocity of the gas molecules colliding with the sides of the container increases, the (1) density of the gas will increase (2) pressure of the gas will increase (3) density of the gas will decrease (4) pressure of the gas will decrease

79. The absolute temperature of a fixed mass of ideal gas is tripled while its volume remains constant. The ratio of the final pressure of the gas to its initial pressure is (1) 1 to 1 (2) 1.5 to 1 (3) 3 to 1 (4) 9 to 1

Note that question 80 has only three choices.

80. Compared to the boiling point of pure water, the boiling point of a salt-water solution is (1) lower (2) higher (3) the same

Group 3–Electromagnetic Applications

If you choose this group, be sure to answer questions 81-90.

81. A galvanometer with a low-resistance shunt in parallel with its moving coil is (1) a motor (2) a generator (3) a voltmeter (4) an ammeter

82. In the diagram below, a solenoid that is free to rotate around an axis at its center, *C*, is placed between the poles of a permanent magnet.

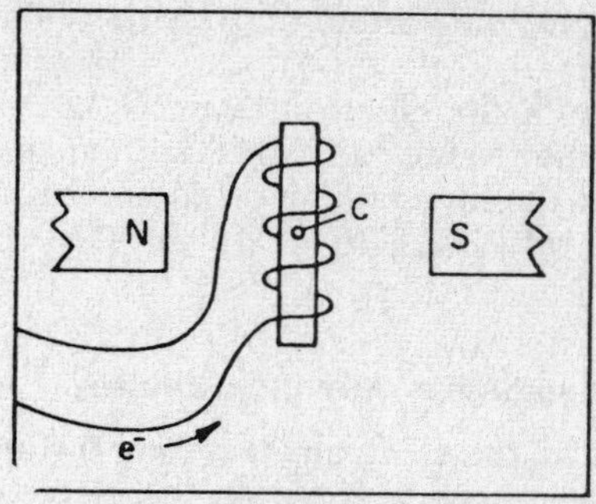

As an electron current starts through the solenoid in the direction shown, the solenoid will (1) remain motionless (2) vibrate back and forth (3) start turning clockwise (4) start turning counterclockwise

83. A potential difference of 50. volts is required to operate an electrical device. The potential difference of the source is 120 volts. The table shows the primary and secondary windings for four available transformers. Which transformer is suitable for this application?

Transformer	Primary	Secondary
A	250	600
B	600	250
C	240	150
D	150	240

(1) *A* (2) *B* (3) *C* (4) *D*

Base your answers to questions 84 through 86 on the diagram below which shows an apparatus for demonstrating the effect of a uniform magnetic field on a beam of electrons moving in the direction shown.

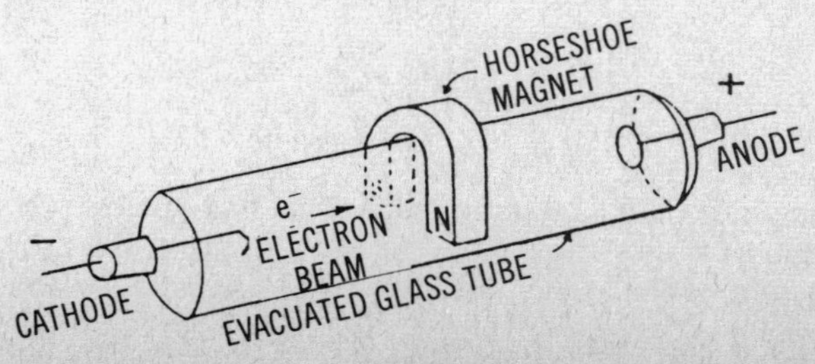

84. As the electron beam enters the magnetic field of the horseshoe magnet, the beam will be deflected (1) toward the south pole of the magnet (2) toward the north pole of the magnet (3) downward, toward the bottom of the tube (4) upward, toward the top of the tube

85. The velocity of the electron beam is 3.0×10^6 meters per second, perpendicular to the 5.0×10^{-3}-tesla magnetic field. What is the magnitude of the force acting on each electron in the beam? (1) 8.0×10^{-22} N (2) 2.4×10^{-15} N (3) 1.7×10^{-9} N (4) 1.5×10^4 N

Note that question 86 has only three choices.

86. If the speed of the electrons traveling through the magnetic field increases, the magnetic force on the electrons will (1) decrease (2) increase (3) remain the same

87. What is the potential difference induced in a wire 0.10 meter long as it moves with a speed of 50. meters per second perpendicular to a magnetic field that has a magnetic flux density of 0.050 tesla? (1) 0.25 V (2) 25 V (3) 250 V (4) 2500 V

88. An ideal transformer has a current of 2.0 amperes and a potential difference of 120 volts across its primary coil. If the current in the secondary coil is 0.50 ampere, the potential difference across the secondary coil is (1) 480 V (2) 120 V (3) 60. V (4) 30. V

89. The diagram below shows conductor *C* between two opposite magnetic poles. Which procedure will produce the greatest induced potential difference in the conductor?

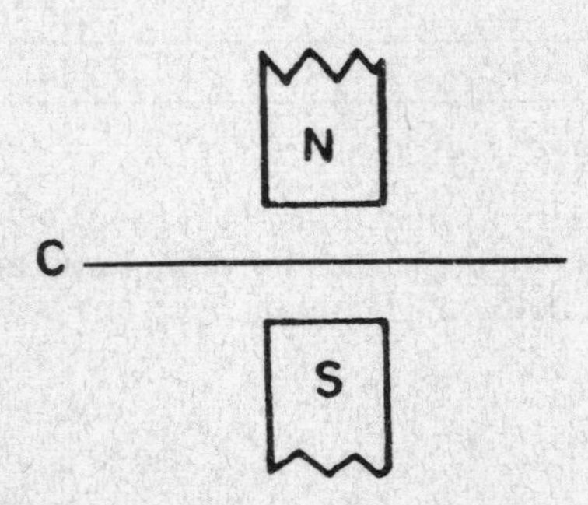

(1) holding the conductor stationary between the poles (2) moving the conductor out of the page (3) moving the conductor toward the right side of the page (4) moving the conductor toward the N-pole

Note that question 90 has only three choices.

90. The loop shown in the diagram below rotates about an axis which is perpendicular to a constant uniform magnetic field.

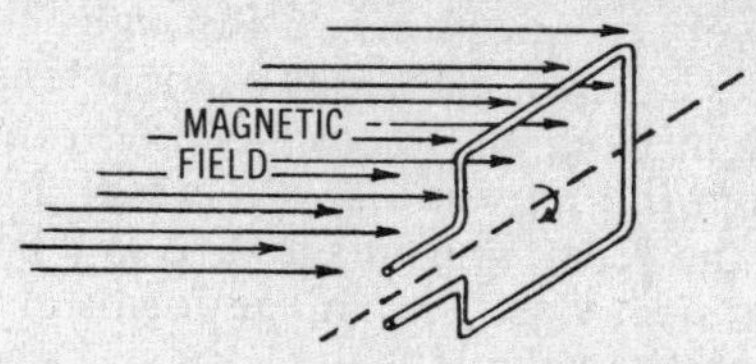

If only the direction of the field is reversed, the magnitude of the maximum induced potential difference will (1) decrease (2) increase (3) remain the same

Group 4–Geometric Optics

If you choose this group, be sure to answer questions 91-100.

91. Which optical device may form an enlarged image? (1) plane mirror (2) glass plate (3) converging lens (4) diverging lens

92. The diagram below represents a spherical mirror with its center of curvature at C and focal point at F. At which position must a point source of light be placed to produce a parallel beam of reflected light?

MIRRORED SURFACE

C A F B

(1) A (2) B (3) C (4) F

93. An object is located 0.12 meter in front of a concave (converging) mirror of 0.16-meter radius. What is the distance between the image and the mirror? (1) 0.07 m (2) 0.20 m (3) 0.24 m (4) 0.48 m

94. In the diagram below, ray XO is incident upon the concave (diverging) lens. Along which path will the ray continue?

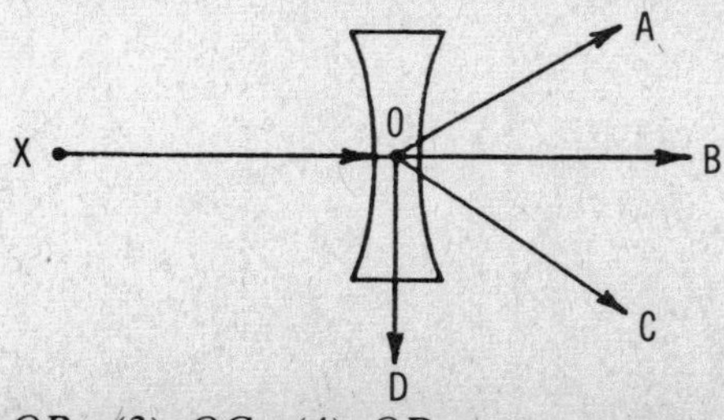

(1) OA (2) OB (3) OC (4) OD

95. An object is placed in front of a convex (diverging) mirror. The image of that object will be (1) nonexistent (2) real and smaller (3) virtual and smaller (4) virtual and larger

Base your answers to questions 96 through 98 on the diagram below which shows a crown glass lens of focal length f in air. Monochromatic red light from an object placed on the left side of the lens passes through the lens and forms a real, inverted image on the right side of the lens. The image size is 0.04 meter and it is smaller than the object size. The image forms at a distance of 0.1 meter from the center of the lens.

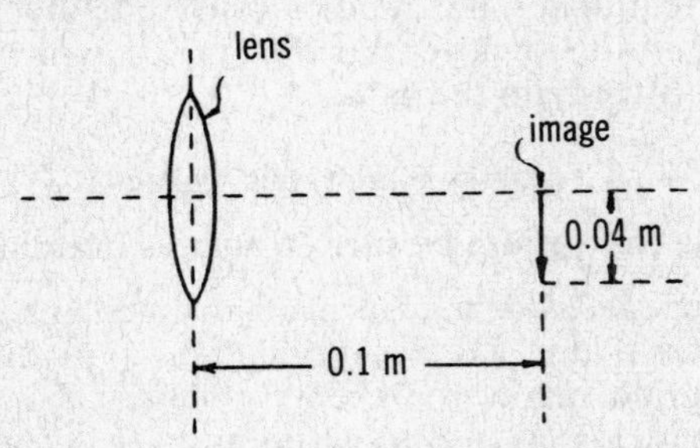

96. If the size of the object is 0.08 meter, then the distance from the object to the center of the lens is (1) 0.1 m (2) 0.2 m (3) 0.3 m (4) 0.5 m

97. The distance from the object to the center of the lens must be (1) greater than $2f$ (2) equal to $2f$ (3) between f and $2f$ (4) less than f

Note that question 98 has only three choices.

98. A flint glass lens of identical curvature is substituted for the crown glass lens. Compared to the focal length of the crown glass lens, the focal length of the flint glass lens is (1) shorter (2) longer (3) the same

99. The diagram below shows two rays of light striking a plane mirror. Which diagram below best represents the reflected rays?

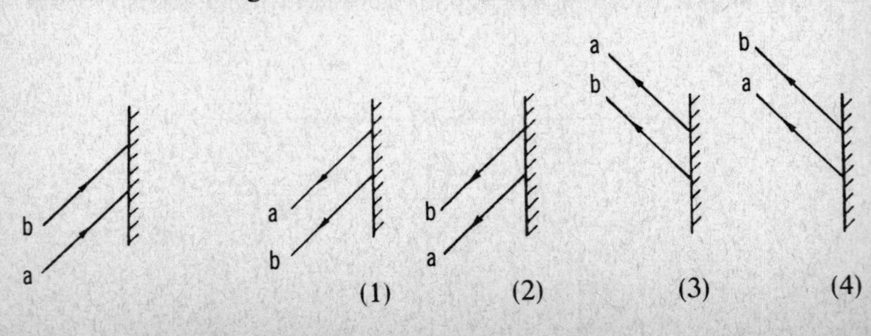

100. A person stands in front of a vertical plane mirror 2.0 meters high as shown in the diagram below. A ray of light reflects off the mirror, allowing him to see his foot. Approximately how far up the mirror from the floor does this ray strike the mirror?

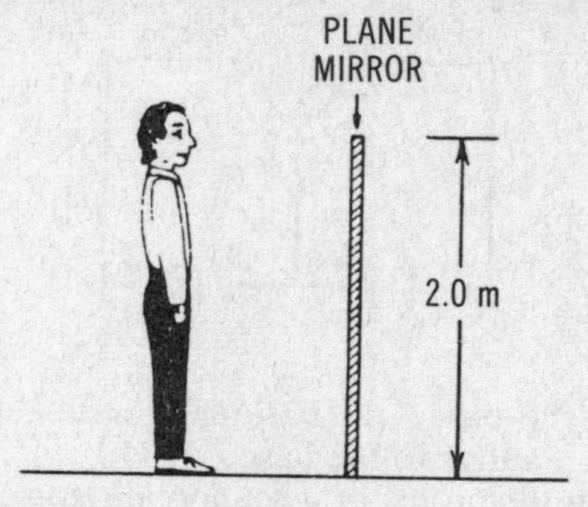

(1) 1.0 m (2) 2.0 m (3) 0.25 m (4) 0 m

Group 5–Solid State

If you choose this group, be sure to answer questions 101-110.

101. Which procedure would cause most of the free electrons in a conductor to move in the same direction? (1) heating the conductor (2) charging the conductor (3) maintaining a potential difference across the conductor (4) holding the conductor between the poles of a magnet

102. Germanium is doped with arsenic to form an *N*-type semiconductor. A majority of the charge carriers in this semiconductor are (1) electrons (2) holes (3) *P*-ions (4) protons

103. The semiconductor material located between the emitter and the collector of a transistor is called the (1) base (2) junction (3) bias (4) rectifier

104. Which device is used for amplifying the flow of electrons in a solid state circuit? (1) diode (2) transistor (3) resistor (4) switch

105. A minute quantity of gallium, which has 3 valence electrons, is added to silicon during the process of crystallization. Which type of semiconductor is formed by this process? (1) *N* (2) *P* (3) *N-P* (4) *P-N*

106. An *N-P-N* transistor is connected as shown in the diagram below. Within the transistor, a forward-biased current will flow from the

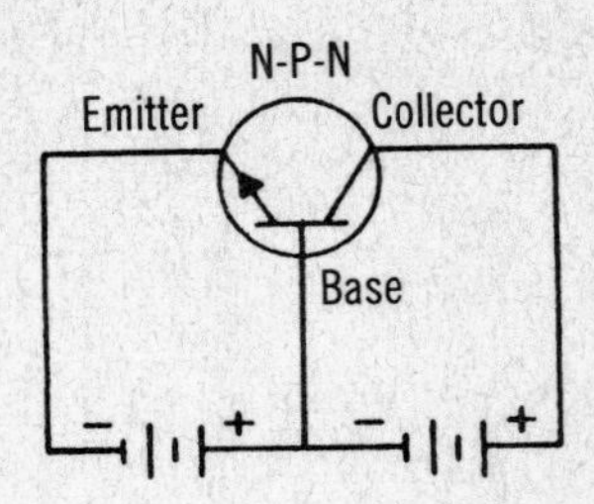

(1) collector to the base (2) base to the collector (3) base to the emitter (4) emitter to the base

107. A hole in the atoms of a semiconductor crystalline lattice structure can be defined as a (1) free electron (2) negative atom (3) region in which an electron is located (4) region from which an electron has vacated

108. A section of *P*-type semiconductor has a potential difference applied across it, as shown in the diagram below. Which statement best describes the flow of charge through the semiconductor?

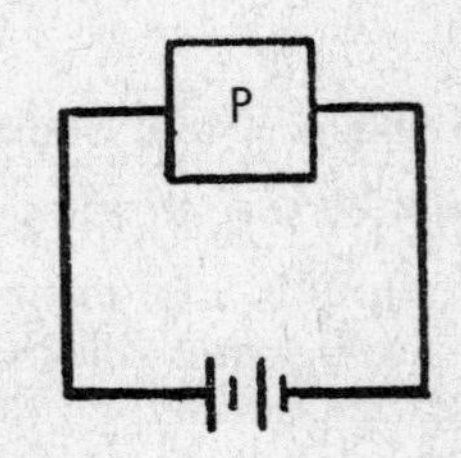

(1) Holes flow toward the positive terminal. (2) Holes flow toward the negative terminal. (3) Protons flow toward the positive terminal. (4) Protons flow toward the negative terminal.

109. A doping agent that adds electrons to a semiconducting material is called (1) an *N*-type semiconductor (2) a *P*-type semiconductor (3) a donor (4) an acceptor

Note that question 110 has only three choices.

110. Compared to the number of free electrons in an insulator of a given size, the number of free electrons in a conductor of the same size is (1) less (2) greater (3) the same

Group 6–Nuclear Energy

If you choose this group, be sure to answer questions 111-120.

111. What is the mass number of an atom with 9 protons, 11 neutrons, and 9 electrons? (1) 9 (2) 18 (3) 20 (4) 29

112. If the mass of one proton is totally converted into energy, it will yield a total energy of (1) 5.1×10^{-19} J (2) 1.5×10^{-10} J (3) 9.3×10^{8} J (4) 9.0×10^{16} J

113. The diagram below represents an inverted test tube over a sample of a radioactive material. Helium has collected in the test tube.

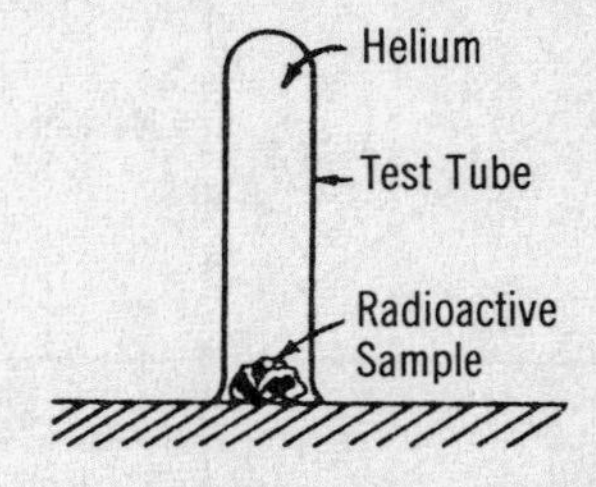

The presence of helium indicates that the sample is most probably undergoing the process of (1) alpha decay (2) beta decay (3) neutron decay (4) gamma emission

114. In the reaction ${}^{24}_{11}Na \rightarrow {}^{24}_{12}Mg + x$, what does x represent? (1) an alpha particle (2) a beta particle (3) a neutron (4) a positron

115. The half-life of an isotope is 14 days. How many days will it take 8 grams of this isotope to decay to 1 gram? (1) 14 (2) 21 (3) 28 (4) 42

116. The uranium isotope ${}^{238}_{92}U$ is used to produce (1) shielding (2) fissionable plutonium (3) control rods (4) heavy water

117. Which reaction is an example of nuclear fusion? (1) ${}^{226}_{88}Ra \rightarrow {}^{222}_{86}Rn + {}^{4}_{2}He + Q$ (2) ${}^{214}_{83}Bi \rightarrow {}^{214}_{84}Po + {}^{0}_{-1}e + Q$ (3) ${}^{235}_{92}U + {}^{1}_{0}n \rightarrow {}^{92}_{36}Kr + {}^{141}_{56}Ba + 3{}^{1}_{0}n + Q$ (4) ${}^{3}_{1}H + {}^{1}_{1}H \rightarrow {}^{4}_{2}He + Q$

118. The function of a moderator in a nuclear reactor is to (1) decrease the speed of the neutrons (2) increase the speed of the neutrons (3) decrease the number of neutrons (4) increase the number of neutrons

119. Neutrons are used in some nuclear reactions as bombarding particles because they are (1) positively charged and are repelled by the nucleus (2) uncharged and are not repelled by the nucleus (3) negatively charged and are attracted by the nucleus (4) uncharged and have negligible mass

Note that question 120 has only three choices.

120. When a nucleus captures an electron, the mass number of the nucleus (1) decreases (2) increases (3) remains the same

Part III

You must answer <u>both</u> questions in this part. Record your answers in the spaces provided on the separate answer paper. [10]

121. Base your answers to parts *a* and *b* on the diagram below of a light ray ($\lambda = 5.9 \times 10^{-7}$ m) in air incident on a rectangular block of Lucite.

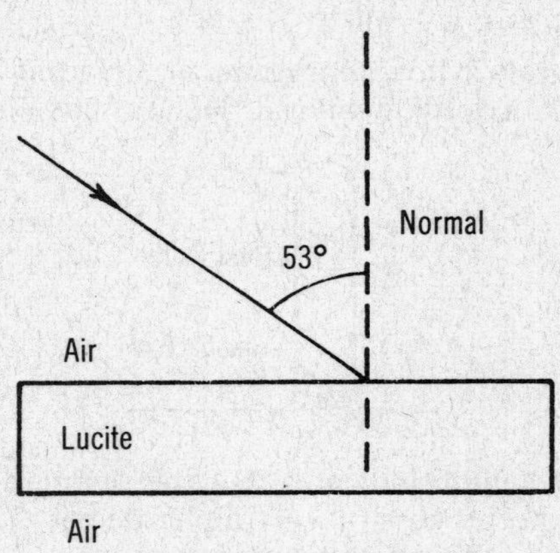

a. Determine the angle of refraction, in degrees, of the light ray as it enters the Lucite from air. [Show all calculations.] [2]

b. On the diagram on the answer paper, *using a protractor and straight edge*, draw the path of the light ray as it travels from air, through the Lucite, and *back into the air*. Label all angles of incidence and refraction with their appropriate numeric values. The diagram above may be used for practice purposes. *Be sure your final answer appears on your answer paper.* [3]

122. Base your answers to parts *a* through *c* on the diagram below which represents a 4.0-kilogram block sliding down a frictionless 30° incline. The top of the incline is 2.5 meters above the ground, and the incline is 5.0 meters long. The uniform acceleration of the block down the incline is 4.9 meters per second squared.

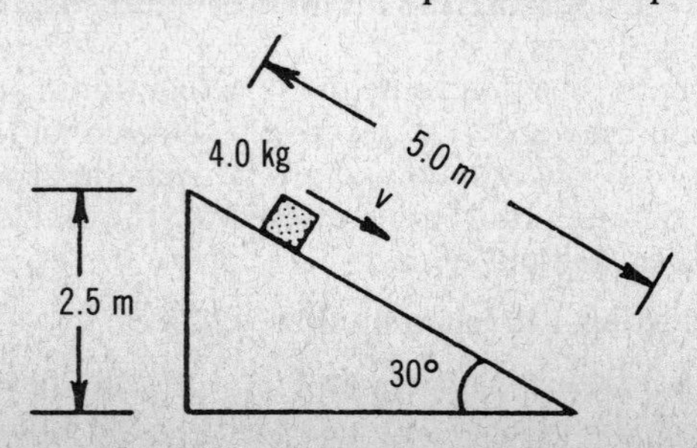

a. On the diagram on the answer paper, draw a vector on the block, representing the weight of the block. The diagram above may be used for practice purposes. *Be sure your final answer appears on your answer paper.* [1]

b. What was the potential energy of the block at the top of the incline? [Show all calculations.] [2]

c. The block was initially at rest at the top of the incline. Determine the speed of the block at the bottom of the incline. [Show all calculations.] [2]

REGENTS HIGH SCHOOL EXAMINATION
PHYSICS
Friday, June 15, 1990

Part I

Answer all 60 questions in this part. [70]

Directions (1-60): For *each* statement or question, select the word or expression that, of those given, best completes the statement or answers the question.

1. A cart starting from rest travels a distance of 3.6 meters in 1.8 seconds. The average speed of the cart is (1) 0.20 m/s (2) 2.0 m/s (3) 0.50 m/s (4) 5.0 m/s

2. An object has a constant acceleration of 2.0 meters per second^2. The time required for the object to accelerate from 8.0 meters per second to 28 meters per second is (1) 20.s (2) 16 s (3) 10. s (4) 4.0 s

3. A ball dropped from a bridge takes 3.0 seconds to reach the water below. How far is the bridge above the water? (1) 15 m (2) 29 m (3) 44 m (4) 88 m

4. A car moving at a speed of 8.0 meters per second enters a highway and accelerates at 3.0 meters per second^2. How fast will the car be moving after it has accelerated for 56 meters? (1) 24 m/s (2) 20. m/s (3) 18 m/s (4) 4.0 m/s

5. A student walks 3 blocks south, 4 blocks west, and 3 blocks north. What is the displacement of the student? (1) 10 blocks east (2) 10 blocks west (3) 4 blocks east (4) 4 blocks west

6. The graph below represents the relationship between distance and time for an object in motion. During which interval is the speed of the object changing?

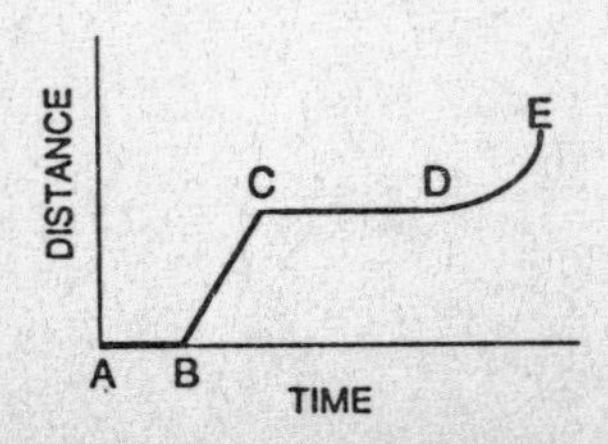

(1) *AB* (2) *BC* (3) *CD* (4) *DE*

7. Which pair of terms are vector quantities? (1) force and mass (2) distance and displacement (3) momentum and acceleration (4) speed and velocity

8. A force of 6.0 newtons north and a force of 8.0 newtons east act concurrently on an object. The magnitude of the resultant of the two forces is (1) 1.3 N (2) 2.0 N (3) 10. N (4) 14 N

9. Which graph best represents the motion of a freely falling body near the Earth's surface?

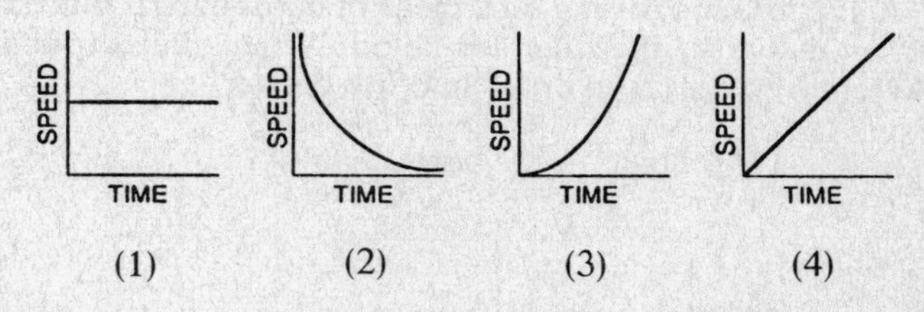

(1) (2) (3) (4)

10. A 50.0-kilogram object in outer space is attracted to a nearby planet with a net force of 400. newtons. What is the magnitude of the object's acceleration? (1) 8.00 m/s² (2) 9.81 m/s² (3) 78.4 m/s² (4) 2,000 m/s²

11. A spring is compressed between two stationary blocks as shown in the diagram below. Block *A* has a mass of 6.0 kilograms. After the spring is released, block *A* moves west at 8.0 meters per second and block *B* moves east at 16 meters per second.

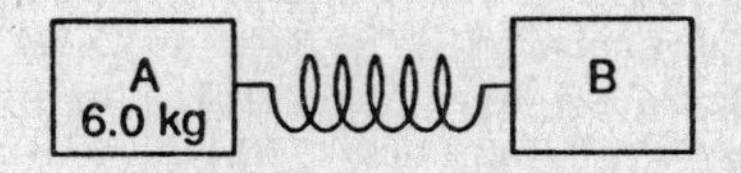

What is the mass of block *B*? [Assume no frictional effects.] (1) 16 kg (2) 12 kg (3) 3.0 kg (4) 6.0 kg

12. Two point masses are located a distance, *D*, apart. The gravitational force of attraction between them can be quadrupled by changing the distance to (1) ½*D* (2) 2*D* (3) ¼*D* (4) 4*D*

13. The table below lists the coefficients of kinetic friction for four materials sliding over steel.

Material	μ_K
aluminum	0.47
brass	0.44
copper	0.36
steel	0.57

A 10.-kilogram block of each of the materials in the table is pulled horizontally across a steel floor at constant velocity. Which block would require the *smallest* applied force to keep it moving at constant velocity? (1) aluminum (2) brass (3) copper (4) steel

14. A 25-kilogram mass travels east with a constant velocity of 40. meters per second. The momentum of this mass is (1) 1.0×10^3 kg•m/s east (2) 9.8×10^3 kg•m/s east (3) 1.0×10^3 kg•m/s west (4) 9.8×10^3 kg•m/s west

15. A copper coin resting on a piece of cardboard is placed on a beaker as shown in the diagram below. When the cardboard is rapidly removed, the coin drops into the beaker.

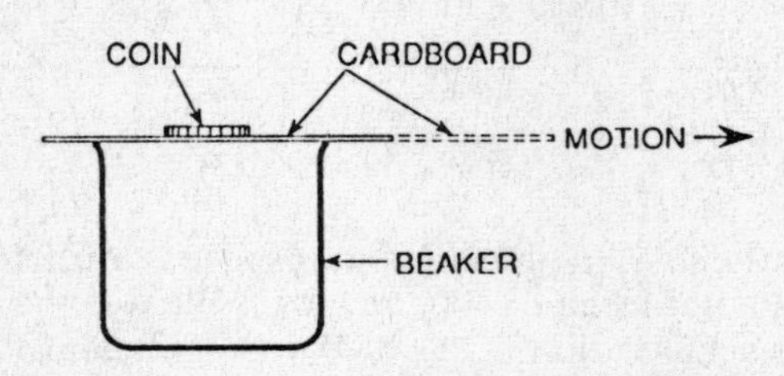

The two properties of the coin which best explain its fall are its weight and its (1) temperature (2) electrical resistance (3) volume (4) inertia

16. A constant braking force of 10 newtons applied for 5 seconds is used to stop a 2.5-kilogram cart traveling at 20 meters per second. The magnitude of the impulse applied to stop the cart is (1) 10 N•s (2) 30 N•s (3) 50 N•s (4) 100 N•s

17. Which is the most likely mass of a high school student? (1) 1 kg (2) 5 kg (3) 60 kg (4) 250 kg

18. The graph below shows the force exerted on a block as a function of the block's displacement in the direction of the force.

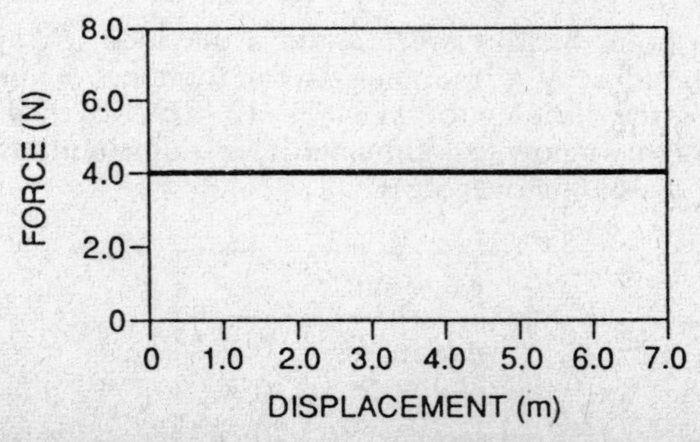

How much work did the force do in displacing the block 5.0 meters? (1) 0 J (2) 20. J (3) 0.80 J (4) 4.0 J

19. A motor has an output of 1,000 watts. When the motor is working at full capacity, how much time will it require to lift a 50-newton weight 100 meters? (1) 5 s (2) 10 s (3) 50 s (4) 100 s

20. When a 5-kilogram mass is lifted from the ground to a height of 10 meters, the gravitational potential energy of the mass is increased by approximately (1) 0.5 J (2) 2 J (3) 50 J (4) 500 J

21. If the speed of an object is doubled, its kinetic energy will be (1) halved (2) doubled (3) quartered (4) quadrupled

22. The graph below represents the relationship between the force applied to a spring and the elongation of the spring.

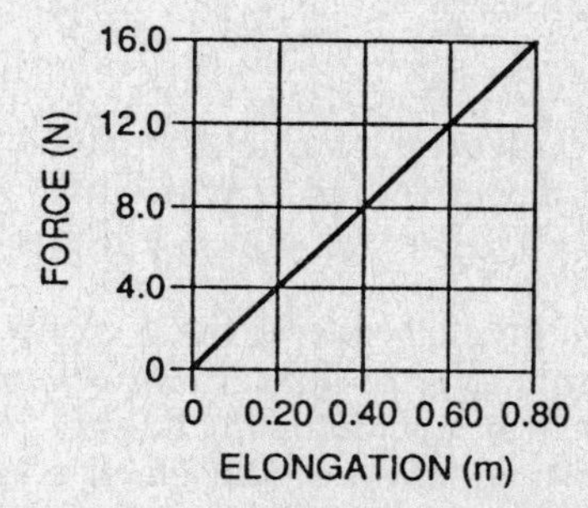

What is the spring constant? (1) 20.N/m (2) 9.8 N/kg (3) 0.80 N•m (4) 0.050 m/N

23. An object 10 meters above the ground has Z joules of potential energy. If the object falls freely, how many joules of kinetic energy will it have gained when it is 5 meters above the ground? (1) Z (2) $2Z$ (3) $Z/2$ (4) 0

24. Sphere A has a charge of $+2 \times 10^{-6}$ coulomb and is brought into contact with a similar sphere, B, which has a charge of -4×10^{-6} coulomb. After it is separated from sphere B, sphere A will have a charge of (1) -1×10^{-6} C (2) -2×10^{-6} C (3) $+2 \times 10^{-6}$ C (4) $+6 \times 10^{-6}$ C

25. Which diagram best represents the charge distribution on a neutral electroscope when a negatively charged rod is held near it?

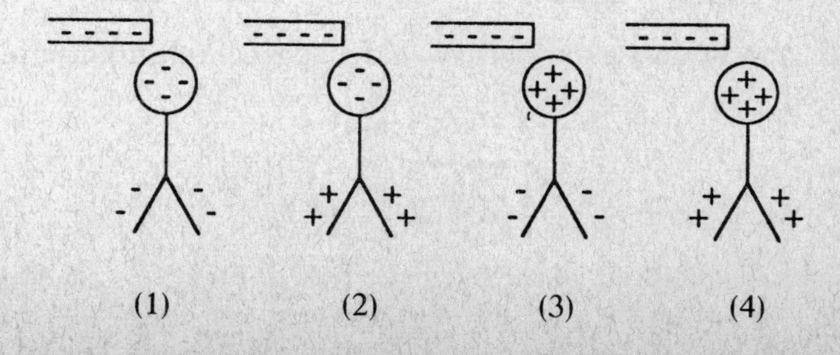

26. An alpha particle with a charge of +2 elementary charges is accelerated by a potential difference of 1.0×10^6 volts. The energy acquired by the particle is (1) 0.50×10^6 eV (2) 2.0×10^6 eV (3) 1.6×10^{-19} eV (4) 3.2×10^{-13} eV

27. The diagram below shows the electric field in the vicinity of two charged conducting spheres, *A* and *B*.

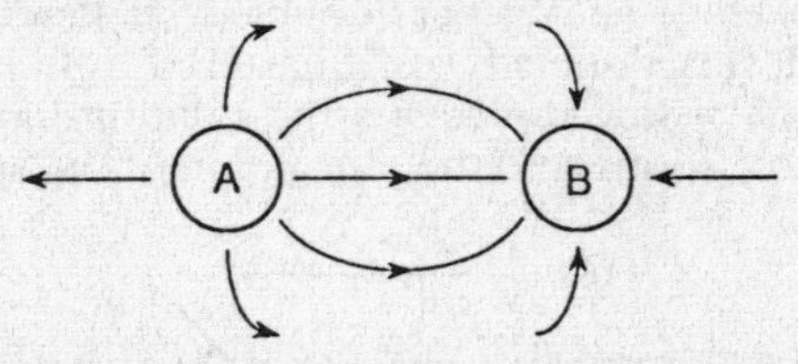

What is the static electric charge on each of the conducting spheres? (1) *A* is negative and *B* is positive. (2) *A* is positive and *B* is negative. (3) Both *A* and *B* are positive. (4) Both *A* and *B* are negative.

28. If the charge on each of two small spheres a fixed distance apart is doubled, the force of attraction between the spheres will be (1) quartered (2) doubled (3) halved (4) quadrupled

29. What is the magnitude of the electric field intensity at a point in the field where an electron experiences a 1.0-newton force? (1) 1.0 N/C (2) 1.6×10^{-19} N/C (3) 6.3×10^{18} N/C (4) 9.1×10^{-31} N/C

30. A copper wire has a resistance of 200 ohms. A second copper wire with twice the cross-sectional area and the same length would have a resistance of (1) 50 Ω (2) 100 Ω (3) 200 Ω (4) 400 Ω

31. In which diagram below is the magnetic flux density at point *P* greatest?

(1) N • P (3) S • P S

(2) N • P N (4) N • P S

32. The wire loop shown below has a clockwise electron current.

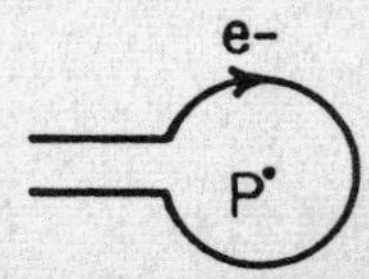

What is the direction of the magnetic field at point *P*? (1) to the right (2) to the left (3) into the page (4) out of the page

33. If 10 coulombs of charge passes a given point in a conductor every 2 seconds, the current at that point is (1) 0.2 A (2) 5 A (3) 10 A (4) 20 A

34. The graph below shows the relationship between current and potential difference for four resistors, *A*, *B*, *C*, and *D*.

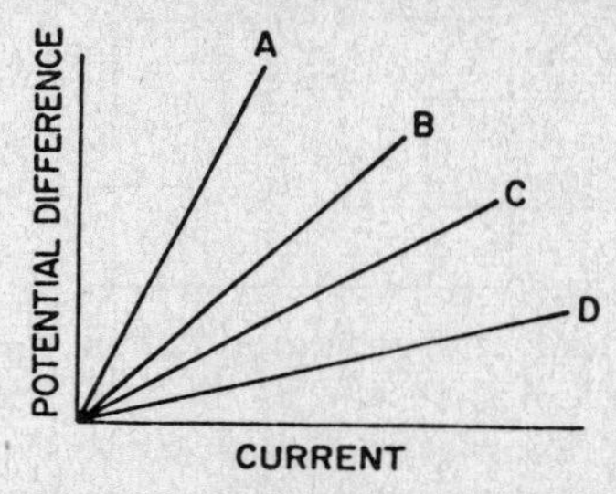

Which resistor has the greatest resistance? (1) A (2) B (3) C (4) D

35. In the circuit shown below, what is the potential difference of the source?

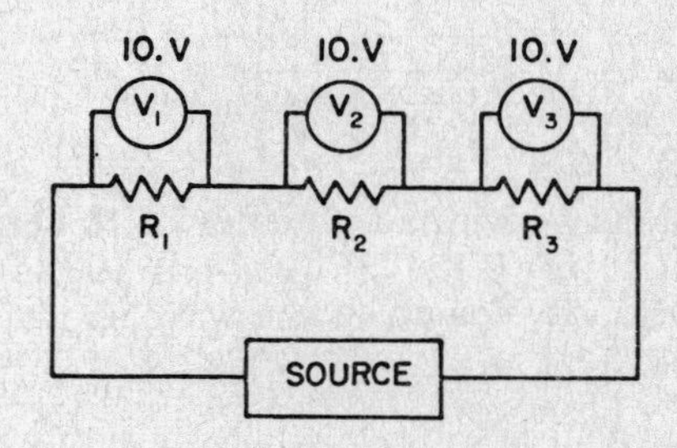

(1) 3.3 V (2) 10. V (3) 30. V (4) 1,000 V

36. The diagram below shows the current in three of the branches of a direct current electric circuit.

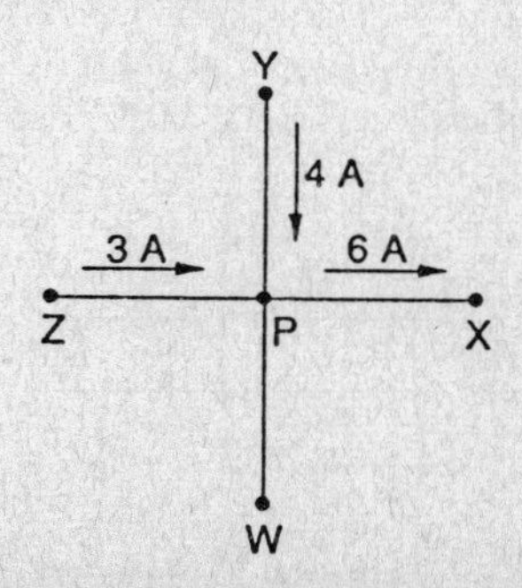

The current in the fourth branch, between junction *P* and point *W*, must be (1) 1 A toward point *W* (2) 1 A toward point *P* (3) 7 A toward point *W* (4) 7 A toward point *P*

37. A 5-ohm and a 10-ohm resistor are connected in series. The current in the 5-ohm resistor is 2 amperes. The current in the 10-ohm resistor is (1) 1 A (2) 2 A (3) 0.5 A (4) 8 A

38. A lamp and an ammeter are connected to a source as shown.

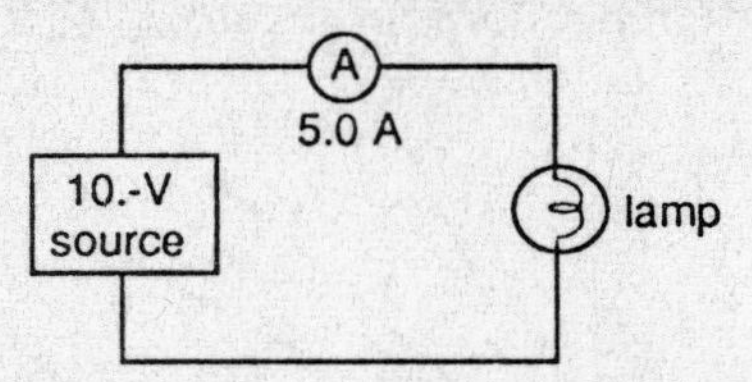

What is the electrical energy expended in the lamp in 3.0 seconds? (1) 50. J (2) 150 J (3) 50. W (4) 150 W

39. The diagram below represents a conductor carrying an electron current in magnetic field *B*. The direction of the magnetic force on the conductor is (1) into the page (2) out of the page (3) toward the top of the page (4) toward the bottom of the page.

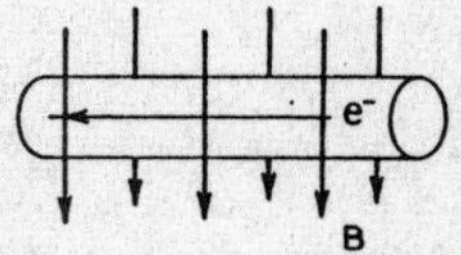

40. A periodic wave with a frequency of 10 hertz would have a period of (1) 1 s (2) 0.1 (3) 10 s (4) 100 s

41. What is the wavelength of the wave shown in the diagram below?

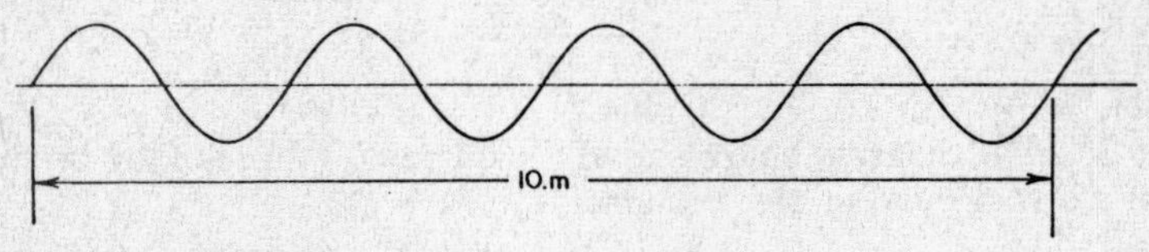

(1) 2.5 m (2) 5.0 m (3) 10. m (4) 4.0 m

42. In the diagram below, *A*, *B*, *C*, and *D* are points near a current-carrying solenoid.

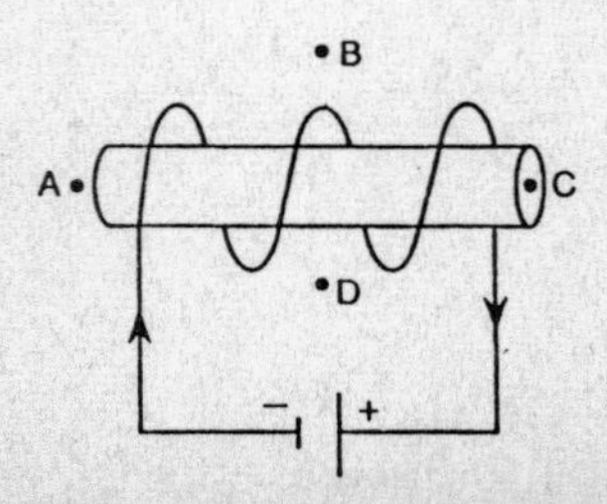

Which point is closest to the north pole of the solenoid? (1) *A* (2) *B* (3) *C* (4) *D*

43. In the diagram below, which wave has the largest amplitude?

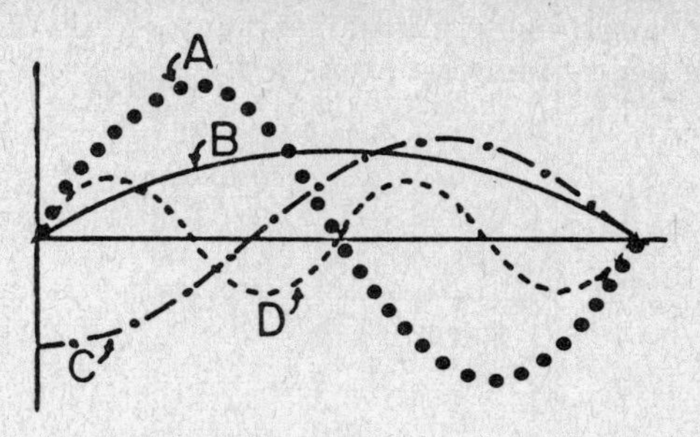

(1) A (2) B (3) C (4) D

44. Which term describes two points on a periodic wave that are moving in the same direction and have the same displacement from their equilibrium positions? (1) dispersed (2) refracted (3) polarized (4) in phase

45. In a vacuum, the wavelength of green light is 5×10^{-7} meter. What is its frequency? (1) 2×10^{-15} Hz (2) 2×10^{-14} Hz (3) 6×10^{14} Hz (4) 6×10^{15} Hz

46. As shown in the diagram below, a transverse wave is moving along a rope.

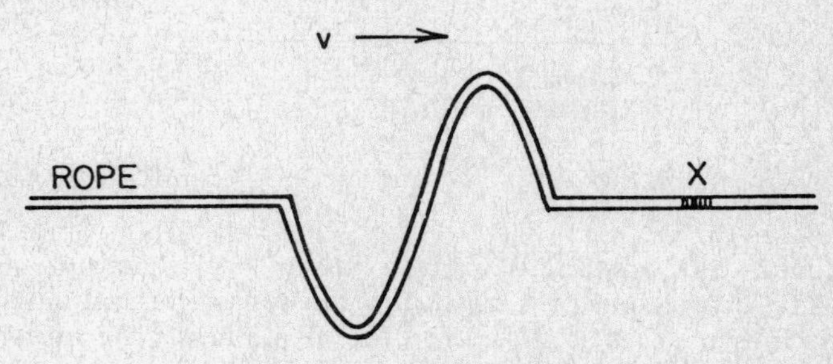

In which direction will segment X move as the wave passes through it? (1) down, only (2) up, only (3) down, then up (4) up, then down

47. A ray of light strikes a mirror at an angle of incidence of 60°. What is the angle of reflection? (1) 0° (2) 30° (3) 60° (4) 90°

48. What is the approximate speed of light in alcohol? (1) 1.4×10^8 m/s (2) 2.2×10^8 m/s (3) 3.0×10^8 m/s (4) 4.4×10^8 m/s

49. A prism disperses white light, forming a spectrum. The best explanation for this phenomenon is that different frequencies of visible light (1) move at different speeds in the prism (2) are reflected inside the prism (3) are absorbed inside the prism (4) undergo constructive interference inside the prism

50. In the diagram below, a ray of light enters a transparent medium from air. If angle X is 45° and angle Y is 30.°, what is the absolute index of refraction of the medium?

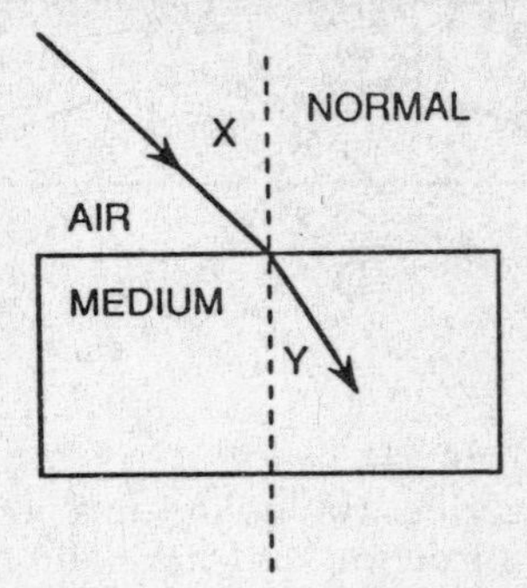

(1) 0.667 (2) 0.707 (3) 1.41 (4) 1.50

51. Which optical medium would have the smallest critical angle (θ_c) in the situation shown in the diagram?

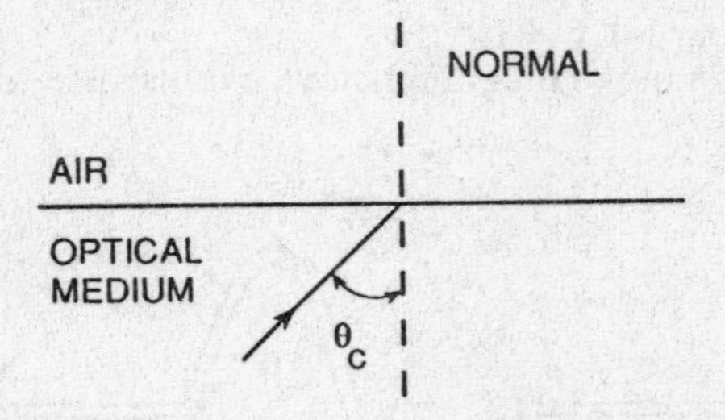

(1) Lucite (2) water (3) Canada balsam (4) diamond

52. The spreading of a wave into the region behind an obstacle is known as (1) diffusion (2) dispersion (3) refraction (4) diffraction

53. An electron in a hydrogen atom drops from the $n = 3$ energy level to the $n = 2$ energy level. The energy of the emitted photon is (1) 1.51 eV (2) 1.89 eV (3) 3.40 eV (4) 4.91 eV

54. Which graph best represents the relationship between the photocurrent in a photoelectric cell and the intensity of the incident light?

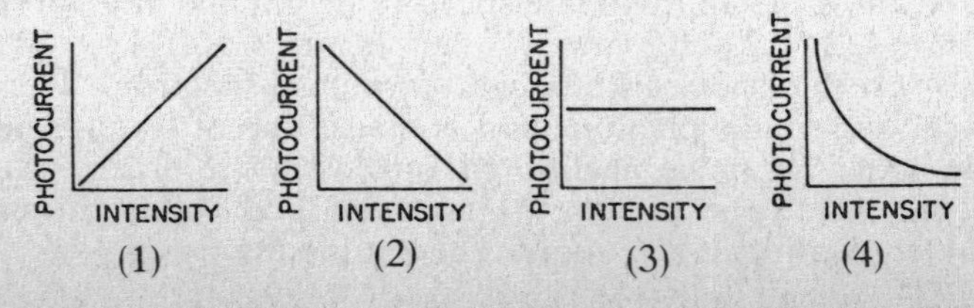

55. The threshold frequency for a photoemissive surface is 4.0 $\times 10^{14}$ hertz. what is the work function of this surface? (1) 1.2×10^{-19} J (2) 2.6×10^{-19} J (3) 6.0×10^{14} J (4) 6.1×10^{47} J

56. The concept that electrons exhibit wave properties can best be demonstrated by the (1) emission of photoelectrons (2) scattering of alpha particles by electrons (3) collisions between photons and electrons (4) production of electron interference patterns

57. According to the Rutherford model of the atom, an atomic nucleus contains (1) all of the atom's electric charge, but none of the atom's mass (2) all of the atom's mass, but none of the atom's electric charge (3) most of the atom's mass and all of the atom's negative charge (4) most of the atom's mass and all of the atom's positive charge

Note that questions 58 through 60 have only three choices.

58. A block is at rest on an inclined plane as shown in the diagram below. As angle θ is increased, the component of the block's weight parallel to the plane (1) decreases (2) increases (3) remains the same

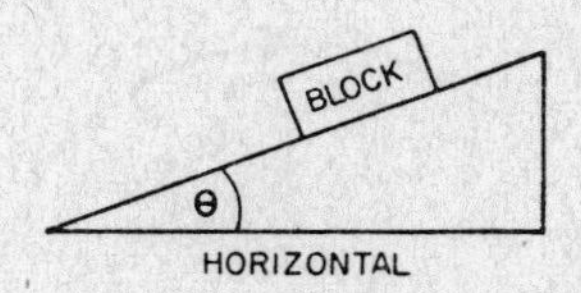

59. Monochromatic light passes through two parallel narrow slits and forms an interference pattern on a screen. As the distance between the two slits is increased, the distance between light bands in the pattern on the screen will (1) decrease (2) increase (3) remain the same

60. As observed from the Earth, the light from a star is shifted toward lower frequencies. This is an indication that the distance between the Earth and the star is (1) decreasing (2) increasing (3) constant

Part II

This part consists of six groups, each containing ten questions. Each group tests an optional area of the course. Choose two of these six groups. Be sure that you answer all ten questions in each group chosen. [20]

Group 1–Motion in a Plane

If you choose this group, be sure to answer questions 61-70.

Base your answers to questions 61 through 64 on the diagram below which shows an object with a mass of 1.0 kilogram attached to a string 0.50 meter long. The object is moving at a constant speed of 5.0 meters per second in a horizontal circular path with center at point *O*.

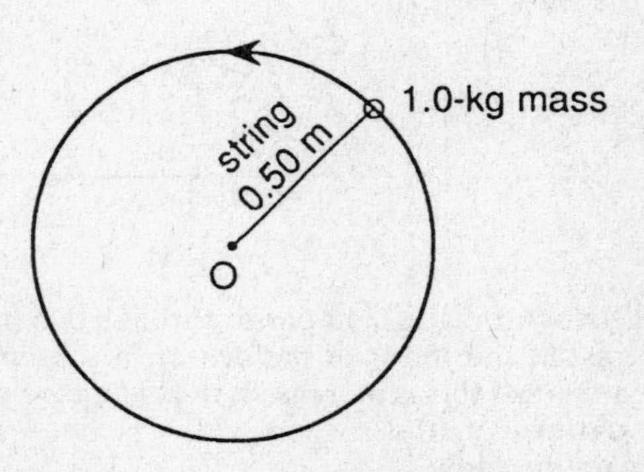

61. What is the magnitude of the centripetal force acting on the object? (1) 2.5 N (2) 10. N (3) 25 N (4) 50. N

62. While the object is undergoing uniform circular motion, its acceleration (1) has a magnitude of zero (2) increases in magnitude (3) is directed toward the center of the circle (4) is directed away from the center of the circle

Note that questions 63 and 64 have only three choices.

63. If the string is cut when the object is at the position shown, the path the object will travel from this position will be (1) toward the center of the circle (2) a curve away from the circle (3) a straight line tangent to the circle

64. If the string is lengthened while the speed of the object remains constant, the centripetal acceleration of the object will (1) decrease (2) increase (3) remain the same

Base your answers to questions 65 through 67 on the diagram below which represents a ball being kicked by a foot and rising at an angle of 30.° from the horizontal. The ball has an initial velocity of 5.0 meters per second. [Neglect friction.]

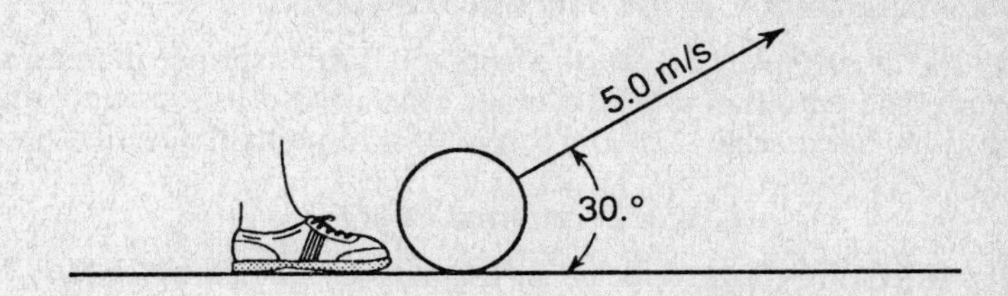

65. What is the magnitude of the horizontal component of the ball's initial velocity? (1) 2.5 m/s (2) 4.3 m/s (3) 5.0 m/s (4) 8.7 m/s

Note that questions 66 and 67 have only three choices.

66. As the ball rises, the vertical component of its velocity (1) decreases (2) increases (3) remains the same

67. If the angle between the horizontal and the direction of the 5.0-meters-per-second velocity decreases from 30.° to 20.°, the horizontal distance the ball travels will (1) decrease (2) increase (3) remain the same

68. In the diagrams below, *P* represents a planet and *S* represents the Sun. Which best represents the path of planet *P* as it orbits the Sun? [The diagrams are not drawn to scale.]

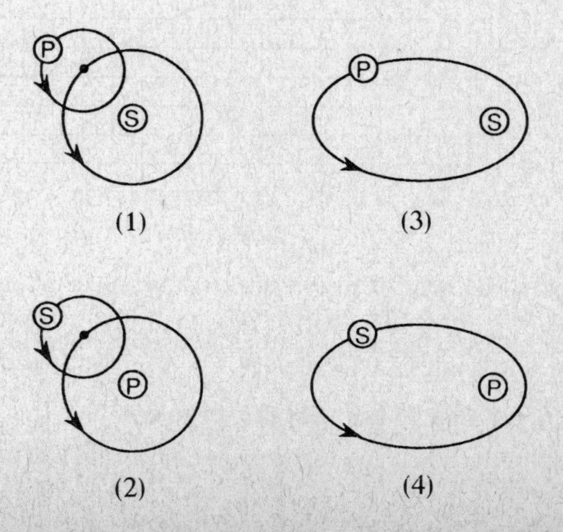

69. The Earth is closest to the Sun during January and farthest from the Sun during July. During which month is the gravitational potential energy of the Earth with respect to the Sun the greatest? (1) January (2) March (3) July (4) September

Note that question 70 has only three choices.

70. As the distance of a satellite from the Earth's surface increases, the time the satellite takes to make one revolution around the Earth (1) decreases (2) increases (3) remains the same

Group 2–Internal Energy

If you choose this group, be sure to answer questions 71-80.

71. The minimum average kinetic energy of the molecules in a substance occurs at a temperature of (1) -273 K (2) 273°C (3) 0°C (4) 0 K

72. How much heat is required to raise the temperature of 1.00 kilogram of liquid alcohol from its melting point to 0°C? (1) 2.43 kJ (2) 196 kJ (3) 284 kJ (4) 476 kJ

Base your answers to questions 73 through 75 on the graph below which represents the variation in temperature as 1.0 kilogram of a gas, originally at 200°C, loses heat at a constant rate of 2.0 kilojoules per minute and eventually becomes a solid at room temperature.

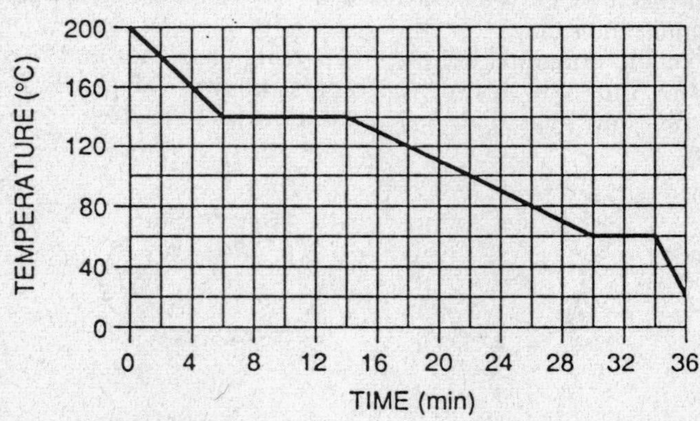

73. What is the heat of fusion of this substance? (1) 18 kJ/kg (2) 9 kJ/kg (3) 8 kJ/kg (4) 4 kJ/kg

74. What is the boiling temperature of this substance? (1) 200°C (2) 140°C (3) 90°C (4) 60°C

Note that question 75 has only three choices.

75. Which phase has the highest specific heat? (1) solid (2) liquid (3) gas

76. Which graph best represents the relationship between volume and absolute temperature for an ideal gas at constant pressure?

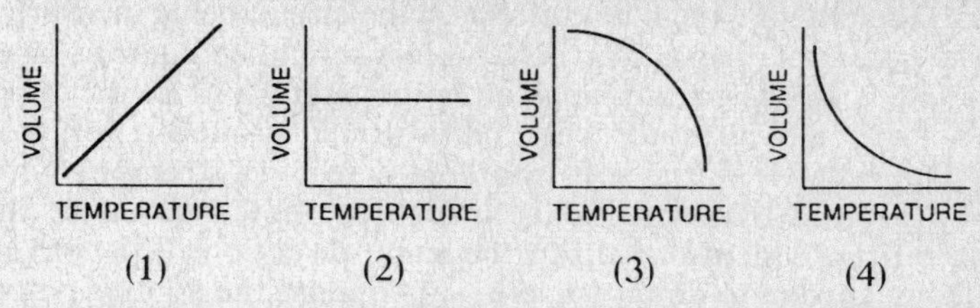

77. In an ideal gas, entropy is a measure of the (1) volume of the molecules (2) mass of the molecules (3) forces of attraction between the molecules (4) disorder of the molecules

Note that questions 78 through 80 have only three choices.

78. As the pressure of a fixed mass of gas is increased at constant temperature, the density of that gas (1) decreases (2) increases (3) remains the same

79. After a hot object is placed in an insulated container with a cold object, the hot object changes temperature and the cold object changes phase. The total amount of internal energy in the system will (1) decrease (2) increase (3) remain the same

80. Compared to the freezing point of pure water, the freezing point of a salt-water solution is (1) lower (2) higher (3) the same

Group 3–Electromagnetic Applications

If you choose this group, be sure to answer questions 81-90.

81. In the diagram below, a loop of wire is situated in uniform magnetic field *B*. The wire carries a constant electron current moving as shown. As viewed from position *F*, how will the loop initially respond to the current?

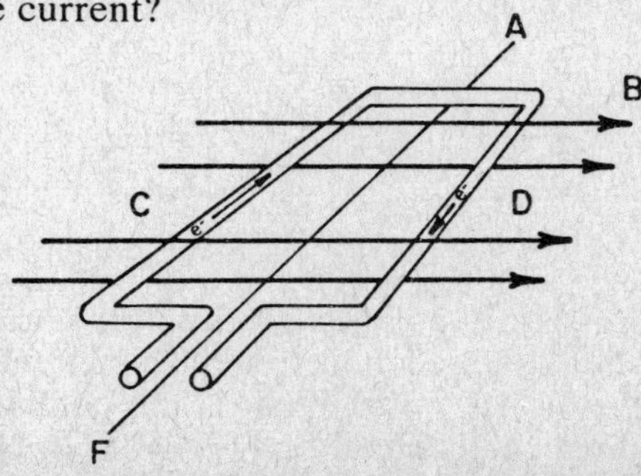

(1) by sliding from *C* to *D* (2) by sliding from *D* to *C* (3) by rotating clockwise about axis *FA* (4) by rotating counter-clockwise about axis *FA*

82. Which statement about ammeters and volt-meters is correct? (1) The internal resistance of both meters should be low. (2) Both meters should have a negligible effect on the circuit being measured. (3) The potential drop across both meters should be made as large as possible. (4) The scale range on both meters must be the same.

83. In a practical motor, the coil is wound around a soft iron core. The purpose of the soft iron core is to (1) strengthen and concentrate the magnetic field through the coil (2) cause the torque on the coil to remain in the same direction (3) convert alternating current to direct current (4) oppose the applied potential difference and reduce the current in the coil

84. What is the purpose of the split-ring commutator in a direct current motor? (1) to eliminate the external magnetic field (2) to increase the current in the armature (3) to maintain the direction of rotation of the armature (4) to decrease the induced back emf

85. An electron is moving at a velocity of 4.0×10^6 meters per second perpendicular to a magnetic field with a flux density of 6.0 teslas. The magnitude of the magnetic force acting on the electron is (1) 1.6×10^{-13} N (2) 6.4×10^{-13} N (3) 3.8×10^{-12} N (4) 2.4×10^{7} N

86. Which device can be used to separate isotopes of an element? (1) a mass spectrometer (2) an electroscope (3) an induction coil (4) two closely spaced double slits

87. Which could *not* be accelerated using an electric field? (1) electron (2) positron (3) alpha particle (4) photon

88. The diagram below shows a wire loop rotating between magnetic poles. During 360° of rotation from the position shown, the induced potential difference changes in (1) direction, only (2) magnitude, only (3) both magnitude and direction (4) neither magnitude nor direction

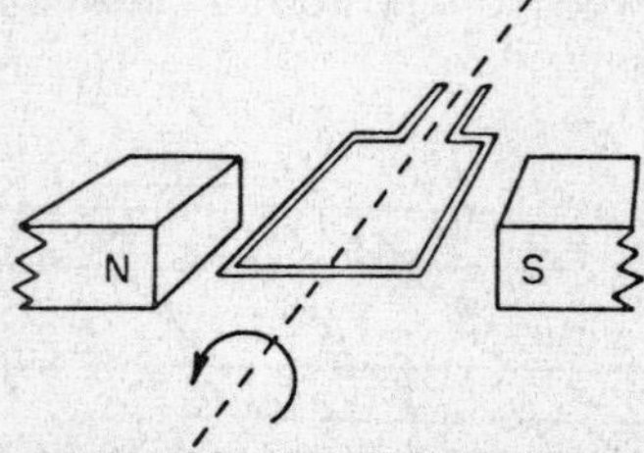

89. A transformer has 150 turns of wire in the primary coil and 1,200 turns of wire in the secondary coil. The potential difference across the primary is 110 volts. What is the potential difference induced across the secondary coil? (1) 14 V (2) 110 V (3) 150 V (4) 880 V

90. Which term best describes the light generated by a laser? (1) diffused (2) coherent (3) dispersive (4) longitudinal

Group 4—Geometric Optics

If you choose this group, be sure to answer questions 91-100.

Base your answers to questions 91 through 94 on the diagram below which shows four rays of light from object *AB* incident upon a spherical mirror whose focal length is 0.04 meter. Point *F* is the principal focus of the mirror, point *C* is the center of curvature, and point *O* is located on the principal axis.

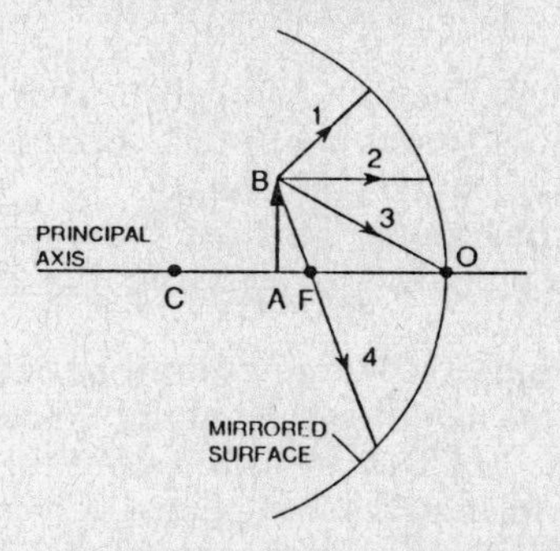

91. Which ray of light will pass through *F* after it is reflected from the mirror? (1) 1 (2) 2 (3) 3 (4) 4

92. If object *AB* is located 0.05 meter from point *O*, its image will be located (1) farther from the mirror than *C* (2) between *C* and *F* (3) between *F* and the mirror (4) behind the mirror

Note that questions 93 and 94 have only three choices.

93. As object *AB* is moved from its present position toward the left, the size of the image produced (1) decreases (2) increases (3) remains the same

94. If the mirror's radius of curvature could be increased, the focal length of the mirror would (1) decrease (2) increase (3) remain the same

95. In the diagram below, a light ray leaves a light source and reflects from a plane mirror.

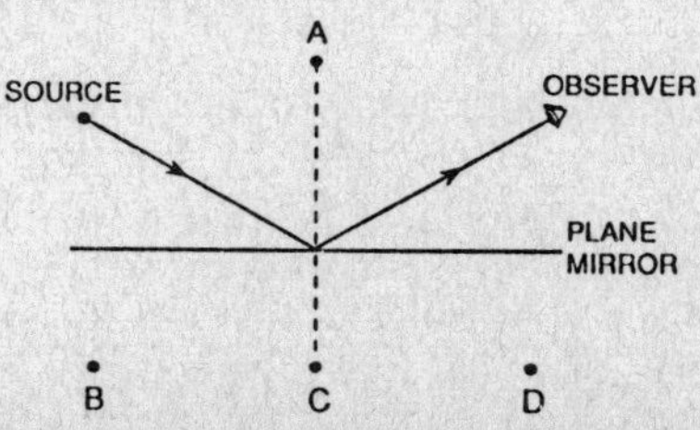

At which point does the image of the source appear to be located? (1) *A* (2) *B* (3) *C* (4) *D*

96. The diagram below shows a thin convex (converging) lens with F as the principal focus.

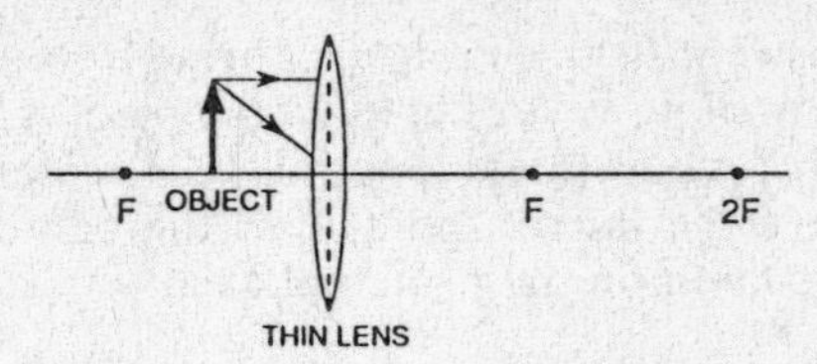

After passing through the lens, the light rays from the arrowhead of the object will (1) converge at F (2) converge at $2F$ (3) emerge as a parallel beam (4) diverge

97. A convex (converging) lens can form images that are (1) real, only (2) virutal, only (3) either real or virtual (4) neither real nor virtual

98. An object is placed 0.40 meter in front of a convex (converging) lens whose focal length is 0.30 meter. What is the image distance? (1) 0.17 m (2) 0.83 m (3) 1.2 m (4) 5.8 m

99. An object 0.16 meter tall is placed 0.20 meter in front of a concave (diverging) lens. What is the size of the image that is formed 0.10 meter from the lens? (1) 0.040 m (2) 0.080 m (3) 0.16 m (4) 0.32 m

100. A student places her eyeglasses directly on a printed page. As she raises them, the lenses cause the image of the print to remain erect while gradually decreasing in size. She should conclude from this that the lenses of the eyeglasses are (1) polarized (2) plane (3) converging (4) diverging

Group 5–Solid State

If you choose this group, be sure to answer questions 101-110.

101. According to accepted atomic models, metals are good conductors because their atoms have (1) more electrons than protons (2) unstable nuclei that emit electrons (3) negative electrons that are attracted to positive protons (4) a number of valence electrons that can move easily

102. What type of semiconductor is represented by the diagram below? (1) P-type (2) N-type (3) N-P-type (4) P-N-type

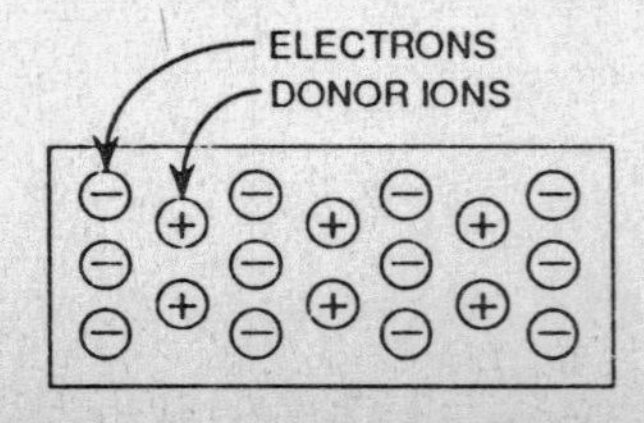

103. Which device could be used to amplify the input signal in a microphone-speaker circuit? (1) diode (2) resistor (3) transistor (4) solenoid

104. Which diagram below correctly represents the basic operating circuit of an *N-P-N* transistor?

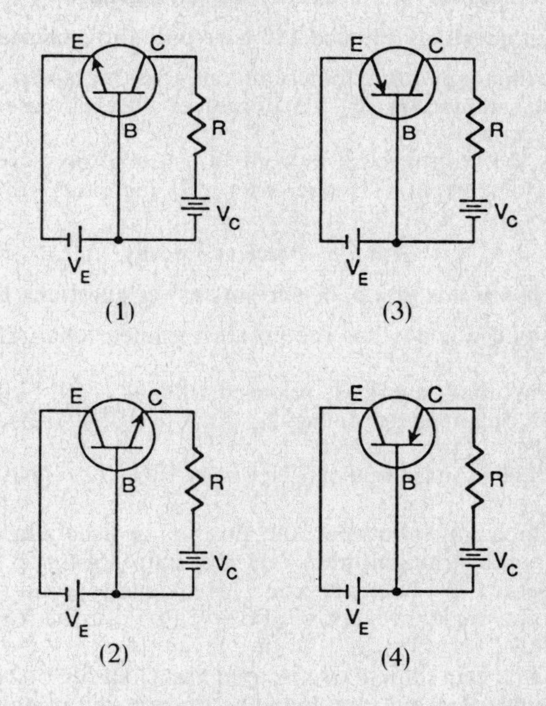

Base your answers to questions 105 through 107 on the diagram below of a circuit containing a semi-conductor device.

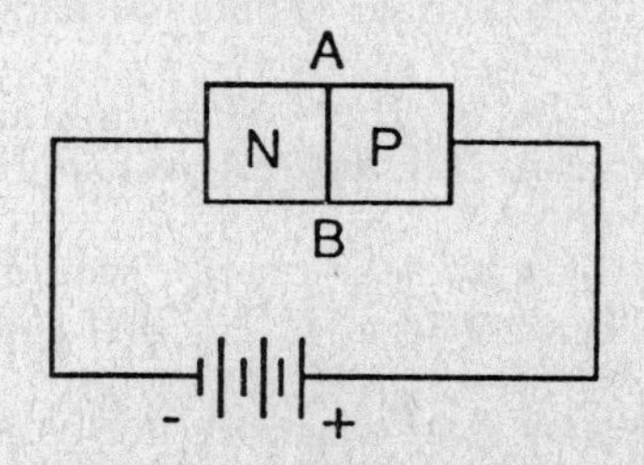

105. Which type of semiconductor device is shown? (1) emitter (2) resistor (3) diode (4) transistor

106. In the diagram, line *AB* identifies the (1) emitter (2) base (3) collector (4) junction

107. As drawn, this device is (1) forward biased (2) reverse biased (3) grounded (4) open

108. Doping material that contains fewer valence electrons per atom than the original semiconductor is classified as (1) a donor (2) an acceptor (3) an insulator (4) an emitter

Note that questions 109 and 110 have only three choices.

109. As the temperature of a semiconductor increases, the resistance of the semiconductor (1) decreases (2) increases (3) remains the same

110. As the emitter-base current in a transistor increases, the base-collector current (1) decreases (2) increases (3) remains the same

Group 6 – Nuclear Energy

If you choose this group, be sure to answer questions 111-120.

111. Which nucleus has the greatest nuclear charge? (1) ${}^{2}_{1}W$ (2) ${}^{8}_{5}X$ (3) ${}^{7}_{3}Y$ (4) ${}^{4}_{2}Z$

112. How much energy is released when 1×10^{-3} kilogram of matter is converted to energy? (1) 3×10^{5} J (2) 3×10^{8} J (3) 9×10^{13} J (4) 9×10^{16} J

113. Which is an isotope of ${}^{237}_{93}\mathrm{Np}$? (1) ${}^{237}_{92}\mathrm{U}$ (2) ${}^{237}_{94}\mathrm{Np}$ (3) ${}^{235}_{92}\mathrm{U}$ (4) ${}^{235}_{93}\mathrm{Np}$

114. Which force between the protons in a helium atom will have the greatest magnitude? (1) gravitational force (2) electrostatic force (3) nuclear force (4) magnetic force

115. In the nuclear reaction ${}^{218}_{84}\mathrm{Po} \rightarrow {}^{214}_{82}\mathrm{Pb} + X$, the X represents (1) ${}^{4}_{2}\mathrm{He}$ (2) ${}^{0}_{-1}\mathrm{e}$ (3) ${}^{0}_{+1}\mathrm{e}$ (4) ${}^{1}_{0}\mathrm{n}$

116. If a certain radioactive isotope has a half-life of 2 days, how much of a 64-kilogram sample of the isotope will remain after 10 days? (1) 1 kg (2) 2 kg (3) 32 kg (4) 4 kg

117. In the Uranium Disintegration Series, when an atom of ${}^{238}_{92}\mathrm{U}$ decays to ${}^{206}_{82}\mathrm{Pb}$, the total number of beta particles emitted is (1) 6 (2) 2 (3) 8 (4) 14

118. The equation ${}^{27}_{13}\mathrm{Al} + {}^{4}_{2}\mathrm{He} \rightarrow {}^{30}_{15}\mathrm{P} + {}^{1}_{0}\mathrm{n}$ is an example of (1) artificial transmutation (2) natural transmutation (3) alpha decay (4) beta decay

119. Which part of a nuclear reactor would most likely contain plutonium? (1) control rod (2) fuel rod (3) moderator (4) shielding

120. Which statement best describes the fission products from nuclear reactors? (1) They are nonradioactive and may be safely discarded. (2) They are nonradioactive and must be treated and/or stored. (3) They are intensely radioactive and may be safely discarded. (4) They are intensely radioactive and must be treated and/or stored.

Part III

You must answer both questions in this part. [10]

121. Base your answers to parts *a* and *b* on the information and the data table below.

An astronaut on a distant planet conducted an experiment to determine the gravitational acceleration on that planet. The data table shows the results of the experiment.

Data Table

Mass (kilograms)	Weight (newtons)
15	106
20.	141
25	179
30.	216
35	249

a. Using the information in the data table, construct a graph on the grid provided on your answer paper, following the directions below. The grid below is provided for practice purposes only. Be sure your final answer appears *on your answer paper.*

(1) Mark an appropriate scale on the axis labeled "Weight (N)." [1]

(2) Plot a weight versus mass graph for the astronaut's data and draw the best-fit line. [2]

b. Using your graph, determine the planet's gravitational acceleration. [Show all calculations, including equations and substitutions with units.] [2]

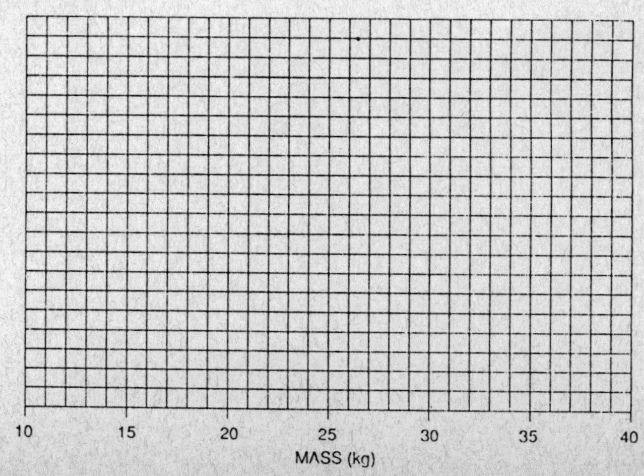

122. Base your answers to parts *a* through *d* on the diagram below which represents a circuit containing a 120-volt power supply with switches S_1 and S_2 and two 60.-ohm resistors.

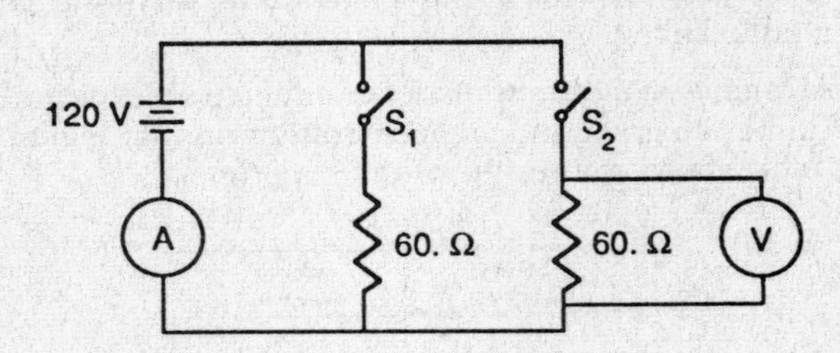

a. If switch S_1 is kept open and switch S_2 is closed, what is the circuit resistance? [1]

b. If switch S_2 is kept open and switch S_1 is closed, how much current will flow through the circuit? [Show all calculations, including equations and substitutions with units.] [2]

c. When both switches are closed, what is the current in the ammeter? [1]

d. When both switches are closed, what is the reading of the voltmeter? [1]

PHYSICS
Friday, June 14, 1991

Part I

Answer all 60 questions in this part. [70]

Directions (1-60): For *each* statement or question, select the word or expression that, of those given, best completes the statement or answers the question. Record your answer on the separate answer sheet in accordance with the directions on the front page of this booklet.

1. Which is a scalar quantity? (1) displacement (2) distance (3) force (4) acceleration

2. What is the approximate mass of a chicken egg? (1) 1×10^1 kg (2) 1×10^2 kg (3) 1×10^{-1} kg (4) 1×10^{-4} kg

3. Compared to the mass of an object at the surface of the Earth, the mass of the object a distance of two Earth radii from the center of the Earth is (1) the same (2) twice as great (3) one-half as great (4) one-fourth as great

4. A runner completed the 100.-meter dash in 10.0 seconds. Her average speed was (1) 0.100 m/s (2) 10.0 m/s (3) 100. m/s (4) 1,000 m/s

5. Which pair of graphs represent the same motion?

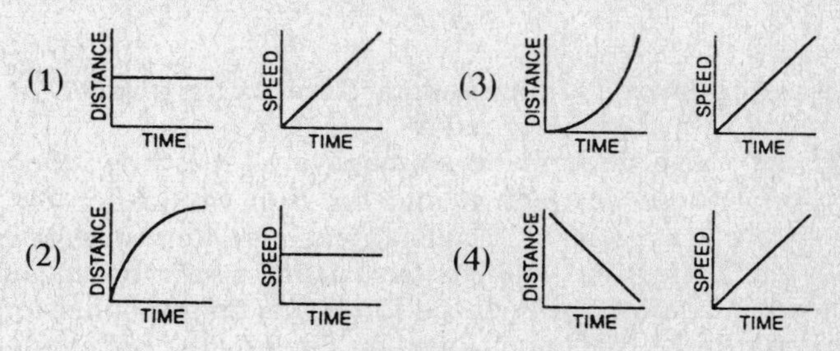

6. A child riding a bicycle at 15 meters per second decelerates at the rate of 3.0 meters per second^2 for 4.0 seconds. What is the child's speed at the end of the 4.0 seconds? (1) 12 m/s (2) 27 m/s (3) 3.0 m/s (4) 7.0 m/s

7. A skier starting from rest skis straight down a slope 50. meters long in 5.0 seconds. What is the magnitude of the acceleration of the skier? (1) 20. m/s^2 (2) 9.8 m/s^2 (3) 5.0 m/s^2 (4) 4.0 m/s^2

8. In the graph below, the acceleration of an object is plotted against the unbalanced force on the object.

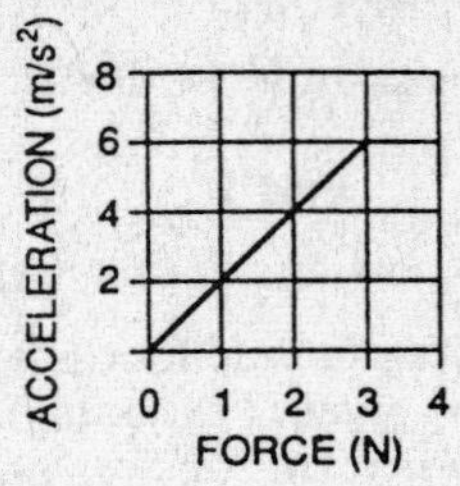

What is the object's mass? (1) 1 kg (2) 2 kg (3) 0.5 kg (4) 0.2 kg

9. What is the gravitational force of attraction between a planet and a 17-kilogram mass that is freely falling toward the surface of the planet at 8.8 meters per second^2? (1) 150 N (2) 8.8 N (3) 1.9 N (4) 0.52 N

10. Two perpendicular forces act on an object as shown in the diagram below.

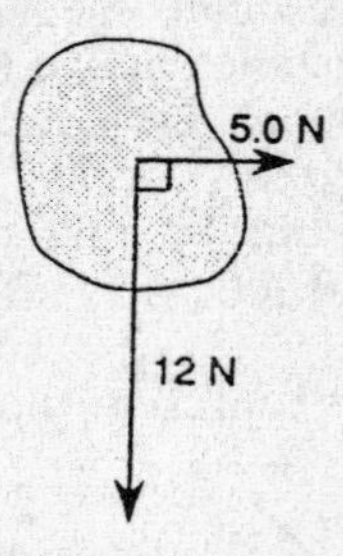

What is the magnitude of the resultant force on the object? (1) 17 N (2) 13 N (3) 7.0 N (4) 5.0 N

11. If the Earth were twice as massive as it is now, then the gravitational force between it and the Sun would be (1) the same (2) twice as great (3) half as great (4) four times as great

12. A 0.025-kilogram bullet is fired from a rifle by an unbalanced force of 200 newtons. If the force acts on the bullet for 0.1 second, what is the maximum speed attained by the bullet? (1) 5 m/s (2) 20 m/s (3) 400 m/s (4) 800 m/s

13. What is an essential characteristic of an object in equilibrium? (1) zero velocity (2) zero acceleration (3) zero potential energy (4) zero kinetic energy

14. A 2.0-kilogram ball traveling north at 4.0 meters per second collides head on with a 1.0-kilogram ball traveling south at 8.0 meters per second. What is the magnitude of the total momentum of the two balls after collision? (1) 0 kg•m/s (2) 8.0 kg•m/s (3) 16 kg•m/s (4) 32 kg•m/s

15. The magnitude of the force that a baseball bat exerts on a ball is 50. newtons. The magnitude of the force that the ball exerts on the bat is (1) 5.0 N (2) 10. N (3) 50. N (4) 250 N

16. Which quantity and unit are correctly paired?

(1) velocity—m/s^2
(2) momentum—$\frac{kg \bullet m}{s^2}$
(3) energy—$\frac{kg \bullet m^2}{s^2}$
(4) work—kg/m

17. A jack exerts a force of 4,500 newtons to raise a car 0.25 meter. What is the approximate work done by the jack? (1) 5.6×10^{-5} J (2) 1.1×10^3 J (3) 4.5×10^3 J (4) 1.8×10^4 J

18. A 6.0×10^2-newton man climbing a rope at a speed of 2.0 meters per second develops power at the rate of (1) 1.2×10^1 W (2) 6.0×10^2 W (3) 3.0×10^2 W (4) 1.2×10^3 W

19. Three people of equal mass climb a mountain using paths *A*, *B*, and *C* shown in the diagram below.

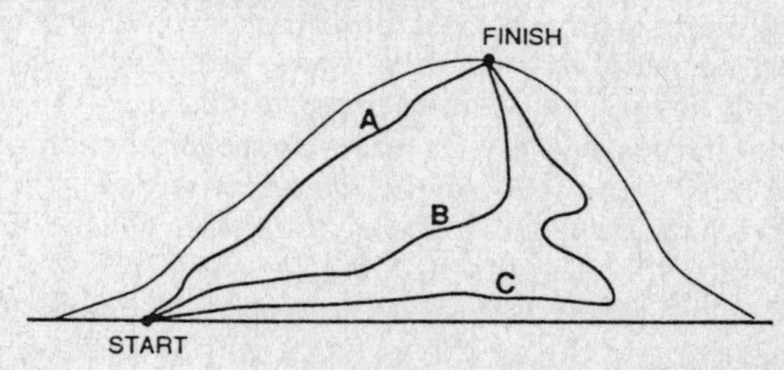

Along which path(s) does a person gain the greatest amount of gravitational potential energy from start to finish? (1) *A*, only (2) *B*, only (3) *C*, only (4) The gain is the same along all paths.

20. Which graph best represents the relationship between the elongation of an ideal spring and the applied force?

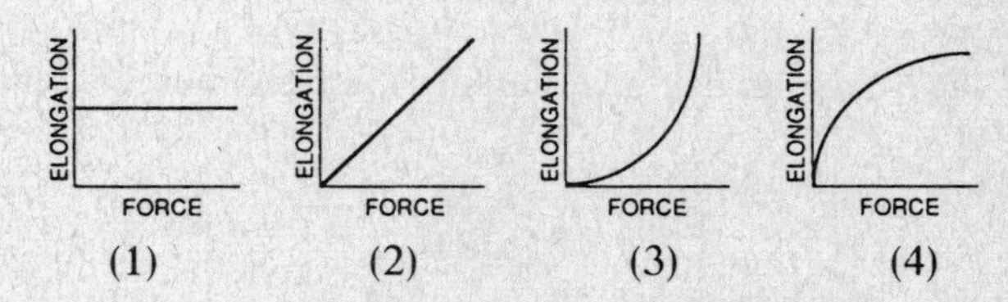

21. A person does 100 joules of work in pulling back the string of a bow. What will be the initial speed of a 0.5-kilogram arrow when it is fired from the bow? (1) 20 m/s (2) 50 m/s (3) 200 m/s (4) 400 m/s

22. As an object falls freely near the Earth's surface, the loss in gravitational potential energy of the object is equal to its (1) loss of height (2) loss of mass (3) gain in velocity (4) gain in kinetic energy

23. Which part of an atom is most likely to be transferred as a body acquires a static electric charge? (1) proton (2) neutron (3) electron (4) positron

24. In the diagram below, *A* is a point near a positively charged sphere.

Which vector best represents the direction of the electric field at point *A*?

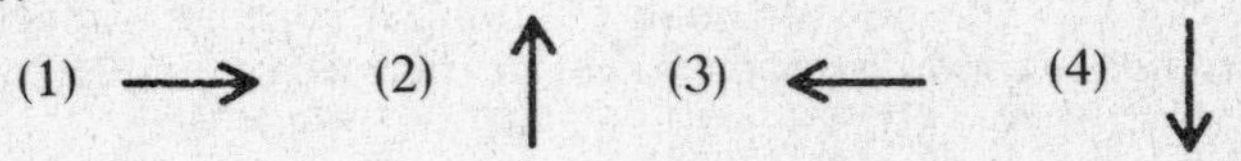

25. When a rod is brought near a neutral electroscope, the leaves diverge. Which statement best describes the charge on the rod? (1) It must be positive. (2) It must be negative. (3) It may be neutral. (4) It may be positive or negative.

26. Two charges that are 2 meters apart repel each other with a force of 2×10^{-5}) newton. If the distance between the charges is decreased to 1 meter, the force of repulsion will be (1) 1×10^{-5} N (2) 5×10^{-6} N (3) 8×10^{-5} N (4) 4×10^{-5} N

27. A metallic sphere is positively charged. The field at the center of the sphere due to this positive charge is (1) positive (2) negative (3) zero (4) dependent on the magnitude of the charge

28. How much energy is needed to move one electron through a potential difference of 1.0×10^2 volts? (1) 1.0 J (2) 1.0×10^2 J (3) 1.6×10^{-19} J (4) 1.6×10^{-17} J

29. An electric iron draws a current of 5 amperes and has a resistance of 20 ohms. The amount of energy used by the iron in 40 seconds is (1) 100 J (2) 500 J (3) 4,000 J (4) 20,000 J

30. The diagram below represents a series circuit containing three resistors.

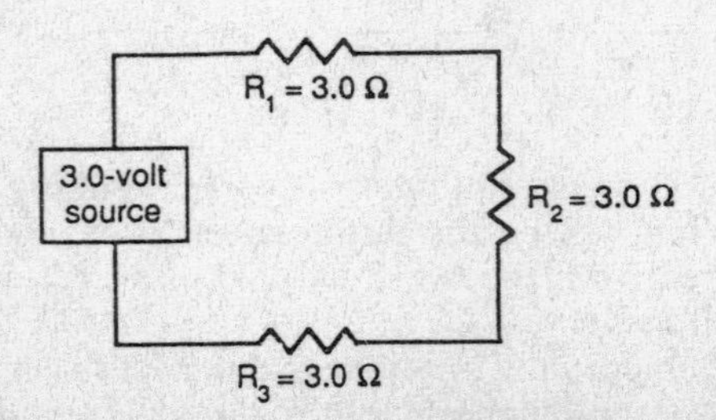

What is the current through resistor R_2? (1) 1.0 A (2) 0.33 A (3) 3.0 A (4) 9.0A

31. In the circuit diagram below, what is the potential difference across the 3.0-ohm resistor?

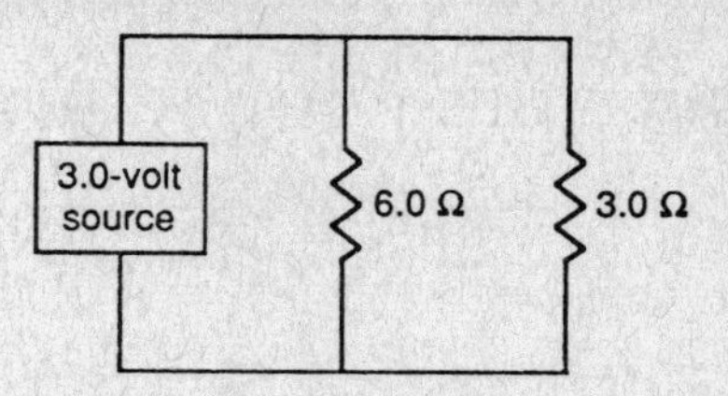

(1) 1.0 V (2) 2.0 V (3) 3.0 V (4) 1.5 V

32. Which circuit segment below has the same total resistance as the circuit segment shown in the diagram at the right?

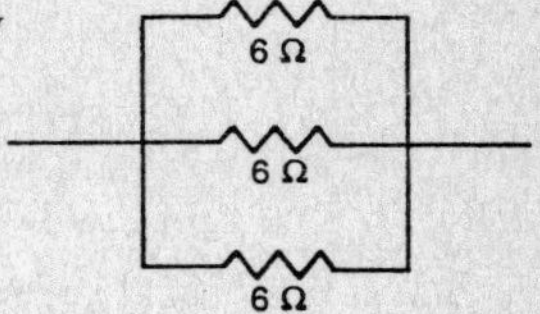

(1) 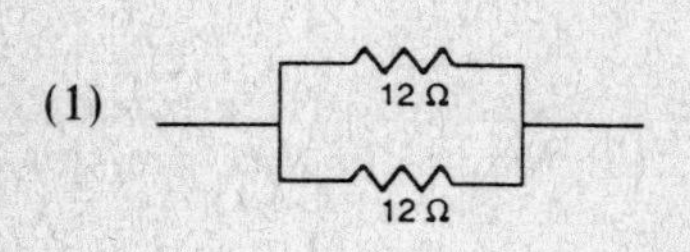

(3) 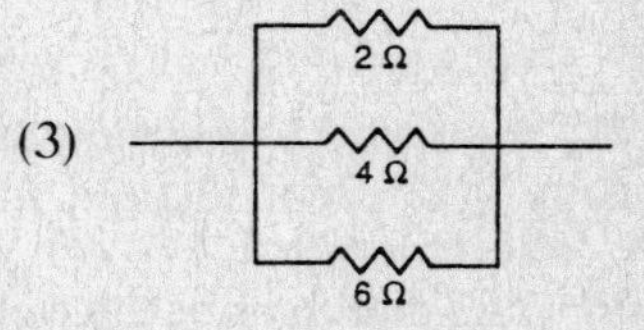

(2) 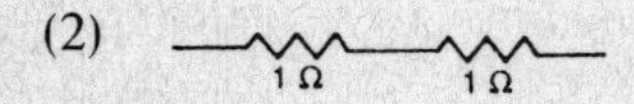

(4) 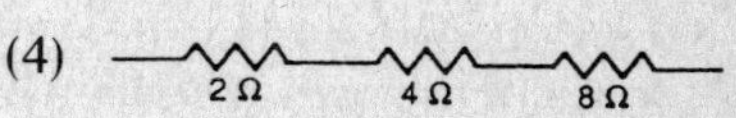

33. While operating at 120 volts, an electric toaster has a resistance of 15 ohms. The power used by the toaster is (1) 8.0 W (2) 120 W (3) 960 W (4) 1,800 W

34. Which circuit shown below could be used to determine the total current and potential difference of a parallel circuit?

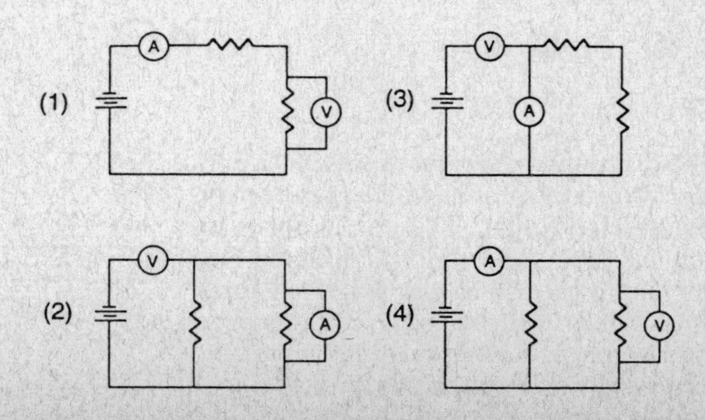

35. A wire carries a current of 6.0 amperes. How much charge passes a point in the wire in 120 seconds? (1) 6.0 C (2) 20. C (3) 360 C (4) 720 C

36. Which diagram best represents the magnetic field between

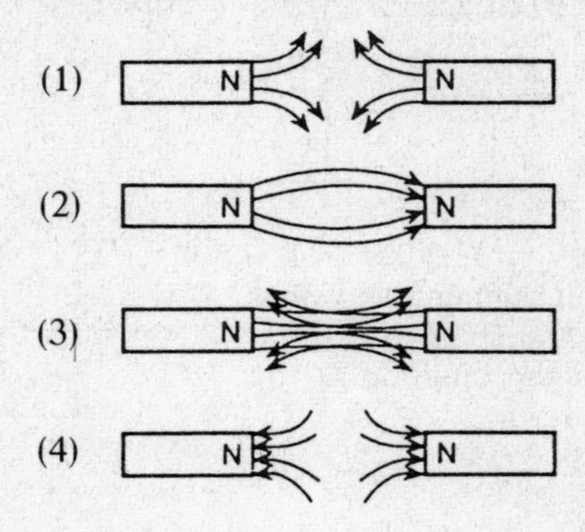

37. A volt is to electric potential as a tesla is to (1) electrical energy (2) electric field intensity (3) magnetic flux density (4) charge density

38. The diagrams at the right show cross sections of conductors with electrons flowing into or out of the page. In which diagram below will the magnetic flux density at point A be greater than the magnetic flux density at point B?

⊙ CURRENT OUT
⊗ CURRENT IN

39. The diagram at the right represents a conductor carrying a current in which the electron flow is from left to right. The conductor is located in a magnetic field which is directed into the page. The direction of the magnetic force on the conductor will be (1) into the page (2) out of the page (3) toward the top of the page (4) toward the bottom of the page

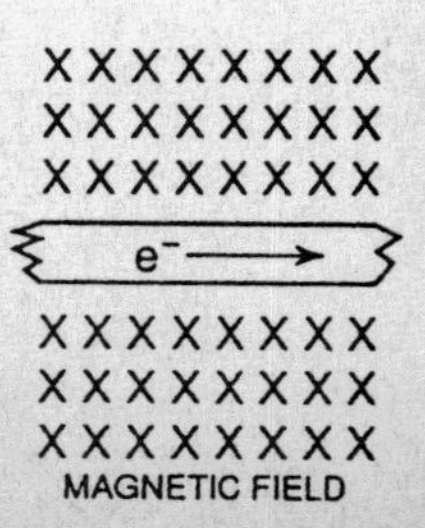

40. The diagram below shows the direction of water waves moving along path XY toward a barrier.

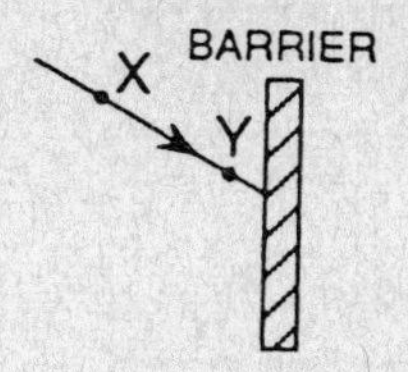

Which arrow represents the direction of the waves after they have reflected from the barrier?

41. What is the frequency of a wave if its period is 0.25 second? (1) 1.0 Hz (2) 0.25 Hz (3) 12 Hz (4) 4.0 Hz

42. The diagram below shows a transverse water wave moving in the direction shown by velocity vector v.

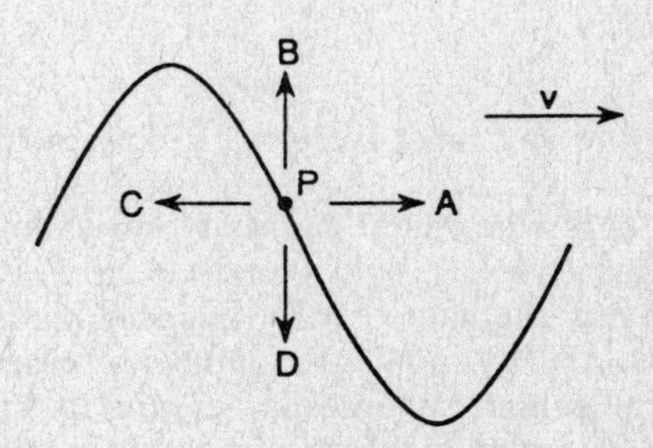

At the instant shown, a cork at point P on the water's surface is moving toward (1) A (2) B (3) C (4) D

43. To the nearest order of magnitude, how many times greater than the speed of sound is the speed of light? (1) 10^4 (2) 10^6 (3) 10^{10} (4) 10^{12}

44. Diagram I shows a glass tube containing undisturbed air molecules. Diagram II shows the same glass tube when a wave passes through it.

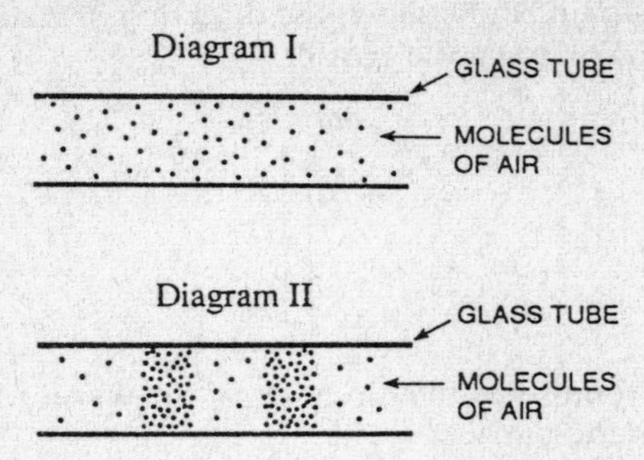

Which type of wave produced the disturbance shown in diagram II? (1) longitudinal (2) torsional (3) transverse (4) elliptical

45. An opera singer's voice is able to break a thin crystal glass if the singer's voice and the glass have the same natural (1) frequency (2) speed (3) amplitude (4) wavelength

46. The diagram below shows a ray of light, *R*, entering glass from air.

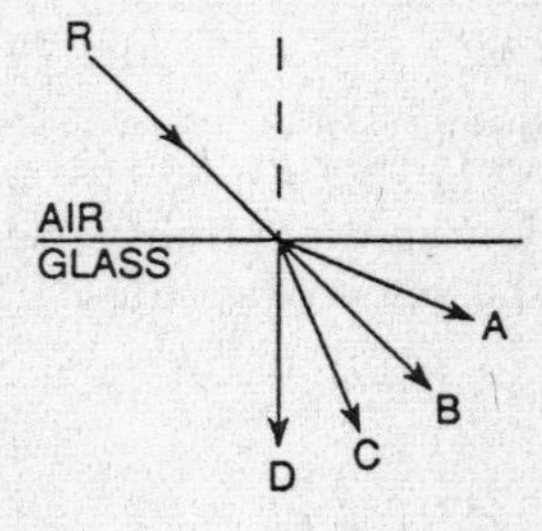

Which path is the ray most likely to follow in the glass? (1) *A* (2) *B* (3) *C* (4) *D*

47. What occurs when light passes from water into flint glass? (1) Its speed decreases, its wavelength becomes smaller, and its frequency remains the same. (2) Its speed decreases, its wavelength becomes smaller, and its frequency increases. (3) Its speed increases, its wavelength becomes larger, and its frequency remains the same. (4) Its speed increases, its wavelength becomes larger, and its frequency decreases.

48. As shown in the diagram below, a beam of light can pass through the length of a curved glass fiber.

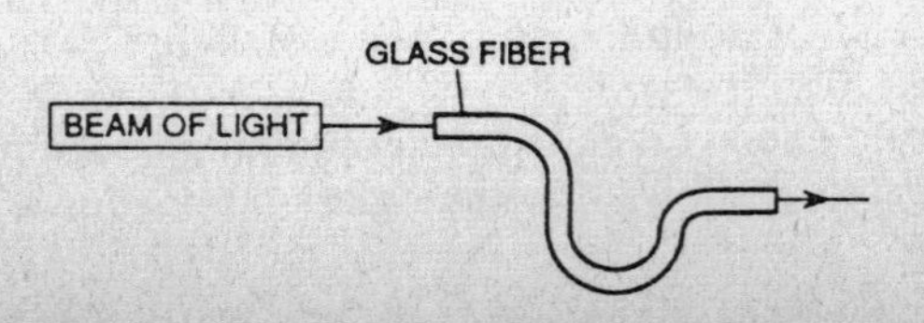

This phenomenon is possible due to the effect of (1) dispersion (2) internal reflection (3) polarization (4) diffraction

49. Compared to visible light, ultraviolet radiation is more harmful to human skin and eyes because ultraviolet radiation has a (1) higher frequency (2) longer period (3) higher speed (4) longer wavelength

50. Polychromatic light passing through a glass prism is separated into its component frequencies. This phenomenon is called (1) diffraction (2) dispersion (3) reflection (4) polarization.

51. Which is a characteristic of light produced by a monochromatic light source? (1) It can be dispersed by a prism. (2) It is a longitudinal wave. (3) Its frequency is constantly changing. (4) It can be polarized.

52. In the diagram below, two speakers are connected to a sound generator. The speakers produce a sound pattern of constant frequency such that a listener will hear the sound very well at *A* and *C*, but not as well at point *B*.

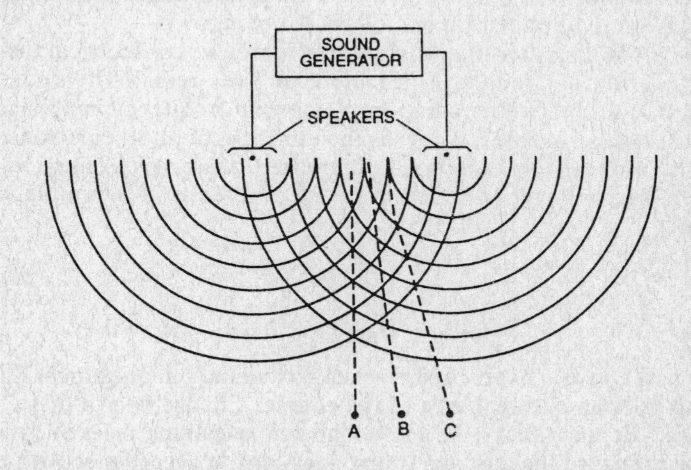

Which wave phenomenon is illustrated by this experiment? (1) interference (2) polarization (3) reflection (4) refraction

53. Blue light has a frequency of approximately 6.0×10^{14} hertz. A photon of blue light will have an energy of approximately (1) 1.1×10^{-48} J (2) 6.0×10^{-34} J (3) 5.0×10^{-7} J (4) 4.0×10^{-19} J

54. An electron in a mercury atom that is changing from the *a* to the *g* level absorbs a photon with an energy of (1) 12.86 eV (2) 10.38 eV (3) 7.90 eV (4) 2.48 eV

55. Which phenomenon can be explained by both the particle model and wave model? (1) reflection (2) polarization (3) diffraction (4) interference

56. What do alpha-particle scattering experiments indicate about an atom's structure? (1) Electrons occupy distinct energy levels. (2) Positive and negative charges are evenly distributed. (3) Negative charge fills the space around the core. (4) Positive charge is concentrated in a small, dense core.

Note that questions 57 through 60 have only three choices.

57. A box initially at rest on a level floor is being acted upon by a variable horizontal force, as shown in the diagram at the right. Compared to the force required to start the box moving, the force required to keep it moving at constant speed is (1) less (2) greater (3) the same

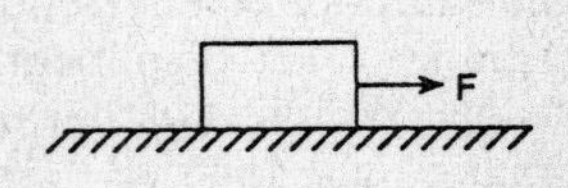

58. A police officer's stationary radar device indicates that the frequency of the radar wave reflected from an automobile is less than the frequency emitted by the radar device. This indicates that the automobile is (1) moving toward the police officer (2) moving away from the police officer (3) not moving

59. If the diameter of a wire were to increase, its electrical resistance would (1) decrease (2) increase (3) remain the same

60. Electromagnetic radiation of constant frequency incident on a photoemissive material causes the emission of photoelectrons. If the intensity of this radiation is increased, the rate of emission of photoelectrons will (1) decrease (2) increase (3) remain the same

Part II

This part consists of six groups, each containing ten questions. Each group tests an optional area of the course. Choose two of these six groups. Be sure that you answer all ten questions in each group chosen. Record the answers to the questions in accordance with the directions on the front page of this booklet. [20]

Group 1—Motion in a Plane

If you choose this group, be sure to answer questions 61-70.

61. Four cannonballs, each with mass *M* and initial velocity *V*, are fired from a cannon at different angles relative to the Earth. Neglecting air friction, which angular direction of the cannon produces the greatest projectile height? (1) 90° (2) 70° (3) 45° (4) 20°

62. A projectile is launched at an angle of 60.° to the horizontal at an initial speed of 10. meters per second. What is the magnitude of the vertical component of its initial speed? (1) 2.5 m/s (2) 4.3 m/s (3) 5.0 m/s (4) 8.7 m/s

63. The diagram below represents a ball undergoing uniform circular motion as it travels clockwise on a string.

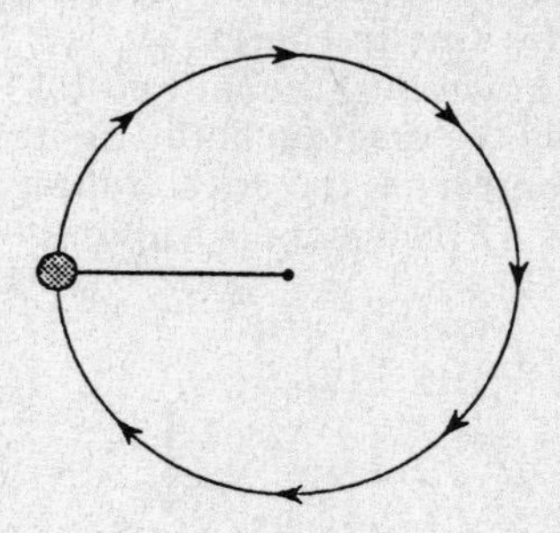

At the moment shown in the diagram, what are the correct directions of both the velocity and centripetal acceleration of the ball?

(1) v ↑ a → (3) v ↓ a ←

(2) v → a ↑ (4) v → a ↓

64. A 3.0-kilogram mass is traveling in a circle of 0.20-meter radius with a speed of 2.0 meters per second. What is its centripetal acceleration? (1) 10. m/s^2 (2) 20. m/s^2 (3) 60. m/s^2 (4) 6.0 m/s^2

65. A car going around a curve is acted upon by a centripetal force, F. If the speed of the car were twice as great, the centripetal force necessary to keep it moving in the same path would be

(1) F (3) $\frac{F}{2}$

(2) $2F$ (4) $4F$

66. The shape of the path of the Earth about the Sun is (1) a circle with the Sun at the center (2) an ellipse with the Sun at one focus (3) an ellipse with the Moon at one focus (4) an ellipse with nothing at either focus

67. The increase in a planet's speed as it approaches the Sun is described by Kepler's second law. What is the best explanation for this empirical law? (1) Since kinetic energy depends on temperature, the planet must move faster as it nears the Sun. (2) Ions from the Sun (solar wind) speed up the planet. (3) The days are shorter in the winter, causing the planet to move faster as it nears the Sun. (4) The gravitational potential energy lost as the planet nears the Sun becomes kinetic energy.

68. An object is thrown into the air and follows the path shown in the diagram at the right. Which vector best represents the acceleration of the object at point *A*? [Neglect air friction.]

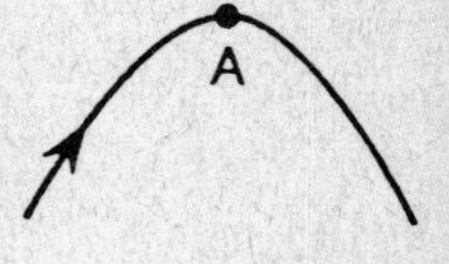

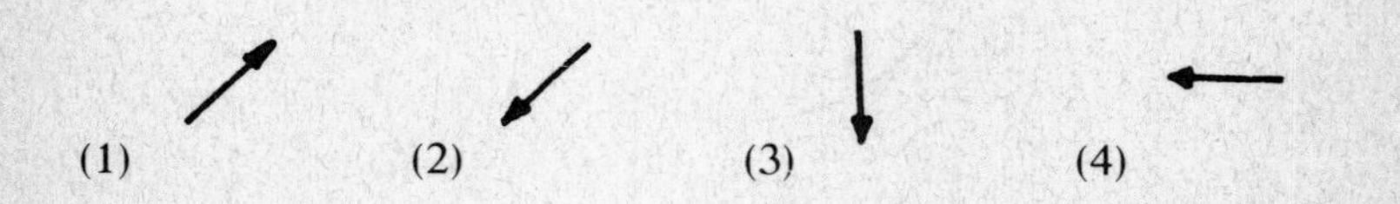

Note that questions 69 and 70 have only three choices.

69. Above a flat horizontal plane, an arrow, *A*, is shot horizontally from a bow at a speed of 50 meters per second, as shown in the diagram below. A second arrow, *B*, is dropped from the same height and at the same instant as *A* is fired.

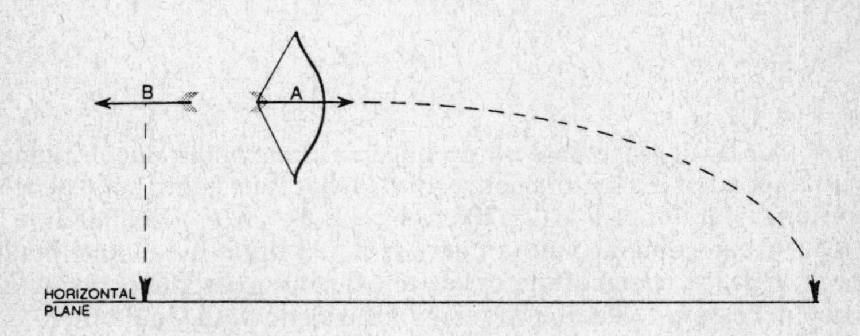

Neglecting air friction, compared to the amount of time *A* takes to strike the plane, the amount of time *B* takes to strike the plane is (1) less (2) greater (3) the same

70. Two satellites, *A* and *B*, are traveling around the Earth in nearly circular orbits. The radius of satellite *A*'s orbit is greater than the radius of satellite *B*'s orbit. Compared to the orbital period of satellite *A*, the orbital period of satellite *B* is (1) less (2) greater (3) the same

Group 2–Internal Energy

If you choose this group, be sure to answer questions 71-80.

Base your answer to question 71 on the graph below which represents the temperature of 2.0 kilograms of a material as a function of the heat added to the substance.

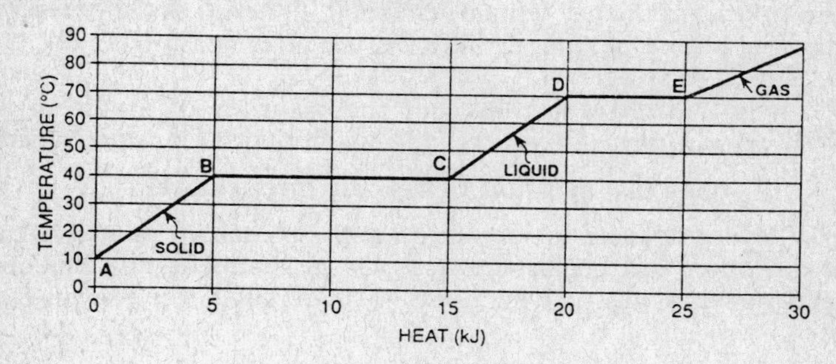

71. During which two intervals shown on the graph is the average potential energy of the molecules of the material increasing? (1) *AB* and *CD* (2) *BC* and *DE* (3) *CD* and *DE* (4) *AB* and *DE*

72. A temperature reading of absolute zero for a system would mean that the system's (1) temperature is -273 K (2) temperature is 273°C (3) kinetic energy is at a maximum (4) internal energy is at a minimum

73. Equal masses of aluminum, silver, tungsten, and zinc, initially at room temperature, are cooled 10° C. Which metal gives off the most heat? (1) aluminum (2) silver (3) tungsten (4) zinc

74. Two objects, *A* and *B*, are in contact with one another. Initially, the temperature of *A* is 300 K and the temperature of *B* is 400 K. Which diagram best represents the net flow of heat in the closed system? [Arrows represent the direction of heat flow.]

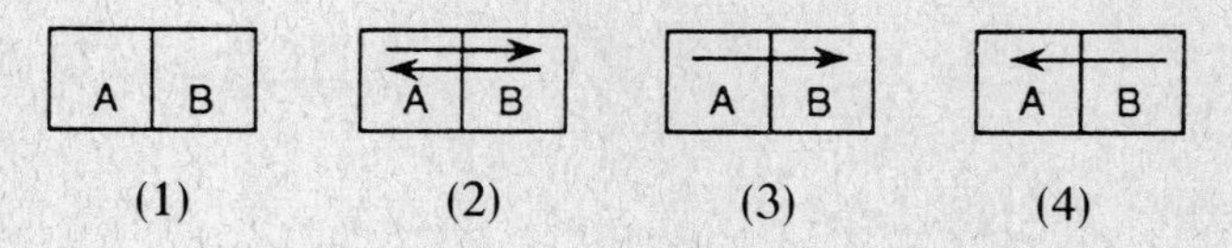

75. What is the final temperature of a 10.-kilogram sample of lead, initially at 0°C, after it has absorbed 39 kilojoules of heat energy? (1) 1.3°C (2) 3.9°C (3) 30.°C (4) 300°C

76. Which substance is a liquid at 0°C? (1) alcohol (2) aluminum (3) ammonia (4) copper

77. Rock salt is thrown on icy pavement to make roadways safer for driving in winter. This process works because the dissolved salt (1) raises the temperature of water (2) raises the freezing point of water (3) lowers the freezing point of water (4) lowers the boiling point of water

78. One kilogram of an ideal gas is heated from 27°C to 327°C. If the volume of the gas remains constant, the ratio of the pressure of the gas before heating to the pressure after heating is (1) 1:1 (2) 1:2 (3) 1:3 (4) 1:4

79. A quantitative measure of the disorder of a system is called (1) entropy (2) fusion (3) equilibrium (4) vaporization

Note that question 80 has only three choices.

80. A force causes an object on a horizontal surface to overcome friction and begin to move. As this happens, the object's internal energy will (1) decrease (2) increase (3) remain the same.

Group 3–Electromagnetic Applications

If you choose this group, be sure to answer questions 81-90.

81. Which type of energy conversion occurs in an electric motor? (1) rotational mechanical energy to electrical energy (2) electrical energy to rotational mechanical energy (3) chemical energy to induced electrical energy (4) induced electrical energy to stored chemical energy

82. In the diagram below, a portion of a wire is being moved upward through a magnetic field.

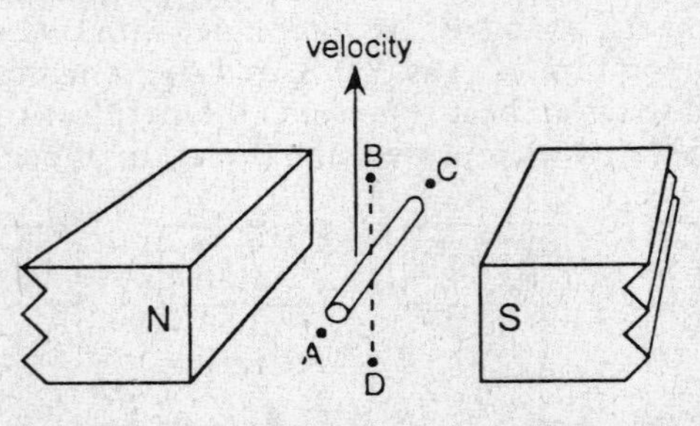

The direction of the induced electron current in the wire is toward point (1) *A* (2) *B* (3) *C* (4) *D*

83. The calculation of the mass of an electron is based on the results of the Millikan oil drop experiment and on the (1) charge on the proton (2) charge-to-mass ratio of the electron (3) mass of an oil drop (4) mass-to-weight ratio of the neutron

84. Electrons are ejected from the filament of a cathode-ray tube when it becomes very hot. This phenomenon is an example of (1) thermionic emission (2) photoelectric emission (3) back electromotive force (4) torque

Base your answers to questions 85 and 86 on the diagram below which represents an electron entering the region between two oppositely charged parallel plates.

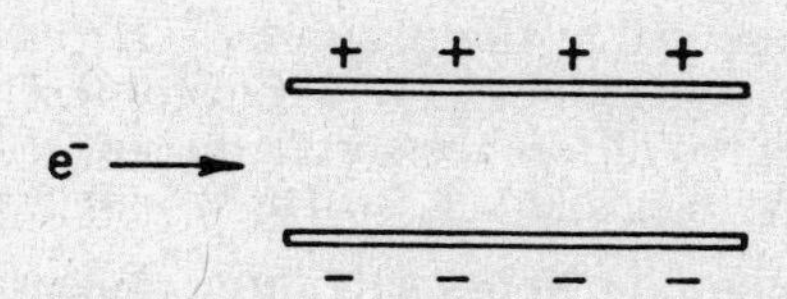

85. In which direction will the electron be deflected by the electric field? (1) toward the bottom of the page (2) toward the top of the page (3) into the page (4) out of the page

86. If the magnitude of the electric field between the plates is 4.0×10^3 volts per meter, the electric force on the electron is (1) 2.5×10^{22} N (2) 4.0×10^3 N (3) 6.4×10^{-16} N (4) 4.0×10^{-23} N

87. In the diagram below, a wire 0.50 meter long is moved at a speed of 2.0 meters per second perpendicularly through a uniform magnetic field with a flux density of 3.0 teslas directed into the page.

SPEED

MAGNETIC FIELD

What is the induced electromotive force? (1) 1.0 V (2) 1.5 V (3) 3.0 V (4) 12 V

Base your answers to questions 88 and 89 on the diagram below which represents a 100% efficient device connected to an alternating current source of 220 volts. The primary coil has 50 turns and the secondary coil has 25 turns. When the device operates, a 0.50-ampere current flows through the primary.

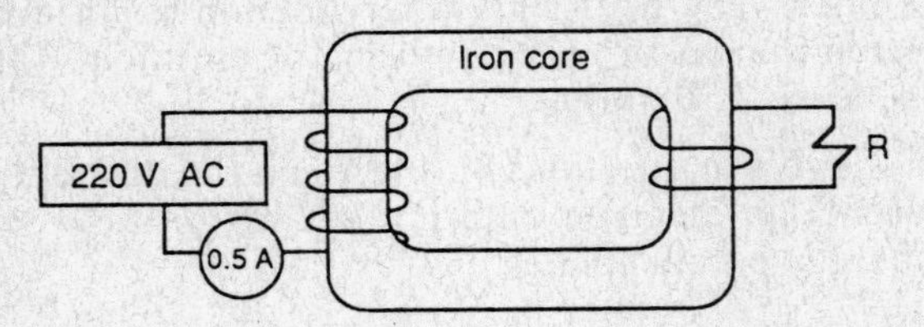

88. The device represented by this diagram is (1) an induction coil (2) a motor (3) a generator (4) a transformer

89. What is the potential difference across resistor *R* in the secondary coil? (1) 110 V (2) 220 V (3) 440 V (4) 2,800 V

Note that question 90 has only three choices.

90. Compared to the internal resistance of a voltmeter, the internal resistance of an ammeter is (1) smaller (2) greater (3) the same.

Group 4–Geometric Optics

If you choose this group, be sure to answer questions 91-100.

91. The image of an object is viewed in a plane mirror. What is the ratio of the object size to the image size? (1) 1:1 (2) 2:1 (3) 1:2 (4) 1:4

92. In the diagram below, how far from the mirror is the light bulb (object) most likely located?

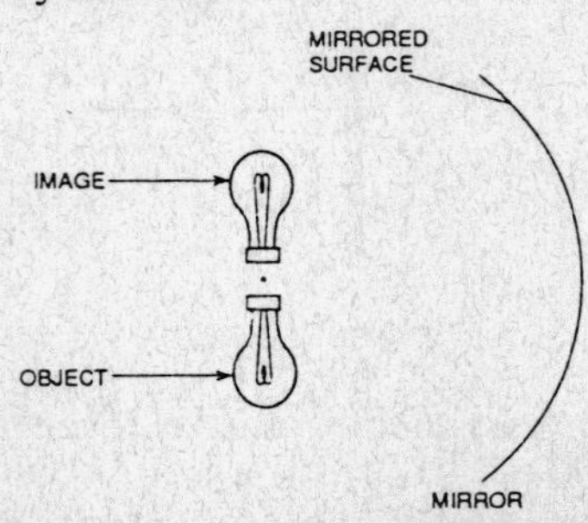

(1) closer than the focal length of the mirror (2) at the principal focus of the mirror (3) at twice the focal length of the mirror (4) farther than twice the focal length of the mirror

93. A light ray is incident upon a cylindrical reflecting surface as shown in the diagram at the right. The ray will most likely be reflected toward letter (1) *A* (2) *B* (3) *C* (4) *D*

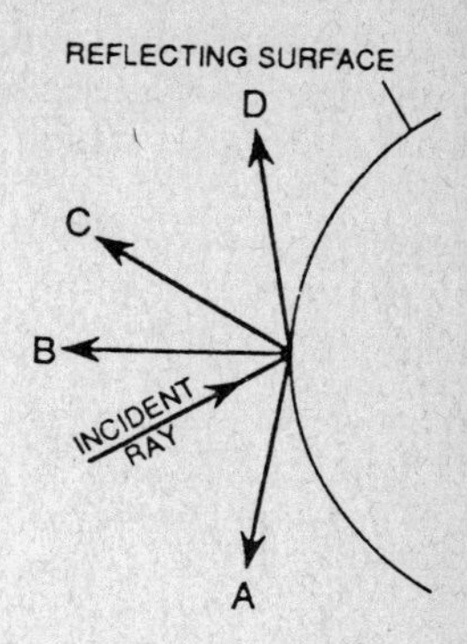

94. A searchlight consists of a high-intensity light source at the focal point of a concave (converging) mirror. The light reflected from the mirror will (1) diverge uniformly (2) converge to a point (3) scatter in all directions (4) form a nearly parallel beam

95. The image produced by a convex (diverging) mirror must be (1) real and erect (2) real and inverted (3) virtual and erect (4) virtual and inverted

96. Which phenomenon of light accounts for the formation of images by a lens? (1) reflection (2) refraction (3) dispersion (4) polarization

Base your answers to questions 97 and 98 on the diagram below which represents a convex lens being used to form the image of an object. The distance from the center of the lens to the object is 0.060 meter. The distance from the center of the lens to the image is 0.120 meter.

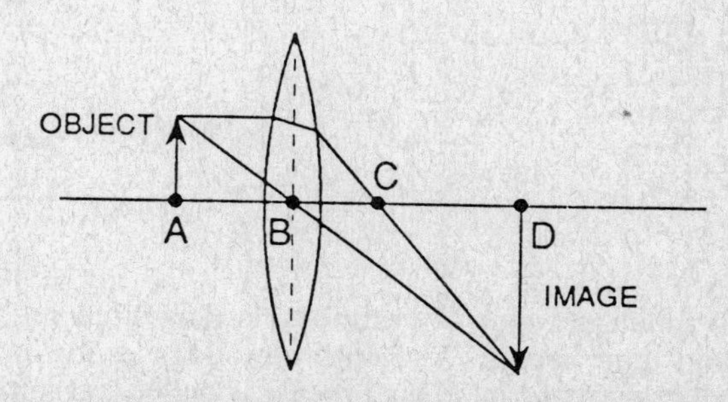

97. If the height of the object is 2.6×10^{-2} meter, the height of the image is (1) 1.3×10^{-2} m (2) 2.0×10^{-2} m (3) 2.6×10^{-2} m (4) 5.2×10^{-2} m

98. The focal length of the lens is (1) 25 m (2) 2.0 m (3) 0.12 m (4) 0.040 m

99. The same frequency of monochromatic light is incident from air upon four lenses having the same curvature, but made of different materials. Which lens has the shortest focal length?

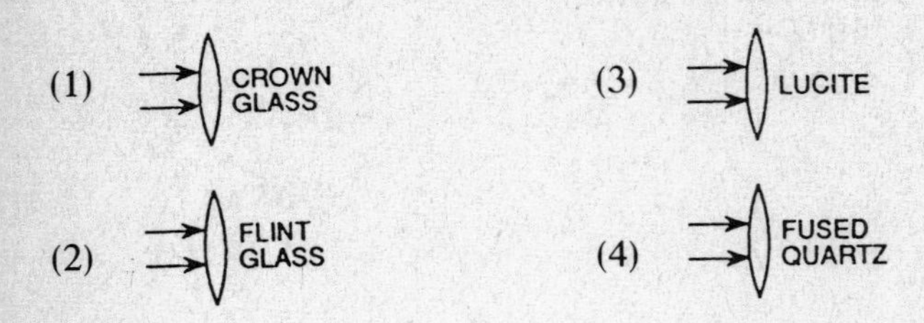

100. A student's solution to an optics problem had a negative value for the focal length. The optical device in the problem was most likely a (1) diverging (concave) lens (2) converging (convex) lens (3) converging (concave) mirror (4) plane mirror

Group 5–Solid State

If you choose this group, be sure to answer questions 101-110.

101. In each diagram below of a band model element, *C* is the conduction band and *V* the valence band. Which diagram best represents a conductor?

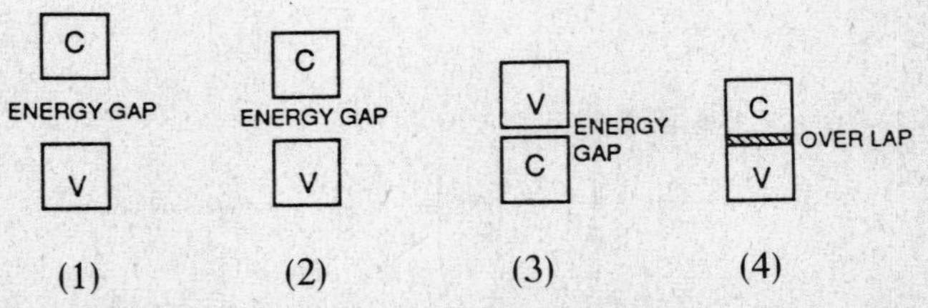

102. A small amount of antimony is deposited on a crystal of silicon and then heated. The antimony diffuses inward to form a semiconducting material. This process is called (1) crystallization (2) conduction (3) solidifying (4) doping

103. Gallium accepts bound valence electrons in a semiconductor. The deficiency of conducting electrons provided by gallium causes an excess of (1) holes (2) neutrons (3) electrons (4) protons

104. How is the semiconductor in the circuit shown below classified?

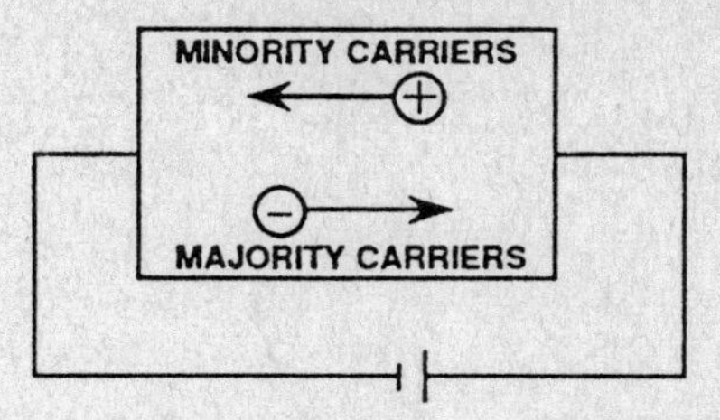

(1) *N*-type (2) *P*-type (3) a diode (4) a transistor

105. In which circuit diagram below is the light bulb most likely to light?

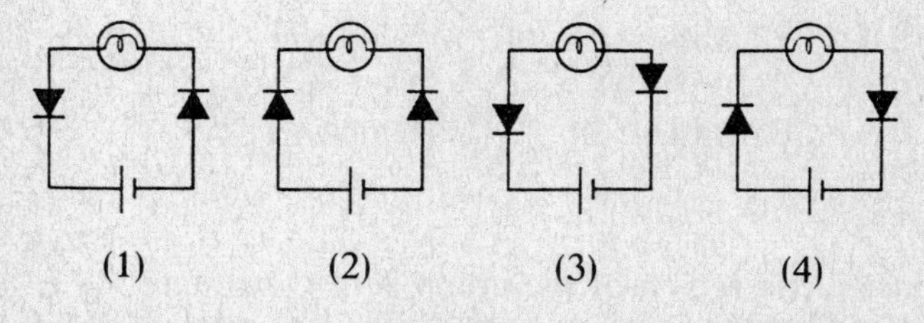

106. Which device is represented by the diagram below?

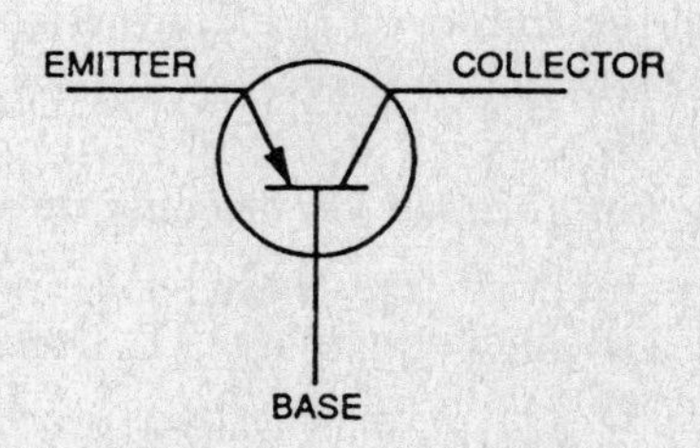

(1) *N-P-N* transistor
(2) *P-N-P* transistor
(3) zener diode
(4) solenoid

107. How is current affected when a semiconductor diode is reverse biased? (1) Current conduction increases because the junction electric field barrier increases. (2) Current conduction increases because the junction electric field barrier decreases. (3) Current conduction decreases because the junction electric field barrier increases. (4) Current conduction decreases because the junction electric field barrier decreases.

108. The graph below represents the relationship between the current and applied potential for a diode.

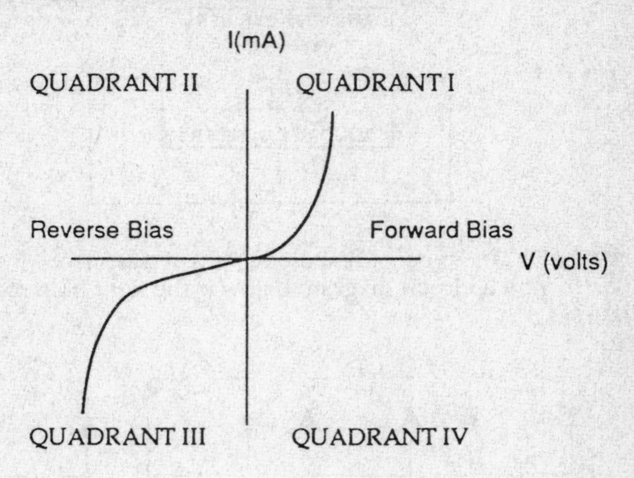

The avalanche region of the graph is in Quadrant (1) I (2) II (3) III (4) IV

109. In an *N-P-N* transistor, the emitter-base combination is forward biased. In this transistor, there is a flow of (1) electrons from the base to the emitter (2) electrons from the emitter to the base (3) holes from the base to the collector (4) holes from the emitter to the base

Note that question 110 has only three choices.

110. As the number of free charges per unit volume of a solid increases, its electrical conductivity (1) decreases (2) increases (3) remains the same

Group 6–Nuclear Energy

If you choose this group, be sure to answer questions 111-120.

111. What is the total number of neutrons in the nucleus of an atom of $^{234}_{91}Pa$? (1) 91 (2) 143 (3) 234 (4) 325

112. Approximately how much energy would be generated if the mass in a nucleus of a $^{2}_{1}H$ atom were completely converted to energy? [The mass of $^{2}_{1}H$ is 2.0 atomic mass units.] (1) 3.2×10^{-19} J (2) 1.5×10^{-10} J (3) 9.3×10^{2} MeV (4) 1.9×10^{3} MeV

113. Which two symbols represent isotopes of the same element?

(1) $^{8}_{2}X$ and $^{8}_{4}X$ (3) $^{3}_{2}X$ and $^{2}_{3}X$

(2) $^{6}_{2}X$ and $^{8}_{2}X$ (4) $^{2}_{2}X$ and $^{4}_{4}X$

114. The half life of $^{234}_{90}Th$ is 24 days. How much of a 128-milligram sample of thorium will remain after 144 days? (1) 5.3 mg (2) 2 mg (3) 21.3 mg (4) 64 mg

115. Which equation is a step in the Uranium Disintegration Series?

(1) $^{234}_{90}Th \rightarrow ^{234}_{91}Pa + ^{0}_{-1}e$ (3) $^{234}_{90}Th \rightarrow ^{226}_{88}Ra + ^{4}_{2}He$

(2) $^{234}_{90}Th \rightarrow ^{230}_{88}Ra + ^{4}_{2}He$ (4) $^{234}_{90}Th \rightarrow ^{226}_{88}Ac + ^{0}_{-1}e$

116. Which particles will *not* increase in kinetic energy in a particle accelerator? (1) alpha particles (2) beta particles (3) protons (4) neutrons

117. Which is an example of electron capture (*K*-capture)?

(1) $^{27}_{13}Al + ^{4}_{2}He \rightarrow ^{30}_{15}P + ^{1}_{0}n$ (3) $^{40}_{19}K + ^{0}_{-1}e \rightarrow ^{40}_{18}Ar$

(2) $^{238}_{92}U + ^{1}_{0}n \rightarrow ^{239}_{92}U \rightarrow ^{239}_{93}Np + ^{0}_{-1}e$ (4) $^{64}_{29}Cu \rightarrow ^{64}_{28}Ni + ^{0}_{+1}e$

118. Which equation represents nuclear fission?

(1) $^{226}_{88}Ra \rightarrow ^{222}_{86}Rn + ^{4}_{2}He$ (3) $^{9}_{4}Be + ^{4}_{2}He \rightarrow ^{12}_{6}C + ^{1}_{0}n$

(2) $^{24}_{11}Na \rightarrow ^{24}_{12}Mg + ^{0}_{-1}e$ (4) $^{235}_{92}U + ^{1}_{0}n \rightarrow ^{141}_{56}Ba + ^{92}_{36}Kr + 3^{1}_{0}N + Q$

119. The critical mass of nuclear reactor materials is defined as the mass of (1) shielding material needed to reflect neutrons back into the reactor (2) moderating material needed to control the core's temperature (3) fissionable material necessary for a chain reaction to take place (4) control rods needed to absorb excess neutrons

120. Which factor associated with radioactive waste has the biggest impact on its storage time? (1) half-life (2) quantity (3) temperature (4) density

Part III

You must answer all questions in this part. Record your answers in the spaces provided on the separate answer paper. Pen or pencil may be used. [10]

121. Base your answers to parts *a* through *c* on the information below. Write your answers in the spaces provided on the separate answer paper.

A student pulls a cart across a horizontal floor by exerting a force of 50. newtons at an angle of 35° to the horizontal.

a On the diagram on your answer paper, *using a protractor and a straightedge*, construct a scaled vector showing the 50.-newton force acting on the cart at the appropriate angle. The force *must* be drawn to a scale of 1.0 centimeter = 10. N. Label the 50.-newton force and the 35° angle on your diagram. *Be sure your final answer appears on your answer paper with the correct labels (numbers and units).* [2]

b Construct the horizontal component of the force vector to scale on your diagram, and label it *H*. [1]

c What is the magnitude of the horizontal component of the force? [1]

122. Base your answers to parts *a* and *b* on the information and diagram below.

The sonar of a stationary ship sends a signal with a frequency of 5.0×10^3 hertz down through water. The speed of the signal is 1.5×10^3 meters per second. The echo from the bottom is detected 4.0 seconds later.

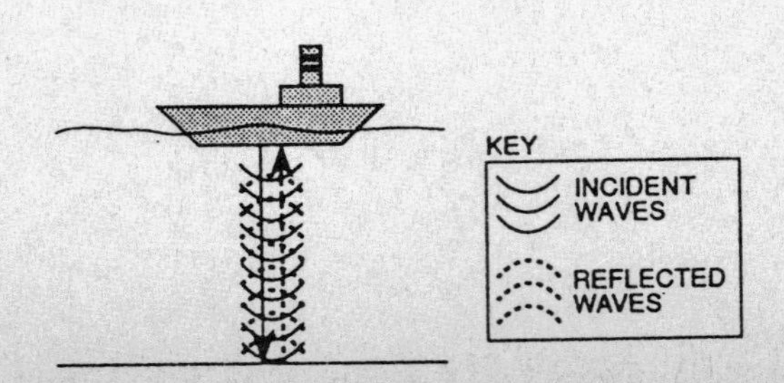

a What is the wavelength of the sonar wave? [Show all calculations, including the equation and substitution with units.] [2]
b What is the depth of the water under the ship? [Show all calculations, including the equation and substitution with units.] [2]

Base your answer to question 123 on the information and graph below.

A student performed an experiment measuring the maximum kinetic energy of emitted photoelectrons as the frequency of light shining on a photoemissive surface was increased. A graph of the student's data appears below.

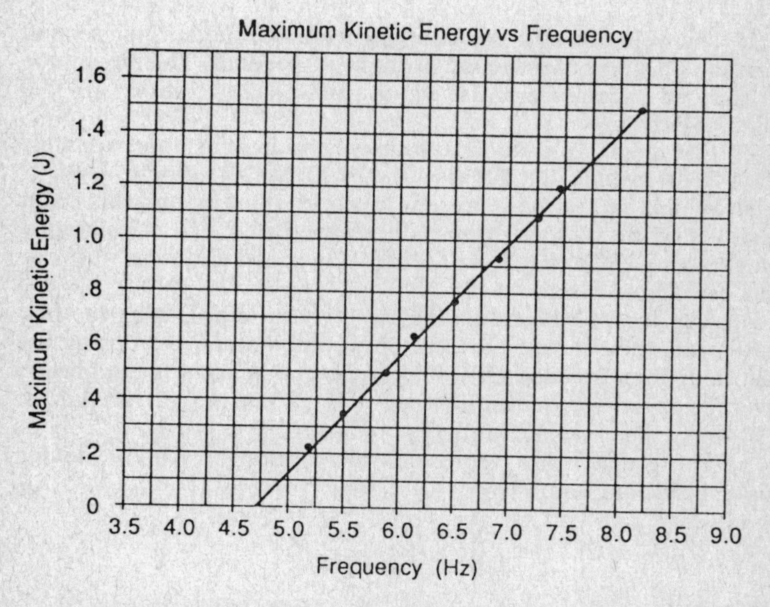

123. On the graph *on your answer paper*, draw a line representing the relationship between maximum kinetic energy and frequency when a photoemissive surface having a larger work function is used. [2]

PHYSICS
Wednesday, June 17, 1992

Part I

Answer all 55 questions in this part. [65]

Directions (1-55): For *each* statement or question, select the word or expression that, of those given, best completes the statement or answers the question. Record your answer on the separate answer sheet in accordance with the directions on the front page of this booklet.

1. The velocity of a car changes from 60. meters per second north to 45 meters per second north in 5.0 seconds. The magnitude of the car's acceleration is (1) 9.8 m/s^2 (2) 15 m/s^2 (3) 3.0 m/s^2 (4) 53 m/s^2

2. The height of a doorknob above the floor is approximately (1) 1×10^2 m (2) 1×10^1 m (3) 1×10^0 m (4) 1×10^{-2} m

3. Which two terms represent a vector quantity and the scalar quantity of the vector's magnitude, respectively? (1) acceleration and velocity (2) weight and force (3) speed and time (4) displacement and distance

4. A group of bike riders took a 4.0-hour trip. During the first 3.0 hours, they traveled a total of 50. kilometers, but during the last hour they traveled only 10. kilometers. What was the group's average speed for the entire trip? (1) 15 km/hr (2) 30. km/hr (3) 40. km/hr (4) 60. km/hr

5. The graph below represents the motion of a body moving along a straight line.

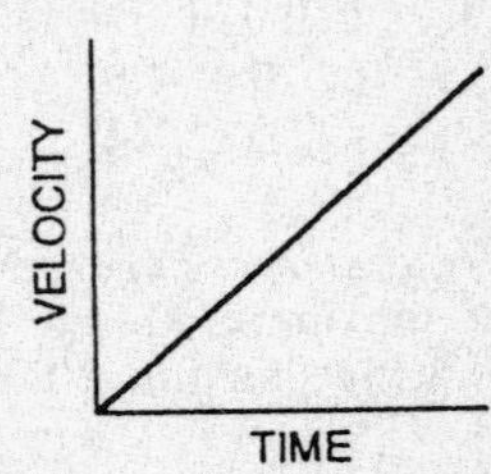

According to the graph, which quantity related to the motion of the body is constant? (1) speed (2) velocity (3) acceleration (4) displacement

6. A student walks 1.0 kilometer due east and 1.0 kilometer due south. Then she runs 2.0 kilometers due west. The magnitude of the student's resultant displacement is closest to (1) 0 km (2) 1.4 km (3) 3.4 km (4) 4.0 km

7. A locomotive starts from rest and accelerates at 0.12 meter per second^2 to a speed of 2.4 meters per second in 20. seconds. This motion could best be described as (1) constant acceleration and constant velocity (2) increasing acceleration and constant velocity (3) constant acceleration and increasing velocity (4) increasing acceleration and increasing velocity

8. A clam dropped by a sea gull takes 3.0 seconds to hit the ground. What is the sea gull's approximate height above the ground at the time the clam was dropped? (1) 15 m (2) 30. m (3) 45 m (4) 90. m

9. Which graph best represents the relationship between mass and acceleration due to gravity for objects near the surface of the Earth? [Neglect air resistance.]

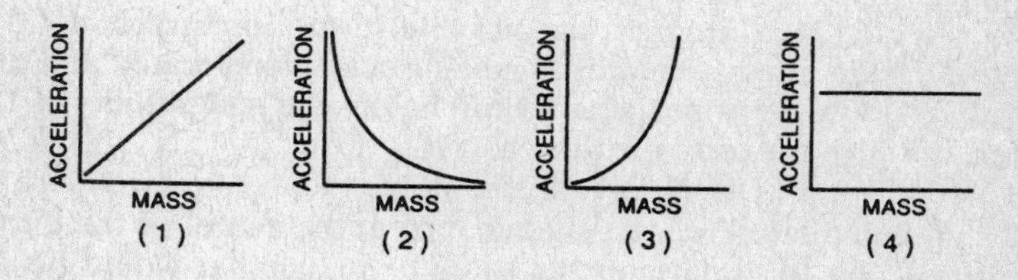

10. Two concurrent forces have a maximum resultant of 45 newtons and a minimum resultant of 5.0 newtons. What is the magnitude of each of these forces? (1) 0.0 N and 45 N (2) 5.0 N and 9.0 N (3) 20. N and 25 N (4) 0.0 N and 50. N

11. The diagram below represents a car resting on a hill.

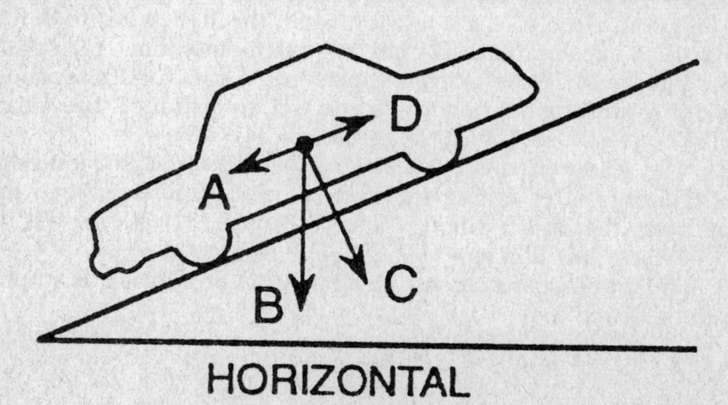

Which vector best represents the weight of the car? (1) *A* (2) *B* (3) *C* (4) *D*

12. In the diagram below, box M is on a frictionless table with forces F_1 and F_2 acting as shown.

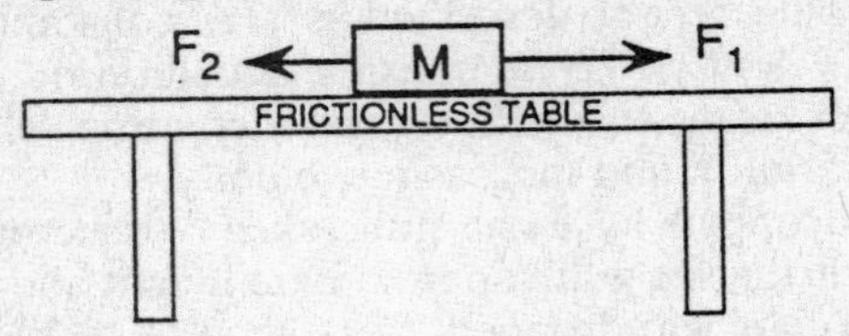

If the magnitude of F_1 is greater than the magnitude of F_2, then the box is (1) moving with a constant speed in the direction of F_1 (2) moving with a constant speed in the direction of F_2 (3) accelerating in the direction of F_1 (4) accelerating in the direction of F_2

13. A 1.2×10^3-kilogram automobile in motion strikes a 1.0×10^{-4}-kilogram insect. As a result, the insect is accelerated at a rate of 1.0×10^2 meters per second^2. What is the magnitude of the force the insect exerts on the car? (1) 1.0×10^{-2} N (2) 1.2×10^{-2} N (3) 1.0×10^1 N (4) 1.2×10^3 N

14. When a satellite is a distance d from the center of the Earth, the force due to gravity on the satellite is F. What would be the force due to gravity on the satellite when its distance from the center of the Earth is $3d$?

(1) F (2) $\frac{F}{9}$ (3) $\frac{F}{3}$ (4) $9F$

15. Two rocks weighing 5 newtons and 10 newtons, respectively, fall freely from rest near the Earth's surface. After 3 seconds of free-fall, compared to the 5-newton rock, the 10-newton rock has greater (1) acceleration (2) height (3) momentum (4) speed

16. A force of 20. newtons is exerted on a cart for 10. seconds. How long must a 50.-newton force act to produce the same impulse? (1) 10. s (2) 2.0 s (3) 5.0 s (4) 4.0 s

17. A 20.-kilogram object strikes the ground with 1,960 joules of kinetic energy after falling freely from rest. How far above the ground was the object when it was released? (1) 10. m (2) 14 m (3) 98 m (4) 200 m

18. The graph below shows the elongation of a spring as a function of the force.

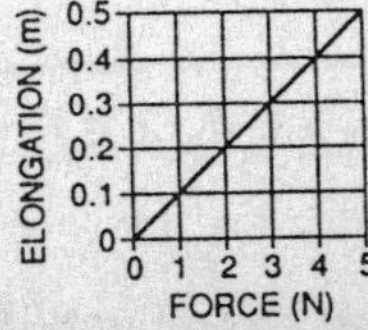

What is the value of the spring constant? (1) 0.1 m/N (2) 0.1 N/m (3) 10 m/N (4) 10 N/m

19. Which graph below best represents the relationship between the potential energy stored in a spring (PE) and the change in the length of the spring from its equilibrium position (X)?

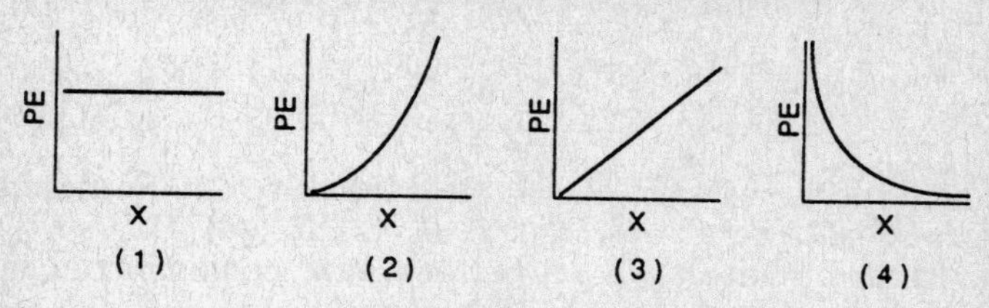

20. An object with +10 elementary charges is grounded and becomes neutral. What is the best explanation for this occurrence? (1) The object gained 10 electrons from the ground (2) the object lost 10 electrons to the ground (3) the object gained 10 protons from the ground (4) the object lost 10 protons to the ground

21. Two identical spheres carry charges of +0.6 coulomb and −0.2 coulomb, respectively. If these spheres touch, the resulting charge on the first sphere will be (1) +0.8 C (2) +0.2 C (3) −0.3 C (4) +0.4 C

22. The diagram represents two charges q_1 and q_2, separated by distance d.

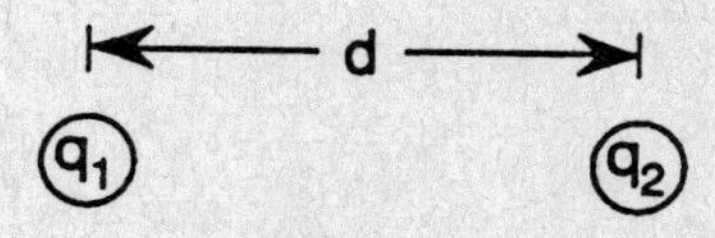

Which change would produce the greatest increase in the electrical force between the two charges? (1) doubling charge q_1, only (2) doubling d, only (3) doubling d and charge q_1, only (4) doubling d and charges q_1 and q_2

23. A helium ion with +2 elementary charges is accelerated by a potential difference of 5.0×10^3 volts. What is the kinetic energy acquired by the ion? (1) 3.2×10^{-19} eV (2) 2.0 eV (3) 5.0×10^3 eV (4) 1.0×10^4 eV

24. Which is a unit of electrical power? (1) volt/ampere (2) ampere/ohm (3) ampere^2/ohm (4) volt^2/ohm

25. Two equal positive point charges, *A* and *B*, are positioned as shown below.

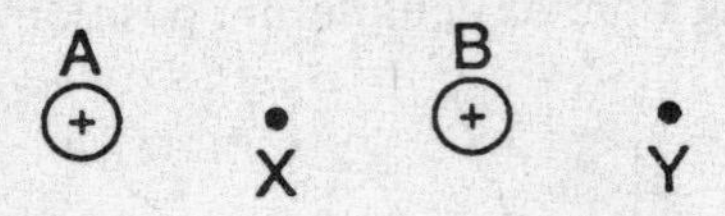

At which location is the electric field intensity due to these two charges equal to zero? (1) *A* (2) *B* (3) *X* (4) *Y*

26. In the circuit shown below, how many coulombs of charge will pass through resistor *R* in 2.0 seconds?

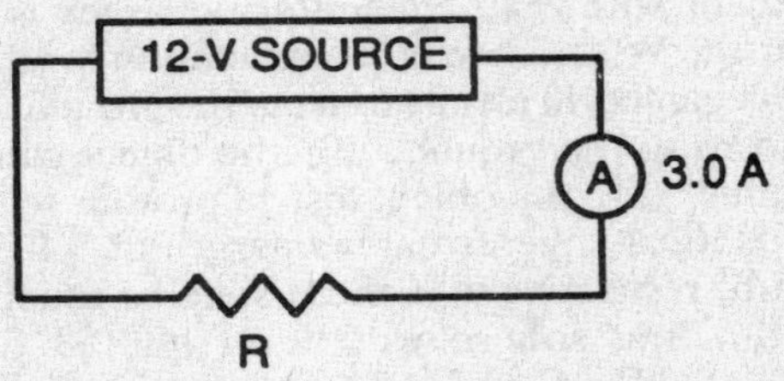

(1) 36 C (2) 6.0 C (3) 3.0 C (4) 4.0 C

27. Which graph below best represents how the resistance (*R*) of a series of copper wires of uniform length and temperature varies with cross-sectional area (*A*)?

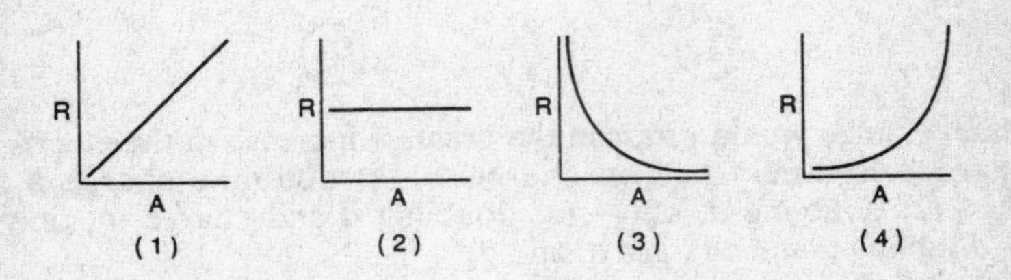

28. A physics student is given three 12-ohm resistors with instruction to create the circuit that would have the lowest possible resistance. The correct circuit would be a (1) series circuit with a total resistance of 36 Ω (2) series circuit with a total resistance of 4 Ω (3) parallel circuit with a total resistance of 36 Ω (4) parallel circuit with a total resistance of 4 Ω

29. Ammeters A_1, A_2, and A_3 are placed in a circuit as shown below.

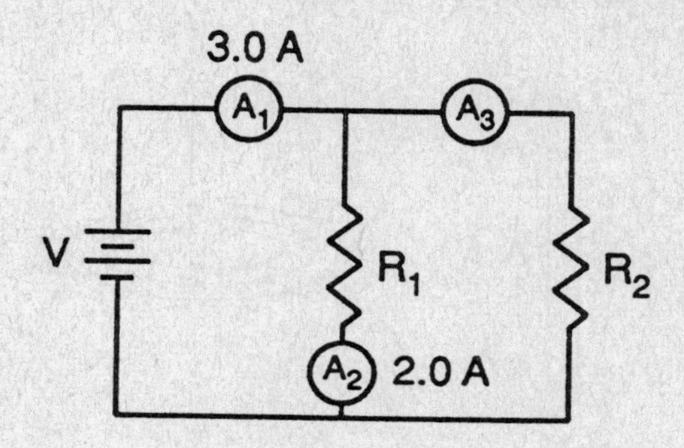

What is the reading on ammeter A_3? (1) 1.0 A (2) 2.0 A (3) 3.0 A (4) 5.0 A

30. What is the approximate amount of electrical energy needed to operate a 1,600-watt toaster for 60. seconds? (1) 27 J (2) 1,500 J (3) 1,700 J (4) 96,000 J

31. An electric motor uses 15 amperes of current in a 440-volt circuit to raise an elevator weighing 11,000 newtons. What is the average speed attained by the elevator? (1) 0.0027 m/s (2) 0.60 m/s (3) 27 m/s (4) 6,000 m/s

32. Which phenomenon does *not* occur when a sound wave reaches the boundary between air and a steel block? (1) reflection (2) refraction (3) polarization (4) absorption

33. The speaker in the diagram below makes use of a current-carrying coil of wire.

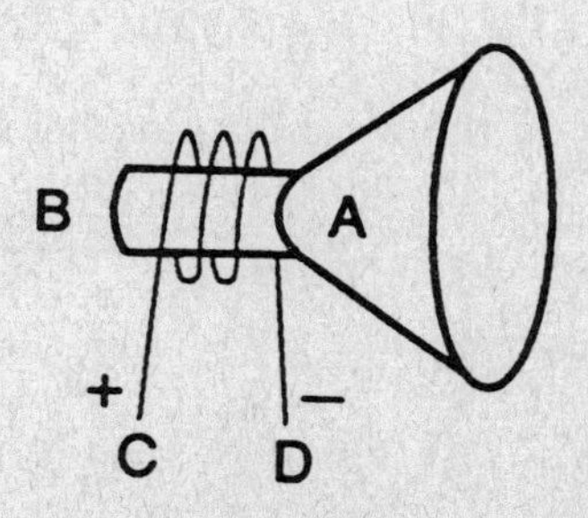

The N-pole of the coil would be closest to (1) *A* (2) *B* (3) *C* (4) *D*

34. An electron current (e^-) moving upward through a straight conductor creates a magnetic field. Which diagram below correctly represents this magnetic field?

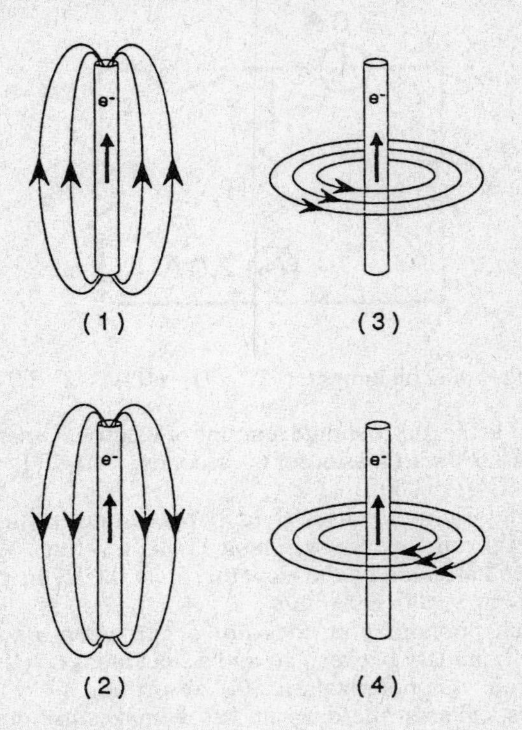

35. What is the period of a periodic wave that has a frequency of 60. hertz? (1) 1.7×10^{-2} s (2) 2.0×10^{4} s (3) 3.0×10^{-3} s (4) 3.3×10^{2} s

36. Which point on the wave diagram below is in phase with point *A*?

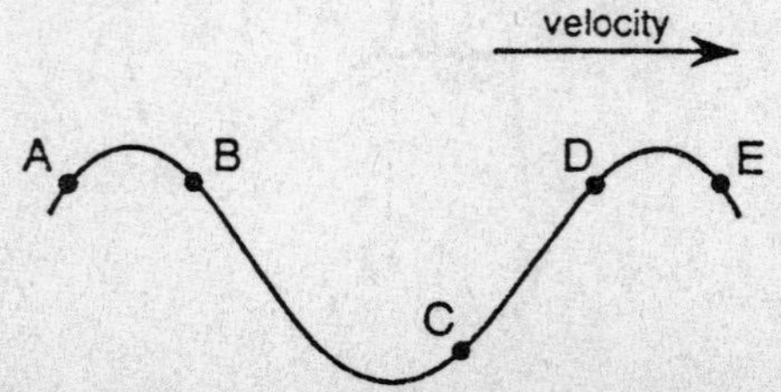

(1) *E* (2) *B* (3) *C* (4) *D*

37. What is the wavelength of a 30.-hertz periodic wave moving at 60. meters per second? (1) 0.50 m (2) 2.0 m (3) 20. m (4) 1,800 m

38. How many nodes are represented in the standing wave diagram below?

(1) 1 (2) 6 (3) 3 (4) 4

39. A car radio is tuned to the frequency being emitted from two transmitting towers. As the car moves at constant speed past the towers, as shown in the diagram below, the sound from the radio repeatedly fades in and out.

This phenomenon can best be explained by (1) refraction (2) interference (3) reflection (4) resonance

40. As a periodic wave travels from one medium to another, which pair of the wave's characteristics cannot change? (1) period and frequency (2) period and amplitude (3) frequency and velocity (4) amplitude and wavelength

41. When light rays from an object are incident upon an opaque, rough-textured surface, no reflected image of the object can be seen. This phenomenon occurs because of (1) regular reflection (2) diffuse reflection (3) reflected angles not being equal to incident angles (4) reflected angles not being equal to refracted angles

42. The diagram below represents a light ray being reflected from a plane mirror. The angle between the incident ray and the reflected ray is 70.°

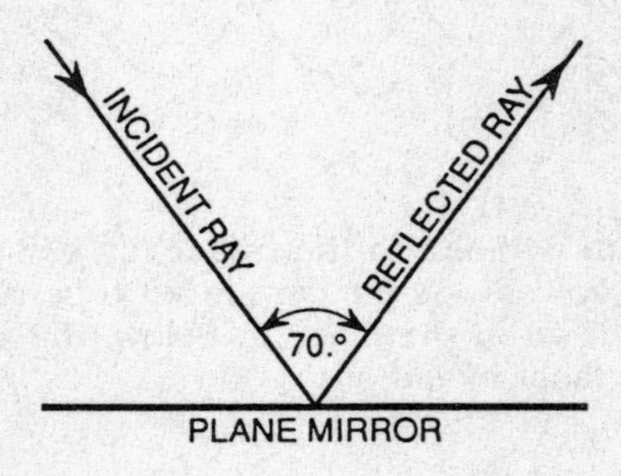

What is the angle of incidence for this ray? (1) 20.° (2) 35° (3) 55° (4) 70.°

43. The diagram below represents a light ray passing from corn oil into medium *X* with no change in velocity.

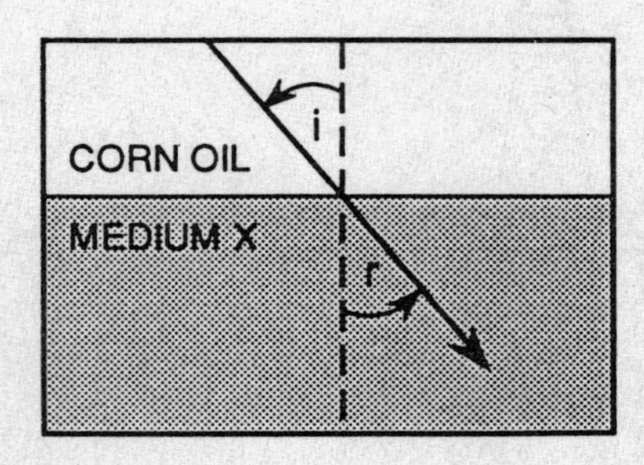

In this diagram, medium *X* could be (1) water (2) diamond (3) glycerol (4) alcohol

44. A ray of light in air is incident on a block of Lucite at an angle of 60.° from the normal. The angle of refraction of this ray in the Lucite is closest to (1) 35° (2) 45° (3) 60.° (4) 75°

Base your answers to questions 45 and 46 on the diagram below which represents a monochromatic light wave passing through the double slits. A pattern of bright and dark bands is formed on the screen.

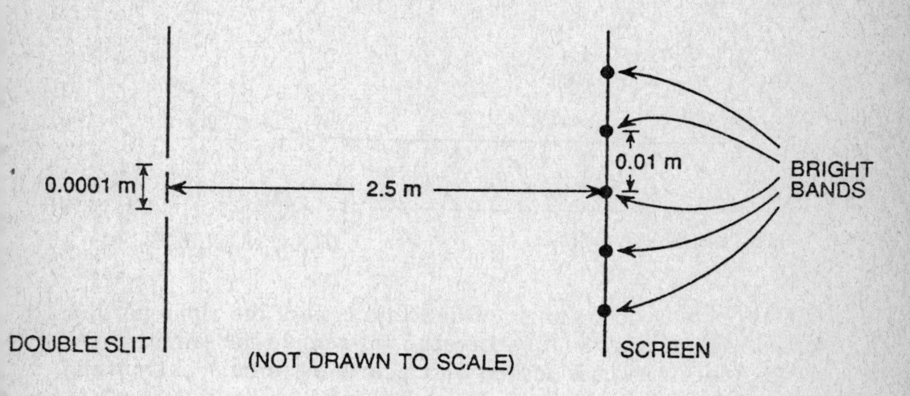

45. What is the color of the light used? (1) violet (2) blue (3) green (4) yellow

Note that question 46 has only three choices.

46. If the original light wave is replaced by a wave of longer wavelength, the space between the bright bands on the screen will (1) decrease (2) increase (3) remain the same

47. The graph below shows the relationship between the frequency of radiation incident on a photosensitive surface and the maximum kinetic energy (KE_{MAX}) of the emitted photoelectrons.

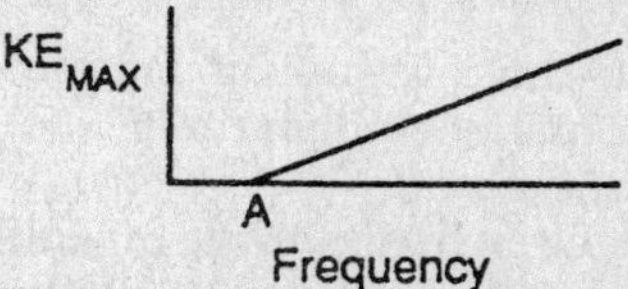

The point labeled A on the graph represents the (1) incident photon intensity (2) photoelectron frequency (3) threshold frequency (4) work function energy

48. A metal has a work function of 1.3×10^{-18} joule. What is the threshold frequency for electromagnetic radiation incident on this metal? (1) 2.0×10^{15} Hz (2) 2.0×10^{14} Hz (3) 8.6×10^{14} Hz (4) 8.6×10^{-52} Hz

49. A photon emitted from an excited hydrogen atom has an energy of 3.02 electronvolts. Which electron energy-level transition would produce this photon? (1) $n = 1$ to $n = 6$ (2) $n = 2$ to $n = 6$ (3) $n = 6$ to $n = 1$ (4) $n = 6$ to $n = 2$

Note that questions 50 through 55 have only three choices.

50. The diagram below represents alpha particle *A* approaching a gold nucleus. *D* is the distance between the path of the alpha particle and the path for a head-on collision

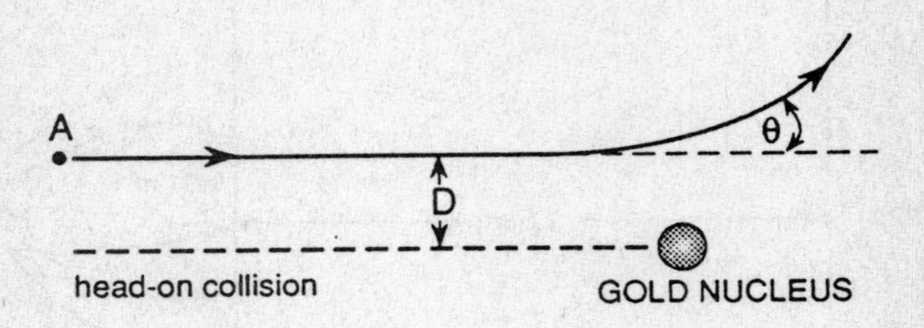

If *D* is decreased, the angle of deflection (θ) of the alpha particle would (1) decrease (2) increase (3) remain the same

51. A lawnmower is pushed with a constant force *F*, as shown in the diagram below.

As angle θ between the lawnmower handle and the horizontal increases, the horizontal component of *F* (1) decreases (2) increases (3) remains the same

52. As the time required to do a given quantity of work decreases, the power developed (1) decreases (2) increases (3) remains the same

53. As the speed of a bicycle moving along a level horizontal surface changes from 2 meters per second to 4 meters per second, the magnitude of the bicycle's gravitational potential energy (1) decreases (2) increases (3) remains the same

54. As shown in the diagram below, pulling a 9.8-newton cart a distance of 0.50 meter along a plane inclined at 15° requires 1.3 joules of work.

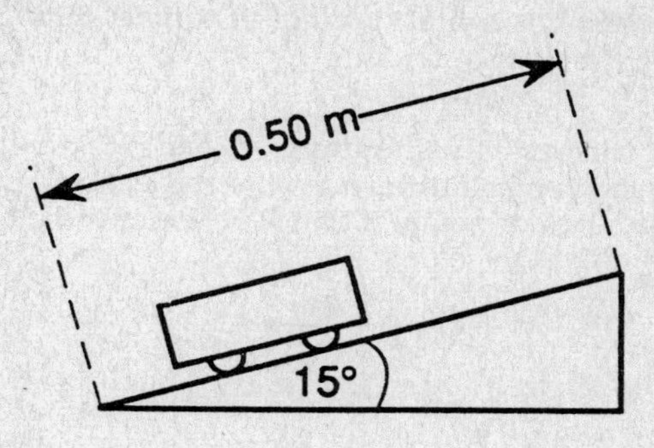

If the cart were raised 0.50 meter vertically instead of being pulled along the inclined plane, the amount of work done would be (1) less (2) greater (3) the same

55. As the momentum of an electron increases, the electron's wavelength (1) decreases (2) increases (3) remains the same

Part II

This part consists of six groups, each containing ten questions. Each group tests an optional area of the course. Choose two of these six groups. Be sure that you answer all ten questions in each group chosen. Record the answers to the questions in accordance with the directions on the front page of this booklet [20]

Group 1—Motion in a Plane

If you choose this group, be sure to answer questions 56-65.

Base your answers to questions 56 through 58 on the information below.

An object is thrown horizontally off a cliff with an initial velocity of 5.0 meters per second. The object strikes the ground 3.0 seconds later.

56. What is the vertical speed of the object as it reaches the ground? [Neglect friction.] (1) 130 m/s (2) 29 m/s (3) 15 m/s (4) 5.0 m/s

57. How far from the base of the cliff will the object strike the ground? [Neglect friction.] (1) 2.9 m (2) 9.8 m (3) 15 m (4) 44 m

58. What is the horizontal speed of the object 1.0 second after it is released? [Neglect friction.] (1) 5.0 m/s (2) 10. m/s (3) 15 m/s (4) 30. m/s

Base your answers to questions 59 and 60 on the diagram below which shows a ball thrown toward the east and upward at an angle of 30° to the horizontal. Point *X* represents the ball's highest point.

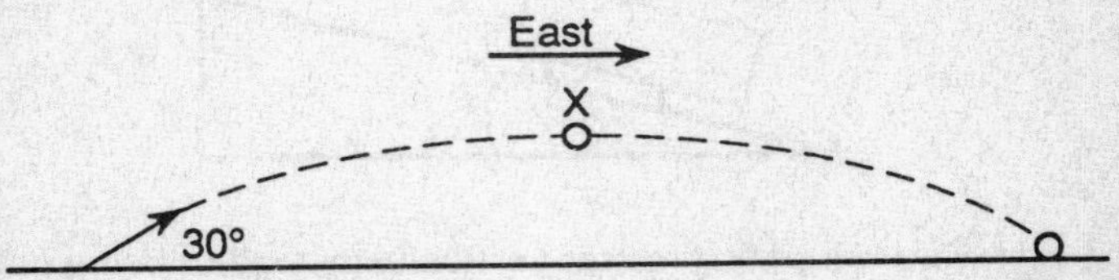

59. What is the direction of the ball's velocity at point *X*? [Neglect friction.] (1) down (2) up (3) west (4) east

60. What is the direction of the ball's acceleration at point *X*? [Neglect friction.] (1) down (2) up (3) west (4) east

61. A batted softball leaves the bat with an initial velocity of 44 meters per second at an angle of 37° above the horizontal. What is the magnitude of the initial vertical component of the softball's velocity? (1) 0 m/s (2) 26 m/s (3) 35 m/s (4) 44 m/s

62. The diagram shows the path of a satellite in an elliptical orbit around the Earth.

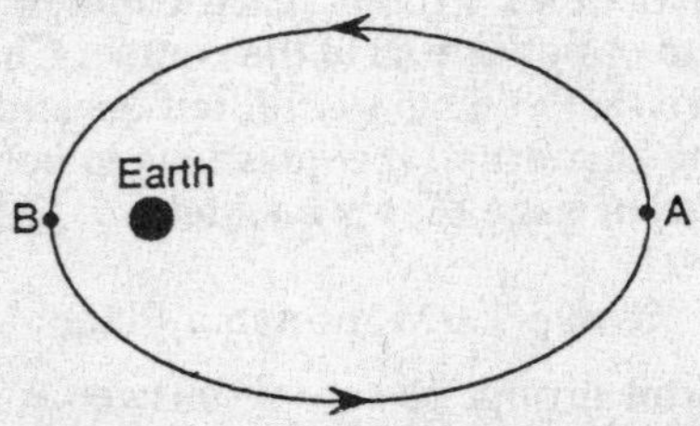

As the satellite moves from point *A* to point *B*, what changes occur in its potential and kinetic energies? (1) both potential energy and kinetic energy increase (2) both potential energy and kinetic energy decrease (3) potential energy increases and kinetic energy decreases (4) potential energy decreases and kinetic energy increases

63. A satellite is in geosynchronous orbit around the Earth. The period of the satellite's orbit is closest to (1) 6 hours (2) 12 hours (3) 24 hours (4) 48 hours

Base your answers to questions 64 and 65 on the diagram below which represents a space station shaped like a wheel and having a radius of 40. meters. The station is rotating clockwise with a speed of 12 meters per second at its outer wall. A 50-kilogram astronaut is standing inside the space station against the outer wall.

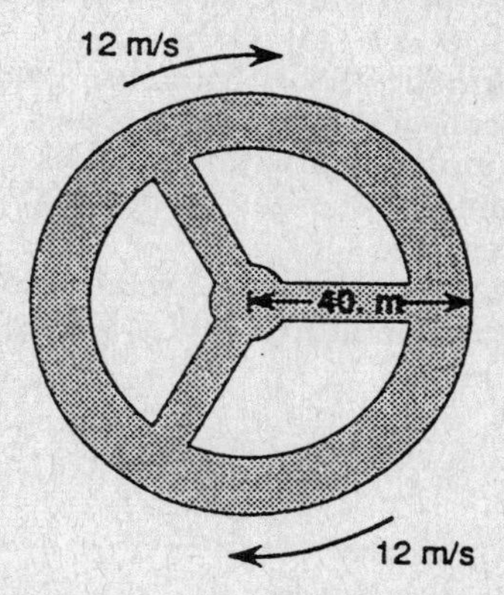

64. What is the apparent weight of the astronaut? (1) 3.6 N (2) 15 N (3) 180 N (4) 490 N

Note that question 65 has only three choices.

65. If the speed of rotation of the space station were doubled, the centripetal acceleration of the astronaut would (1) decrease (2) increase (3) remain the same

Group 2–Internal Energy

If you choose this group, be sure to answer questions 66-75.

66. What is the difference between 15°C and 6°C, expressed in Kelvin? (1) 282 K (2) 273 K (3) 15 K (4) 9 K

67. A certain mass of lead requires 1 kilojoule of heat energy to raise its temperature from 20.°C to 44°C. If 1 kilojoule of heat energy is added to the same mass of copper at 20.°C, what will be the final temperature of the copper? (1) 28°C (2) 32°C (3) 44°C (4) 92°C

68. Maximum average molecular kinetic energy in the solid phase of mercury is reached at a temperature of (1) -39°C (2) 0°C (3) 357°C (4) 396°C

69. If two substances are placed in contact with each other and no net exchange of internal energy occurs between them, the substances must have the same (1) specific heat (2) melting point (3) temperature (4) heat of fusion

70. What is the approximate amount of heat energy needed to change 5.0 kilograms of ice at 0°C to water at 0°C? (1) 21 kJ (2) 340 kJ (3) 1,700 kJ (4) 11,000 kJ

71. According to kinetic theory, pressure exerted by a gas is caused by the (1) collision of gas molecules with each other (2) collision of gas molecules with the walls of the container (3) negligible volume of the gas molecules (4) large forces between gas molecules

72. Which graph best represents the relationship between pressure (P) and absolute temperature (T_k) for a fixed mass of an ideal gas at constant volume?

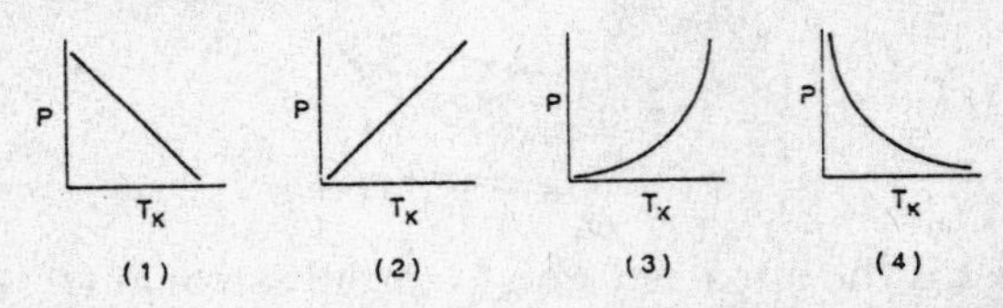

73. When a student drops a beaker, it shatters, spreading randomly shaped pieces of glass over a large area of the floor. According to the second law of thermodynamics, the measure of the disorder of this system is known as (1) vaporization (2) absolute order (3) molecular collision (4) entropy

Note that questions 74 and 75 have only three choices.

74. Compared to the quantity of heat needed to raise the temperature of 10 grams of liquid water 5°C, the heat needed to raise the temperature of 10 grams of ice 5°C is (1) less (2) greater (3) the same

75. The graph below shows temperature versus time for 1 kilogram of water at constant pressure as heat is added to a constant rate.

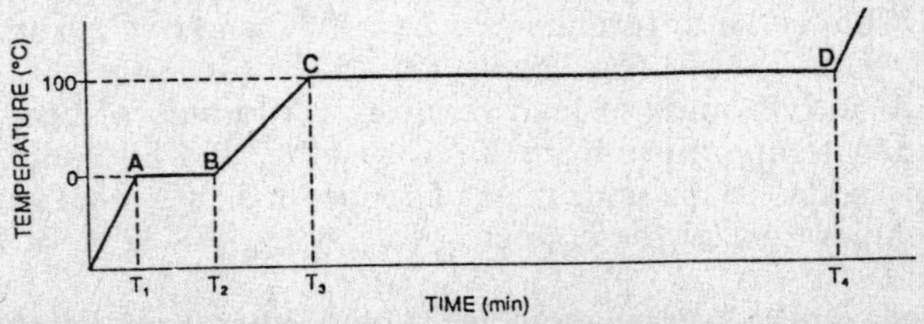

Dissolving a salt in the water will cause interval AB to occur at (1) a lower temperature (2) a higher temperature (3) the same temperature

Group 3—Electromagnetic Applications

If you choose this group, be sure to answer questions 76-85.

76. Which electrical device must have high resistance in order to function properly? (1) electric motor (2) ammeter (3) voltmeter (4) galvanometer

Base your answers to questions 77 and 78 on the diagram below which represents an electron with velocity of 2.0×10^6 meters per second directed into a region between two large, flat charged parallel plates.

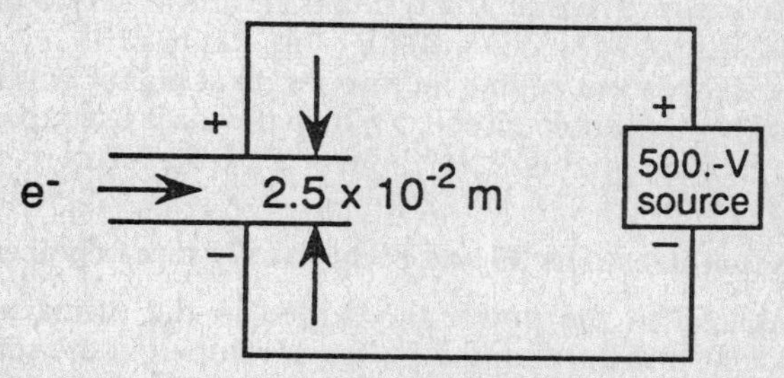

77. The magnitude of the electric field intensity between the plates is (1) 1.0×10^2 N/C (2) 2.0×10^4 N/C (3) 5.0×10^{-5} N/C (4) 2.5×10^{-11} N/C

78. The direction of the acceleration of the electron in the region between the plates is (1) into the page (2) out of the page (3) toward the bottom of the page (4) toward the top of the page

Base your answers to questions 79 and 80 on the diagram below which represents a beam of electrons moving through a uniform magnetic field. The magnetic field is directed into the page.

79. As the beam of electrons moves through the magnetic field, the electrons will be deflected (1) into the page (2) out of the page (3) toward the left (4) toward the right

80. If the speed of an electron in the magnetic field is 6.0×10^6 meters per second and a force of 5.0×10^{-14} newton acts on the electron, what is the flux density of the magnetic field? (1) 5.2×10^{-2} T (2) 8.3×10^{-21} T (3) 3.0×10^{-7} T (4) 1.9×10^1 T

81. The thermionic emission of electrons from the metal filament of an operating light bulb is caused by (1) heat (2) magnetism (3) radio waves (4) sound waves

82. A conductor 0.10 meter long moves with a velocity of 8.0 meters per second perpendicular to a magnetic field measuring 4.0×10^{-2} newton per ampere-meter. What is the magnitude of the electromotive force induced in the conductor (1) 5.0×10^{-3} V (2) 2.0×10^{-1} V (3) 3.2×10^{-2} V (4) 4.0×10^{-3} V

83. Some fluorescent ceiling lights operate at higher voltage than that supplied by household circuits. Which device is used to increase the voltage for these lights? (1) laser (2) transformer (3) motor (4) generator

Note that questions 84 and 85 have only three choices.

84. Compared to the power developed in the primary coil of a 100% efficient transformer, the power developed in the secondary coil is (1) less (2) greater (3) the same

85. The only difference between two motors is the material of their armature cores. Motor *A* has its coil wrapped around a piece of soft iron, and motor *B* has its coil wrapped around a piece of wood. Compared to the force exerted on the armature of motor *A*, the force exerted on the armature of motor *B* is (1) less (2) greater (3) the same

Group 4—Geometric Optics

If you choose this group, be sure to answer questions 86-95.

86. When the calculated image distance for an image formed using a curved mirror has a negative value, the image must be (1) real (2) virtual (3) reduced (4) enlarged

87. The diagram below represents two light rays emerging from a candle flame and being reflected from a plane mirror.

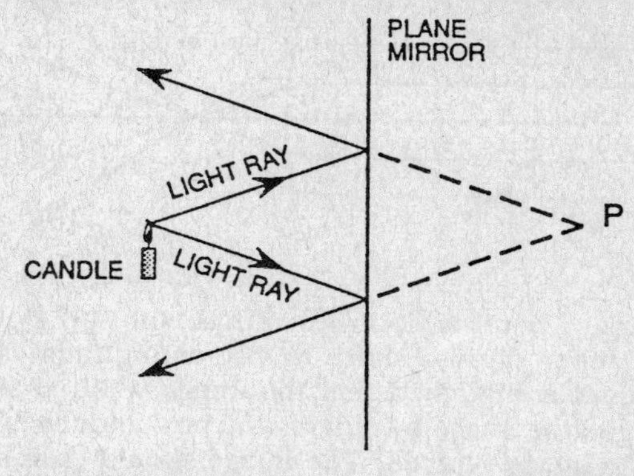

What does point *P* represent? (1) the virtual image point of the candle flame (2) the real image point of the candle flame (3) the focal point of the mirror (4) the center of curvature of the mirror

88. A pencil 0.10 meter long is placed 1.0 meter in front of a concave (converging) mirror whose focal length is 0.50 meter. The image of the pencil is (1) erect and 0.030 meter long (2) erect and 0.10 meter long (3) inverted and 0.030 meter long (4) inverted and 0.10 meter long

89. If the distance between an object and a concave (converging) mirror is more than twice the focal length of the mirror, the image formed will be (1) real and behind the mirror (2) real and in front of the mirror (3) virtual and behind the mirror (4) virtual and in front of the mirror

90. In the diagram below, an object is located in front of a convex (diverging) mirror. *F* is the virtual focal point of the mirror and *C* is its center of curvature. Ray *R* is parallel to the principal axis.

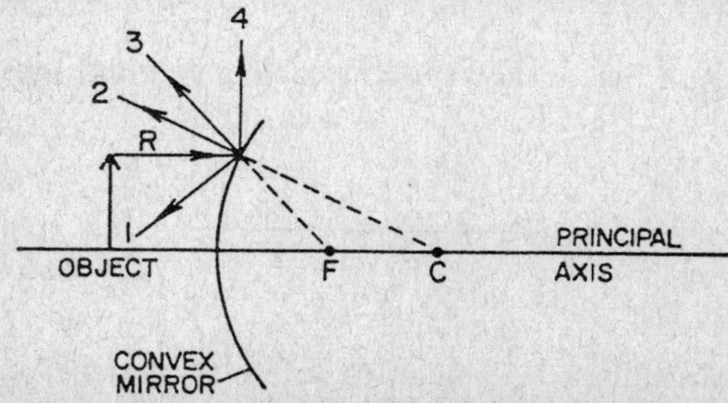

Ray *R* will most likely be reflected along path (1) 1 (2) 2 (3) 3 (4) 4

91. Which phenomenon is represented by the diagram below?

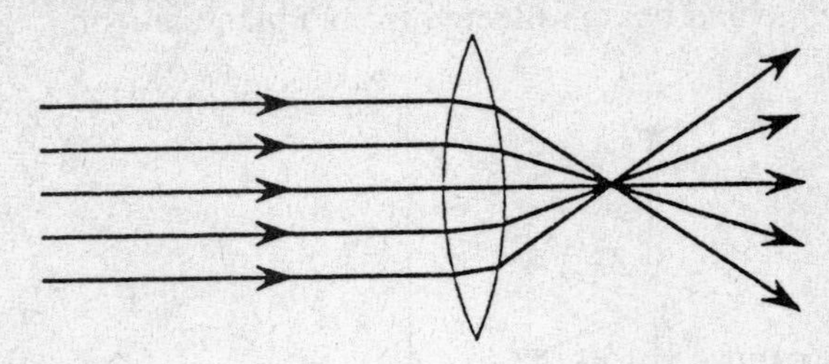

(1) reflection (2) refraction (3) diffraction (4) polarization

92. As a teacher showed slides by projecting them on a fixed screen, a student complained that the image was too small. The teacher enlarged the image by moving the projector away from the screen, but the image blurred. The image should then have been brought into focus by (1) moving the lens closer to the slide (2) moving the lens away from the slide (3) decreasing the amount of light in the room (4) increasing the power of the projector lamp

93. Which ray diagram is *incorrect?*

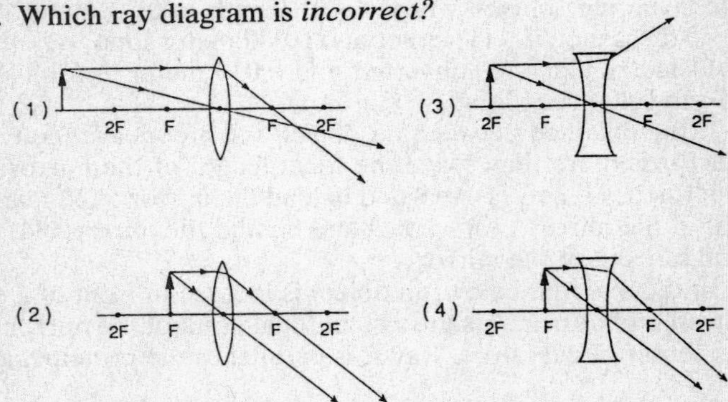

94. The diagram below shows a convex (converging) lens with focal length *f*

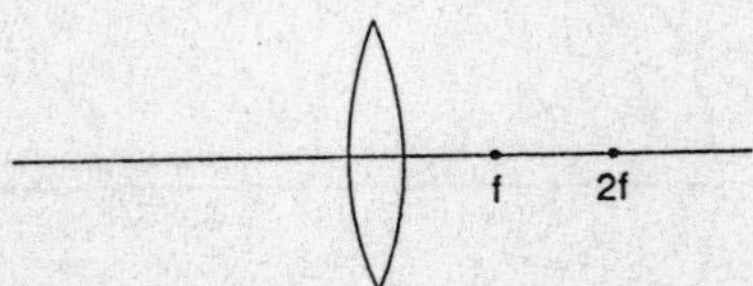

Where should an object be placed to produce a virtual image? (1) at *f* (2) at *2f* (3) between *f* and the lens (4) between *2f* and the *f*

Note that question 95 has only three choices.

95. The diagram below represents two rays of red light passing through a converging lens.

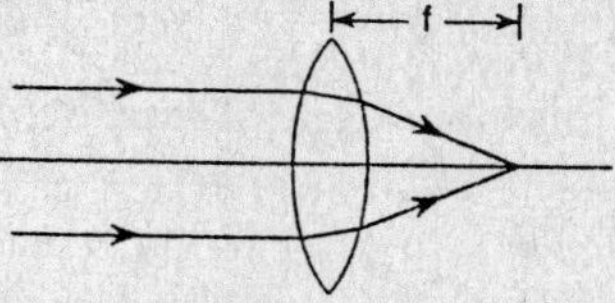

If the two rays were blue light, distance *f* would be. (1) shorter (2) longer (3) the same

Group 5 – Solid State

If you choose this group, be sure to answer questions 96-105.

96. Which is an important factor in determining whether two materials can be used to form a semiconductor material? (1) One material must be able to donate electrons and the other must be able to accept electrons. (2) Both materials must have the same number of electrons in the valence shells. (3) One material must have positive electrons and the other must have negative electrons. (4) Both materials must have more neutrons than protons in their nuclei.

97. Which graph best represents the relationship between the conductivity and the temperature of semiconductors?

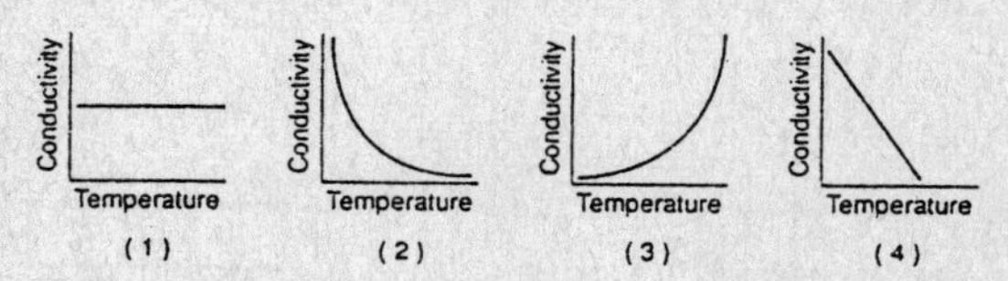

98. In the circuit shown in the diagram below, ammeter *A* reads 4 milliamperes when connected to an *N*-type semiconductor.

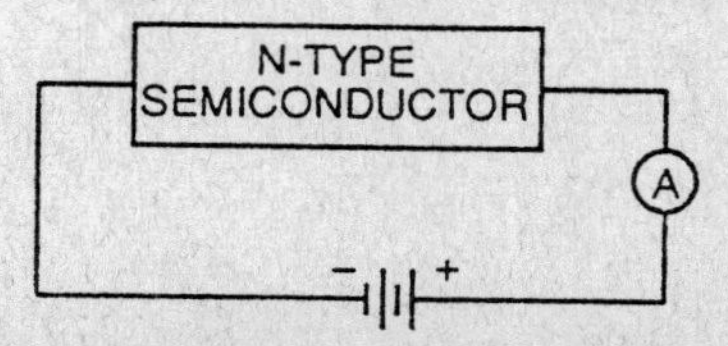

If the connections to the battery are reversed, the reading of ammeter *A* will be (1) 8 mA (2) 2 mA (3) 0 mA (4) 4 mA

99. Why do current carriers have difficulty crossing a *P-N* junction? (1) The junction area has a large positive charge. (2) The junction area has a large negative charge. (3) The junction acts as an electric field barrier. (4) The resistance across the junction is extremely low.

100. In the P-N junction diode shown below, in which direction do both the holes in the P-type material and the electrons in the N-type material move?

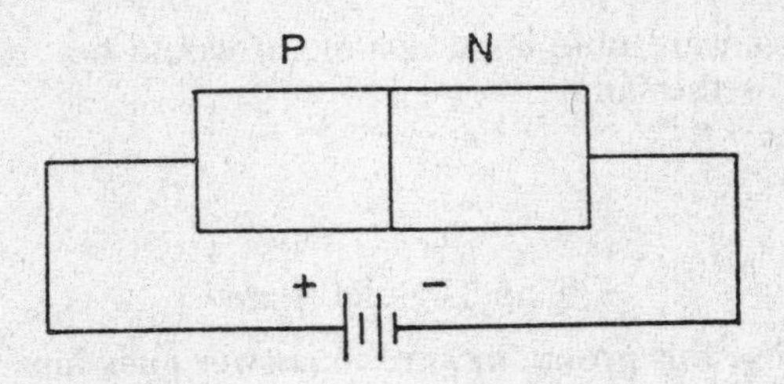

(1) away from the junction (2) toward the junction (3) toward the right (4) toward the left

101. Which diagram shows both a forward and a reverse bias?

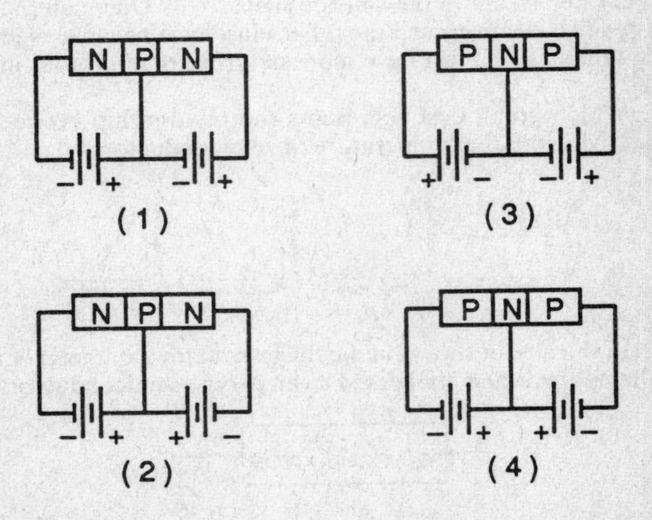

102. If the transistor shown in the diagram at the right is to serve as an amplifier, most of the electrons must pass from (1) *B* to *E* (2) *C* to *B* (3) *C* to *E* (4) *E* to *C*

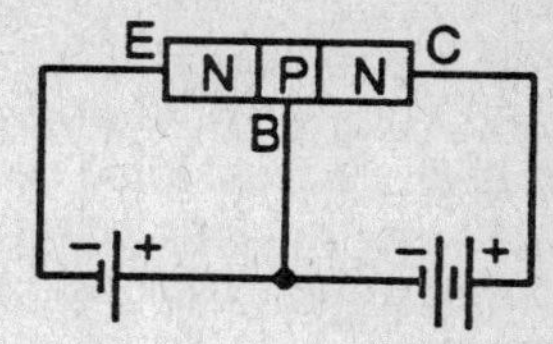

Note that questions 103 through 105 have only three choices.

103. As the amount of doping material used in a diode increases, the potential difference needed to cause the avalanche (1) decreases (2) increases (3) remains the same

104. When a semiconductor is replaced in a circuit by an insulator, the resistance of that section of the circuit (1) decreases (2) increases (3) remains the same

105. Adding small amounts of an impurity such as phosphorus (5 valence electrons) to a semiconductor will cause the net charge of the semiconductor to (1) decrease (2) increase (3) remain the same

Group 6–Nuclear Energy

If you choose this group, be sure to answer questions 106-115.

106. A neutral atom of an isotope of element *X* has 44 electrons and 63 neutrons. What is the mass number of this isotope? (1) 19 (2) 44 (3) 63 (4) 107

107. Compared to electrostatic forces, nuclear forces are (1) weaker and of shorter range (2) weaker and of longer range (3) stronger and of shorter range (4) stronger and of longer range

108. The half-life of $^{223}_{88}$Ra is 11.2 days. If mass M of this radium isotope is present initially, how much $^{222}_{88}$Ra remains at the end of 56 days?

(1) $\frac{1}{2}M$ (2) $\frac{1}{4}M$ (3) $\frac{1}{5}M$ (4) $\frac{1}{32}M$

109. Which nuclear symbol represents an isotope of polonium (Po) that is part of the Uranium Disintegration Series? (1) $^{214}_{83}$Po (2) $^{214}_{84}$Po (3) $^{218}_{82}$Po (4) $^{222}_{84}$Po

110. A decrease in both mass number and atomic number of a nucleus occurs due to the emission of (1) an alpha particle (2) a beta particle (3) a neutron (4) a positron

111. What kind of nuclear reaction is shown below?

$$^{1}_{1}H + ^{3}_{1}H \rightarrow ^{4}_{2}He + Q$$

(1) alpha decay (2) beta decay (3) fusion (4) fission

112. If a proton were absorbed by $^{222}_{86}Rn$, the symbol for the resulting nucleus would be (1) $^{222}_{87}Fr$ (2) $^{223}_{87}Fr$ (3) $^{222}_{85}At$ (4) $^{223}_{86}Rn$

113. In nuclear reactors, the function of a moderator is to decrease the neutrons' (1) binding energy (2) electromagnetic energy (3) potential energy (4) kinetic energy

Note that questions 114 and 115 have only three choices.

114. As two nuclei are moved closer together, the electrostatic force of repulsion between them (1) decreases (2) increases (3) remains the same

115. As a star gives off energy in a thermonuclear reaction, the mass of the star (1) decreases (2) increases (3) remains the same

Part III

You must answer all questions in this part. Record you answers in the spaces provided on the separate answer paper. Pen or pencil may be used. [15]

116. Base your answers to parts a throuch c on the information below.

A student performs a laboratory activity in which a 15-newton force acts on a 2.0-kilogram mass. The work done over time is summarized in the table below.

DATA TABLE

Time (s)	Work (J)
0	0
1.0	32
2.0	59
3.0	89
4.0	120

a Using the information in the data table, construct a graph on the grid provided *on your answer paper,* following the directions below. The grid below is provided for practice purposes only. *Be sure your final answer appears on your answer paper.* (1) Develop an appropriate scale for work, and plot the points for a *work*-versus-*time* graph. [1] (2) Draw the best-fit line. [1]
b Calculate the value of the slope of the graph constructed in part *a*. (Show all calculations, including equations and substitutions with units.) [2]
c Based on your graph, how much time did it take to do 75 joules of work? [1]

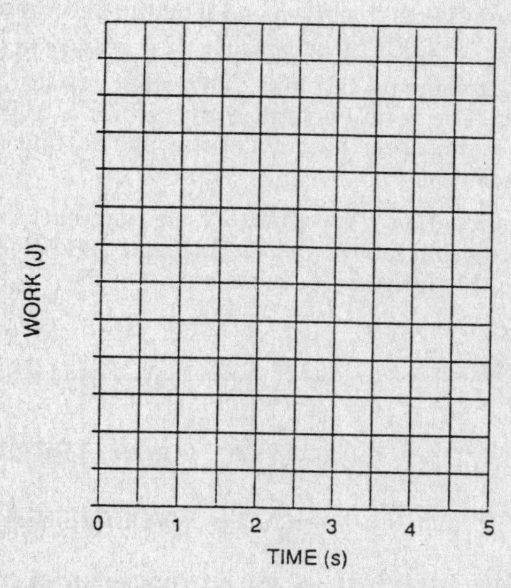

117. Base your answers to parts *a* through *c* on the diagram and information below.

Two railroad carts, *A* and *B*, are on a frictionless, level track. Cart *A* has a mass of 2.0×10^3 kilograms and a velocity of 3.0 meters per second toward the right. Cart *B* has a velocity of 1.5 meters per second toward the left. The magnitude of the momentum of cart *B* is 6.0×10^3 kilogram-meters per second. When the two carts collide, they lock together

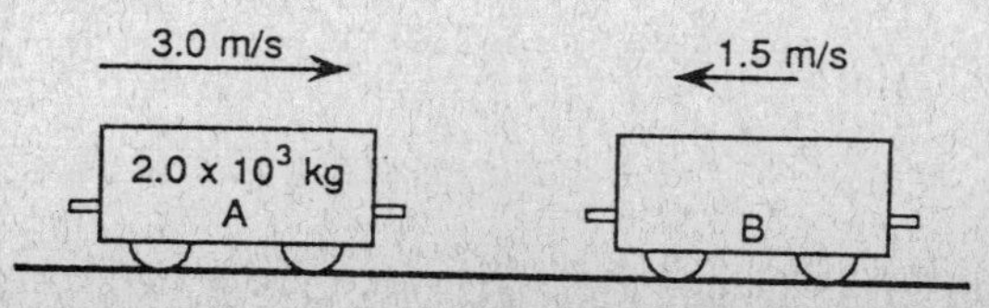

a What is the magnitude of the momentum of cart *A* before the collision? (Show all calculations, including equations and substitutions with units.) [2]
b On the diagram on your answer sheet, construct a scaled vector that represents the momentum of cart *A* before the collision. The momentum vector *must* be drawn to a scale of 1.0 centimeter = 1,000 kilogram-meters per second. *Be sure your final answer appears on your answer paper with correct labels (numbers and units).* [1]
c In one or more *complete sentences,* describe the momentum of the two carts after the collision and justify your answer based on the initial momenta of both carts. [2]

118. Base your answers to parts *a* and *b* on the information below.

Two resistors are connected in parallel to a 12-volt battery. One resistor, R_1, has a value of 18 ohms. The other resistor, R^2, has a value of 9 ohms. The total current in the circuit is 2 amperes. A student wishes to measure the current through R_1 and the potential difference across R_2.

a Using the symbols below for a battery, an ammeter, a voltmeter, and resistors, draw and label a circuit diagram that will enable the student to make the desired measurements. [3]

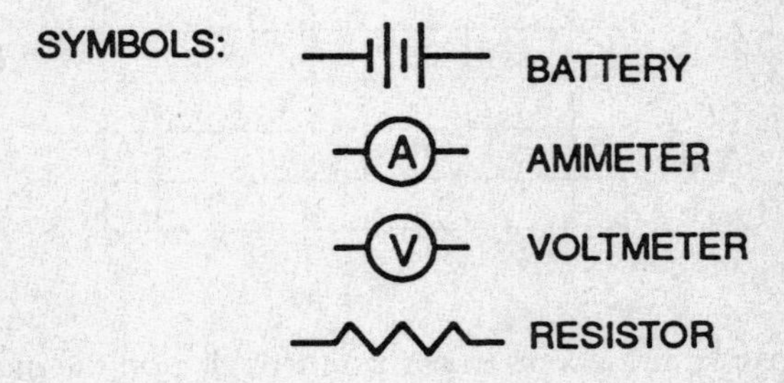

b Calculate the value of the current in resistor R_1. (Show all calcuations, including equations and substitutions with units.) [2]

117 *a*

b

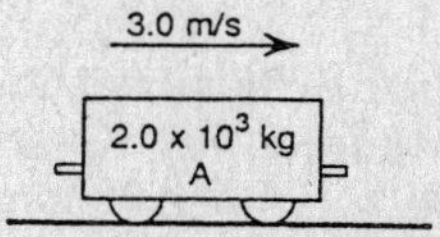

3.0 m/s

2.0×10^3 kg
A

c ______________________________

____

____

____

____

____

118 *a*

b

I do hereby affirm, at the close of this examination, that I had no unlawful knowledge of the questions or answers prior to the examination, and that I have neither given nor received assistance in answering any of the questions during the examination.

____
Signature

PHYSICS
Thursday, June 17, 1993

Part I

Answer all 55 questions in this part. [65]

Directions (1-55): For *each* statement or question, select the word or expression that, of those given, best completes the statement or answers the question. Record your answer on the separate answer paper in accordance with the directions of the front page of this booklet.

1. A car travels a distance of 98 meters in 10. seconds. What is the average speed of the car during this 10.-second interval? (1) 4.9 m/s (2) 9.8 m/s (3) 49 m/s (4) 98 m/s

2. Which measurement of an average classroom door is closest to 1 meter? (1) thickness (2) width (3) height (4) surface area

3. A boat initially traveling at 10. meters per second accelerates uniformly at the rate of 5.0 meters per second^2 for 10. seconds. How far does the boat travel during this time? (1) 50. m (2) 250 m (3) 350 m (4) 500 m

4. The graph below represents the relationship between distance and time for an object.

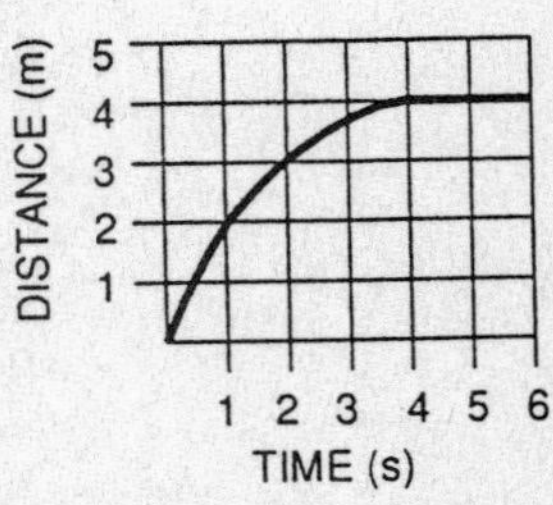

What is the instantaneous speed of the object at t = 5.0 seconds? (1) 0 m/s (2) 2.0 m/s (3) 5.0 m/s (4) 4.0 m/s

5. An object accelerates uniformly from rest to a speed of 50. meters per second in 5.0 seconds. The average speed of the object during the 5.0-second interval is (1) 5.0 m/s (2) 10. m/s (3) 25 m/s (4) 50. m/s

6. A 5-newton ball and a 10-newton ball are released simultaneously from a point 50 meters above the surface of the Earth. Neglecting air resistance, which statement is true? (1) The 5-N ball will have a greater acceleration than the 10-N ball. (2) The 10-N ball will have a greater acceleration than the 5-N ball. (3) At the end of 3 seconds of free-fall, the 10-N ball will have a greater momentum than the 5-N ball. (4) At the end of 3 seconds of free-fall, the 5-N ball will have a greater momentum than the 10-N ball.

7. In the diagram below, the weight of a box on a plane inclined at 30.° is represented by the vector *W*.

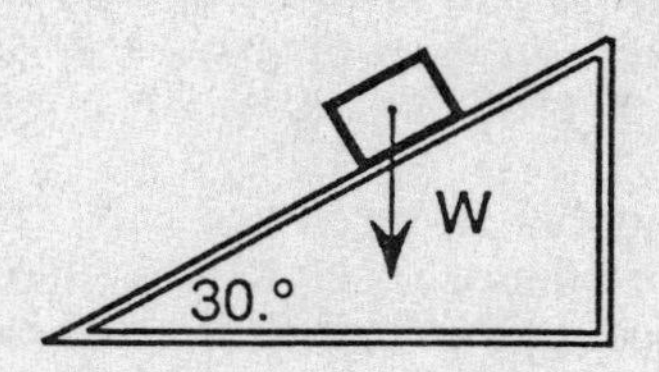

What is the magnitude of the component of the weight (*W*) that acts parallel to the incline? (1) *W* (2) 0.50*W* (3) 0.87*W* (4) 1.5*W*

8. The diagram at the right represents a force acting at point *P*. Which pair of concurrent forces would produce equilibrium when added to the force acting at point *P*?

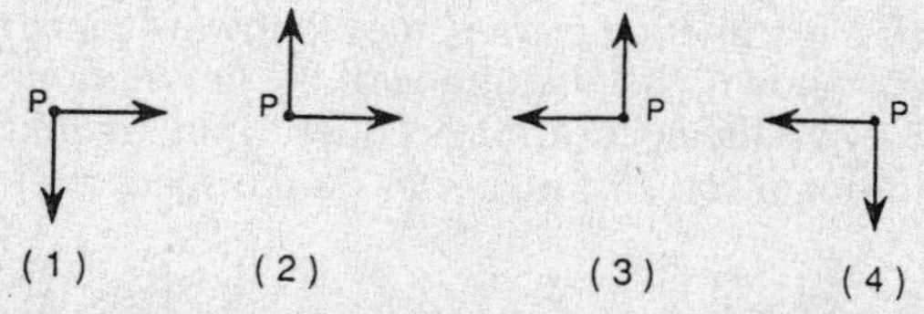

9. A boat heads directly eastward across a river at 12 meters per second. If the current in the river is flowing at 5.0 meters per second due south, what is the magnitude of the boat's resultant velocity? (1) 7.0 m/s (2) 8.5 m/s (3) 13 m/s (4) 17 m/s

10. A bird feeder with two birds has a total mass of 2.0 kilograms and is supported by wire as shown in the diagram below.

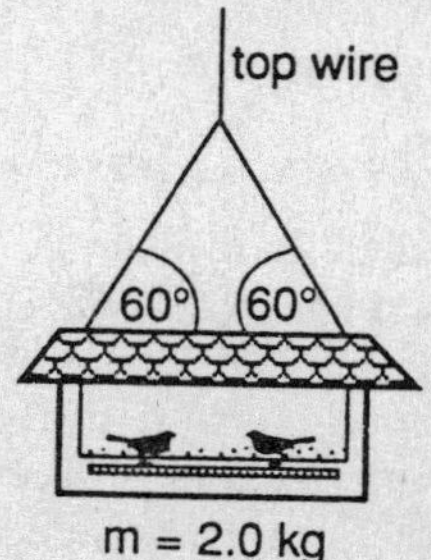

The force in the top wire is approximately (1) 10. N (2) 14 N (3) 20. N (4) 39 N

11. A 50.-kilogram woman wearing a seat belt is traveling in a car that is moving with a velocity of +10. meters per second. In an emergency, the car is brought to a stop in 0.50 second. What force does the seat belt exert on the woman so that she remains in her seat? (1) -1.0×10^3 N (2) -5.0×10^2 N (3) -5.0×10^1 N (4) -2.5×10^1 N

12. A 0.10-kilogram ball dropped vertically from a height of 1.0 meter above the floor bounces back to a height of 0.80 meter. The mechanical energy lost by the ball as it bounces is approximately (1) 0.080 J (2) 0.20 J (3) 0.30 J (4) 0.78 J

13. A student rides a bicycle up a 30.° hill at a constant speed of 6.00 meters per second. The combined mass of the student and bicycle is 70.0 kilograms. What is the kinetic energy of the student-bicycle system during this ride? (1) 210. J (2) 420. J (3) 1,260 J (4) 2,520 J

Base your answers to questions 14 and 15 on the information and diagram below.

Spacecraft *S* is traveling form planet P_1 toward planet P_2. At the position shown, the magnitude of the gravitational force of planet P_1 on the spacecraft is equal to the magnitude of the gravitational force of planet P_2 on the spacecraft.

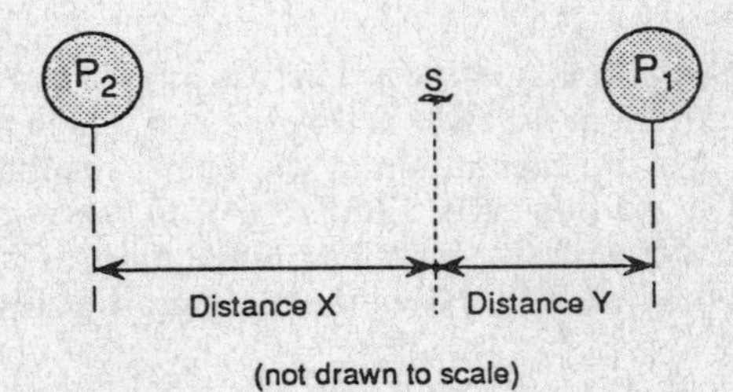

Note that questions 14 and 15 have only three choices.

14. If distance *X* is greater than distance *Y*, then the mass of P_1 must be (1) less than the mass of P_2 (2) greater than the mass of P_2 (3) equal to the mass of P_2

15. As the spacecraft moves from the position shown toward planet P_2, the ratio of the gravitational force of P_2 on the spacecraft to the gravitational force of P_1 on the spacecraft will (1) decrease (2) increase (3) remain the same

16. The graph at the right shows the relationship between weight and mass for a series of objects. The slope of this graph represents (1) change of position (2) normal force (3) momentum (4) acceleration due to gravity

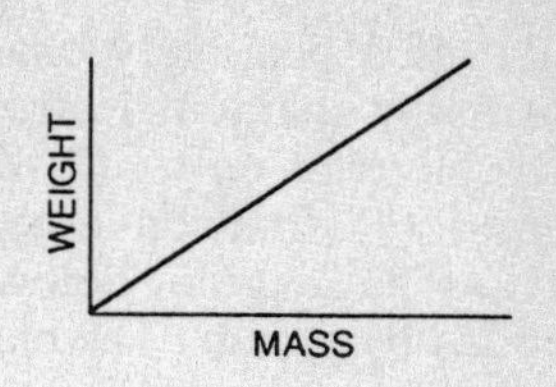

17. Each diagram below shows a different block being pushed by a force across a surface at a constant velocity.

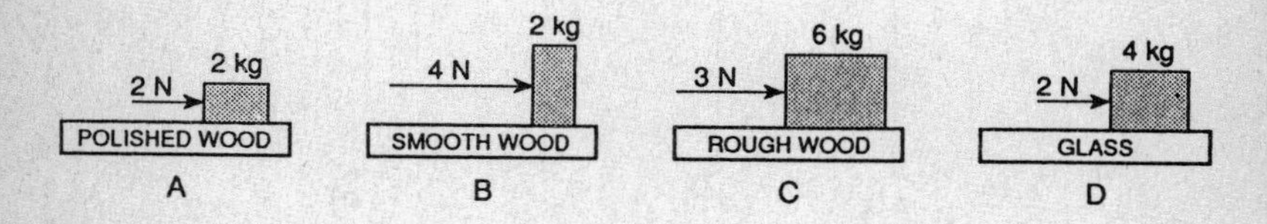

In which two diagrams is the force of friction the same? (1) *A* and *B* (2) *B* and *D* (3) *A* and *D* (4) *C* and *D*

18. A student running up a flight of stairs increases her speed at a constant rate. Which graph best represents the relationship between work and time for the student's run up the stairs?

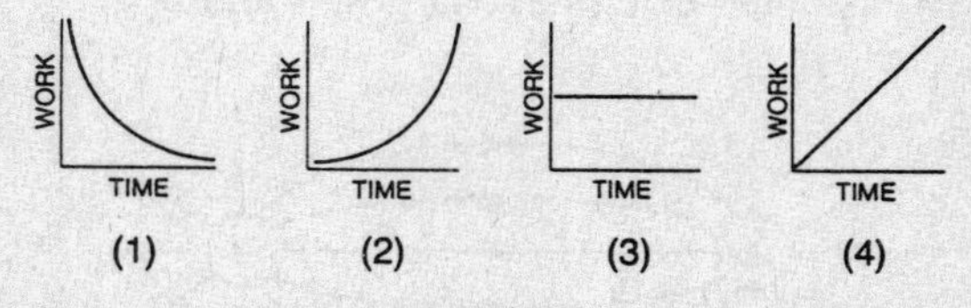

19. A net force of 5.0 newtons moves a 2.0-kilogram object a distance of 3.0 meters in 3.0 seconds. How much work is done on the object? (1) 1.0 J (2) 10. J (3) 15 J (4) 30. J

20. Which graph best represents the relationship between the elongation of a spring whose elastic limit has not been reached and the force applied to it?

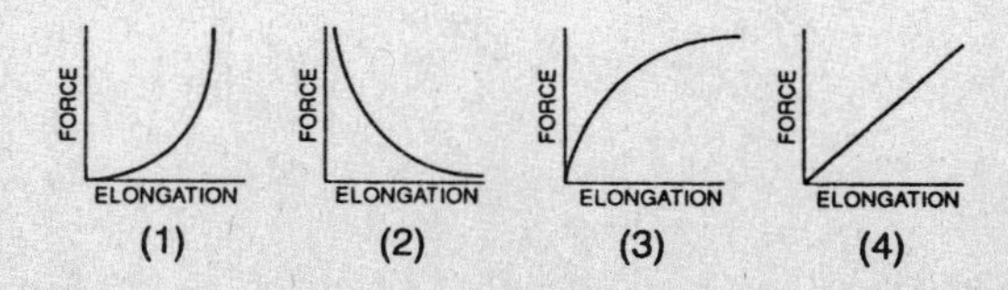

21. If a positively charged rod is brought near the knob of a positively charged electroscope, the leaves of the electroscope will (1) converge, only (2) diverge, only (3) first diverge, then converge (4) first converge, then diverge

22. The diagram below shows four charged metal spheres suspended by strings. The charge of each sphere is indicated.

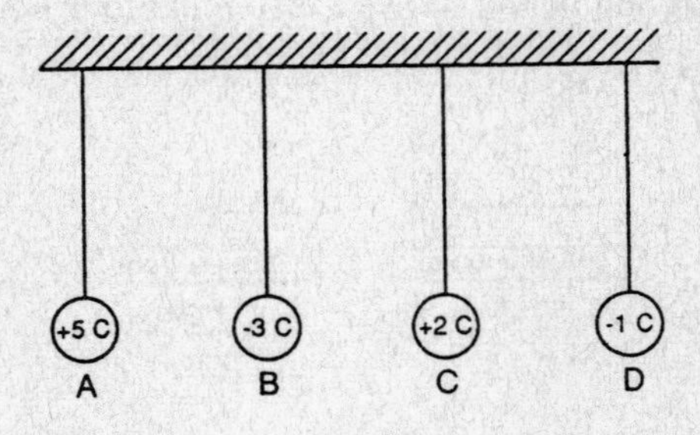

If spheres *A*, *B*, *C*, and *D* simultaneously come into contact, the net charge on the four spheres will be (1) +1 C (2) +2 C (3) +3 C (4) +4 C

23. The diagram below represents the electric field lines in the vicinity of two isolated electrical charges, *A* and *B*.

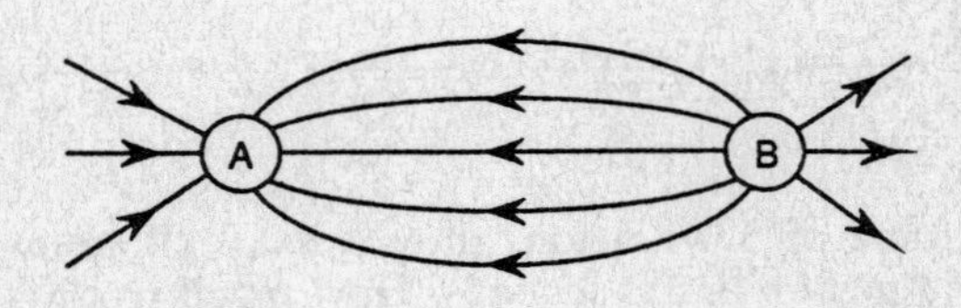

Which statement identifies the charges of *A* and *B*? (1) *A* is negative and *B* is positive. (2) *A* is positive and *B* is negative. (3) *A* and *B* are both positive. (4) *A* and *B* are both negative.

Base your answers to questions 24 through 26 on the diagram below which represents a frictionless track. A 10-kilogram block starts from rest at point *A* and slides along the track.

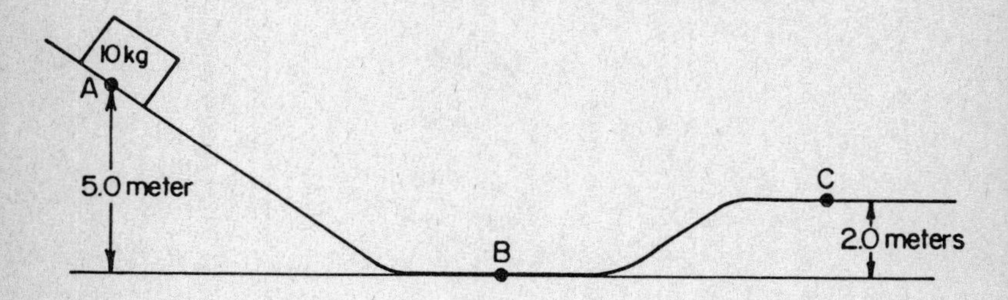

24. As the block moves from point *A* to point *B*, the total amount of gravitational potential energy changed to kinetic energy is approximately (1) 5 J (2) 20 J (3) 50 J (4) 500 J

25. What is the approximate speed of the block at point *B*? (1) 1 m/s (2) 10 m/s (3) 50 m/s (4) 100 m/s

26. What is the approximate potential energy of the block at point *C*? (1) 20 J (2) 200 J (3) 300 J (4) 500 J

27. If the potential difference between two oppositely charged parallel metal plates is doubled, the electric field intensity at a point between them is (1) halved (2) unchanged (3) doubled (4) quadrupled

28. Moving a point charge of 3.2×10^{-19} coulomb between points *A* and *B* in an electric field requires 4.8×10^{-19} joule of energy. What is the potential difference between these two points? (1) 0.67 V (2) 2.0 V (3) 3.0 V (4) 1.5 V

29. The slope of the line on the graph at the right represents (1) resistance of a material (2) electric field intensity (3) power dissipated in a resistor (4) electrical energy

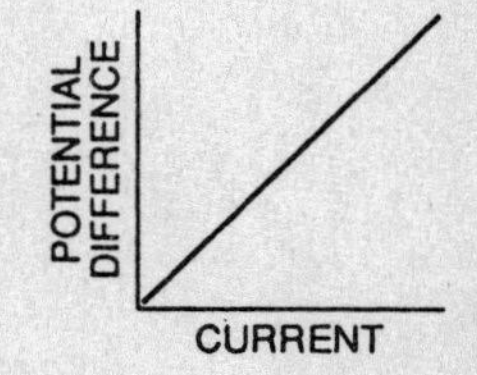

30. In the diagrams below, ℓ represents a unit length of copper wire and A represents a unit cross-sectional area. Which copper wire has the *smallest* resistance at room temperature?

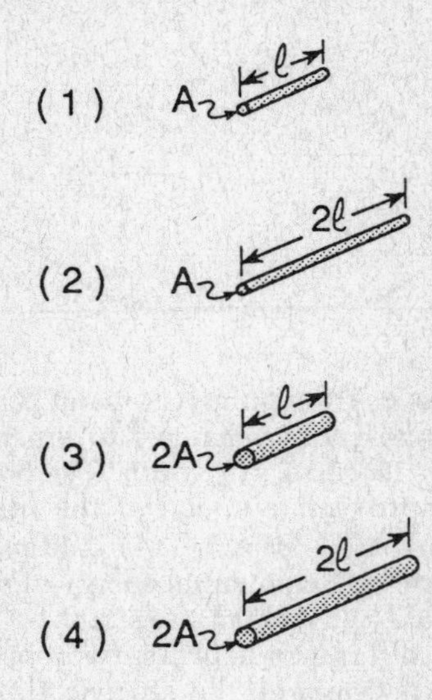

31. Two resistors are connected to a source of voltage as shown in the diagram below.

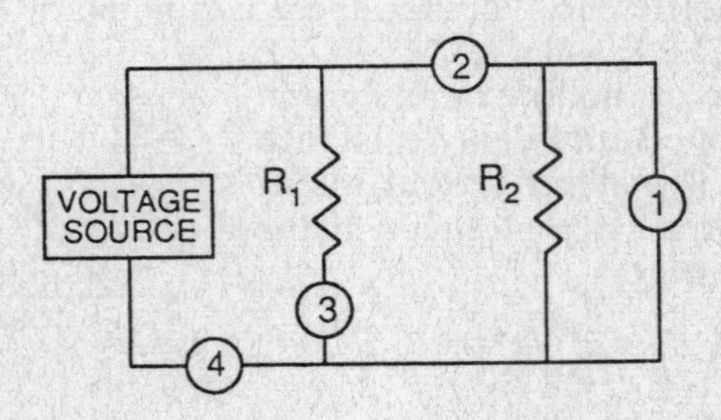

At which position should an ammeter be placed to measure the current passing only through resistor R_1? (1) 1 (2) 2 (3) 3 (4) 4

32. A toaster dissipates 1,500 watts of power in 90. seconds. The amount of electric energy used by the toaster is approximately (1) 1.4×10^5 J (2) 1.7×10^1 J (3) 5.2×10^8 J (4) 6.0×10^{-2} J

33. In the diagram below, a steel paper clip is attached to a string which is attached to a table. The clip remains suspended beneath a magnet.

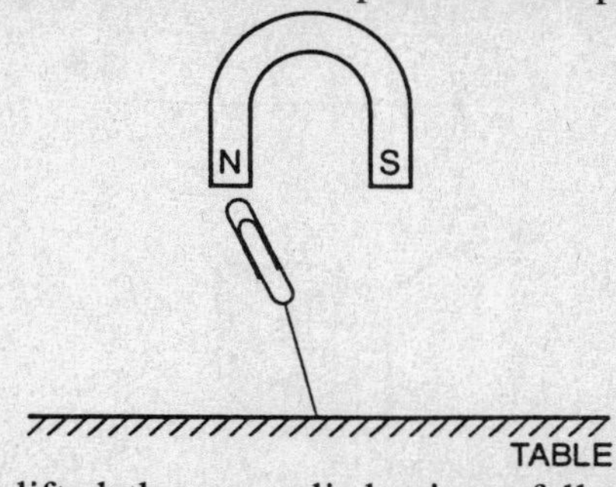

As the magnet is lifted, the paper clip begins to fall as a result of (1) an increase in the potential energy of the clip (2) an increase in the gravitational field strength near the magnet (3) a decrease in the magnetic properties of the clip (4) a decrease in the magnetic field strength near the clip

34. The diagram below shows the magnetic field that results when a piece of iron is placed between unlike magnetic poles.

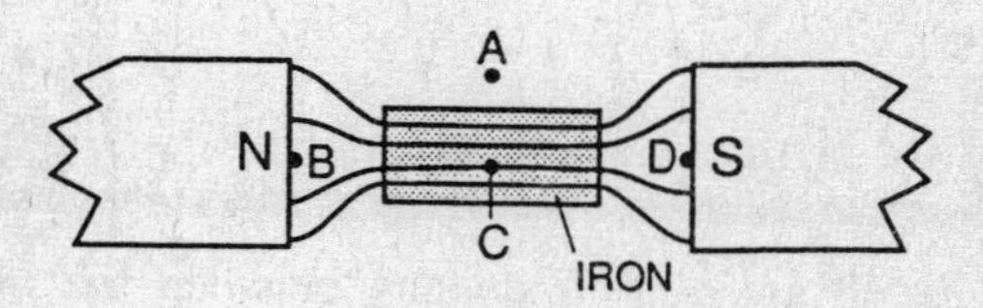

At which point is the magnetic field strength greatest? (1) *A* (2) *B* (3) *C* (4) *D*

35. A wire carrying an electron current (e^-) is placed between the poles of a magnet, as shown in the diagram below.

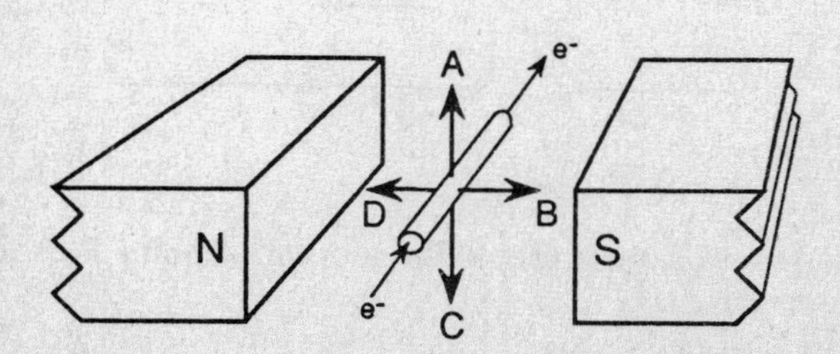

Which arrow represents the direction of the magnetic force on the current? (1) *A* (2) *B* (3) *C* (4) *D*

36. The diagram below shows a coil of wire connected to a battery.

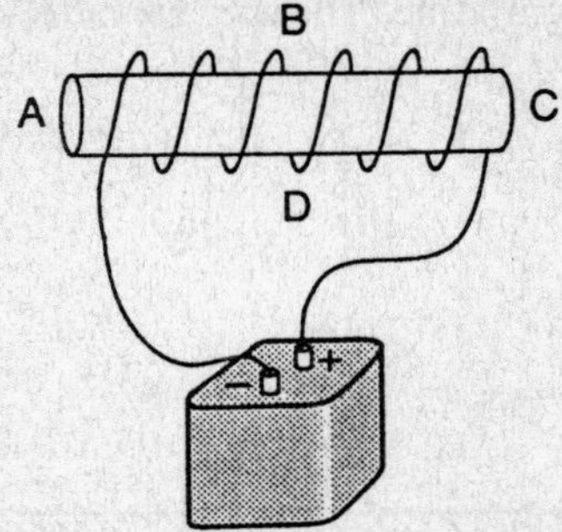

The *N*-pole of this coil is closest to (1) *A* (2) *B* (3) *C* (4) *D*

37. The diagram below shows radar waves being emitted from a stationary police car and reflected by a moving car back to the police car.

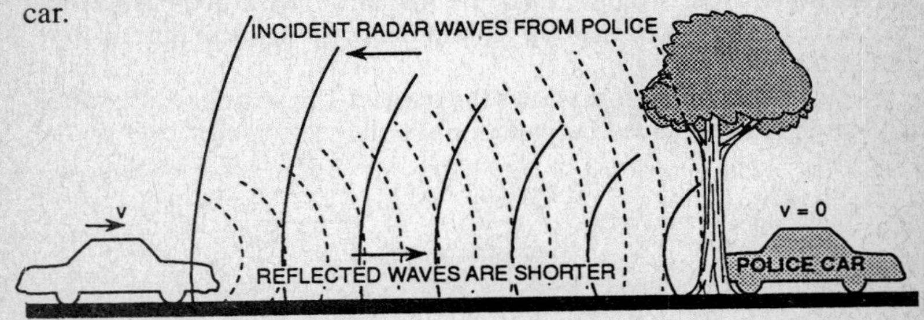

The difference in apparent frequency between the incident and reflected waves is an example of (1) constructive interference (2) refraction (3) the Doppler effect (4) total internal reflection

38. The diagram below shows a transverse pulse moving to the right in a string.

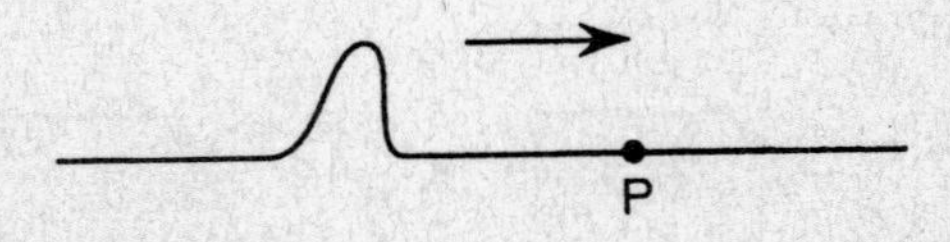

Which diagram best represents the motion of point *P* as the pulse passes point *P*?

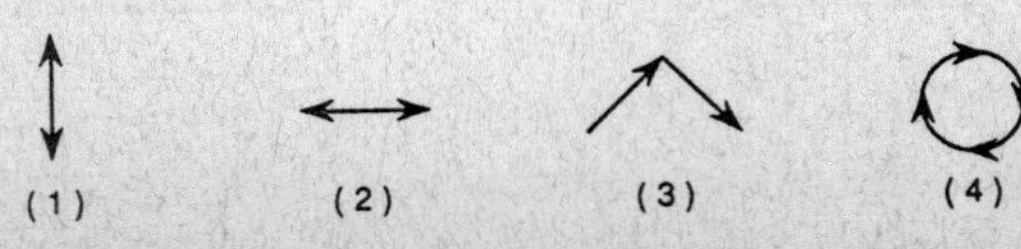

39. Light is to brightness as sound is to (1) color (2) loudness (3) period (4) speed

40. The periodic wave in the diagram below has a frequency of 40. hertz.

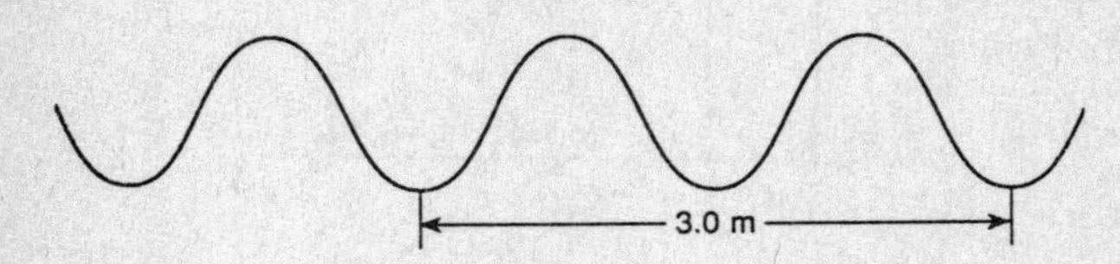

What is the speed of the wave? (1) 13 m/s (2) 27 m/s (3) 60. m/s (4) 120 m/s

41. Two waves have the same frequency. Which wave characteristic must also be identical for both waves? (1) phase (2) amplitude (3) intensity (4) period

42. A typical microwave oven produces radiation at a frequency of 1.0×10^{10} hertz. What is the wavelength of this microwave radiation? (1) 3.0×10^{-1} m (2) 3.0×10^{-2} m (3) 3.0×10^{10} m (4) 3.0×10^{18} m

43. Two wave sources operating in phase in the same medium produce the circular wave patterns shown in the diagram below. The solid lines represent wave crests and the dashed lines represent wave troughs.

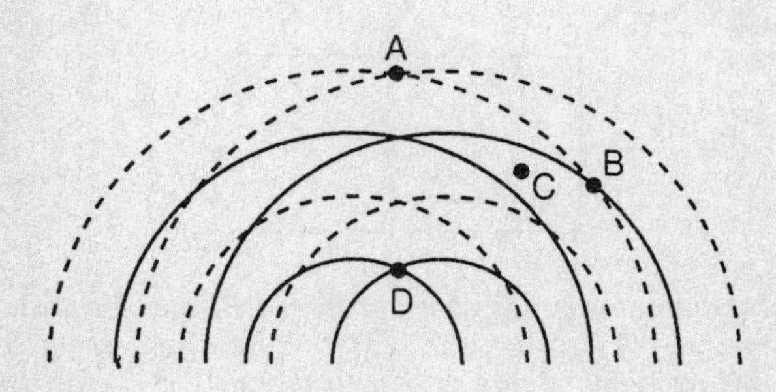

Which point is at a position of maximum destructive interference? (1) *A* (2) *B* (3) *C* (4) *D*

44. The distance between successive antinodes in the standing wave pattern shown at the right is equal to (1) 1 wavelength (2) 2 wavelengths (3) $\frac{1}{2}$ wavelength (4) $\frac{1}{3}$ wavelength

45. The diagram below shows a ray of light passing from medium *X* into air.

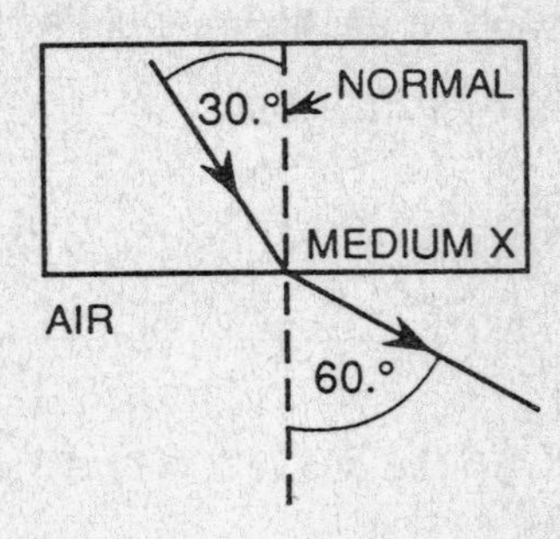

What is the absolute index of refraction of medium *X*? (1) 0.50 (2) 2.0 (3) 1.7 (4) 0.58

46. In the diagram below, a ray of monochromatic light (*A*) and a ray of polychromatic light (*B*) are both incident upon an air-glass interface.

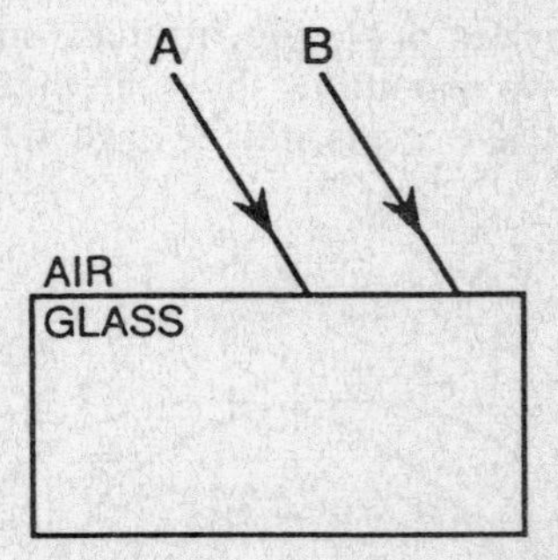

Which phenomenon could occur with ray *B*, but *not* with ray *A*? (1) dispersion (2) reflection (3) polarization (4) refraction

47. If the critical angle for a substance is 44°, the index of refraction of the substance is equal to (1) 1.0 (2) 0.69 (3) 1.4 (4) 0.023

48. The diagram below shows a beam of light entering and leaving a "black box."

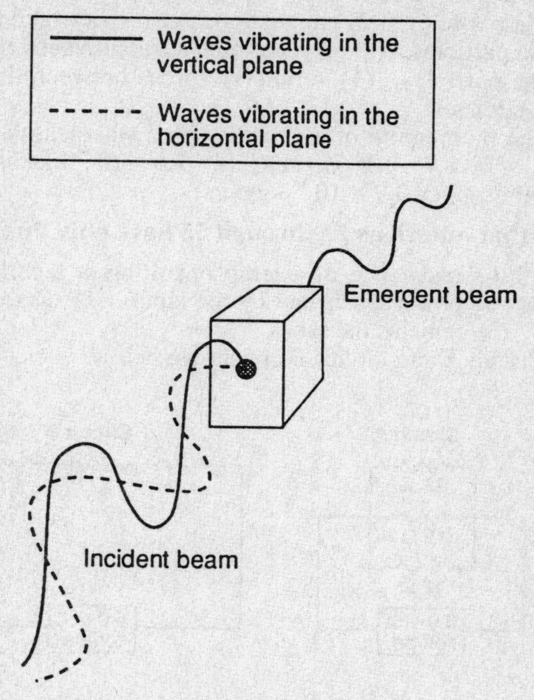

The box most likely contains a (1) prism (2) converging lens (3) double slit (4) polarizer

49. Which graph best represents the relationship between the intensity of light that falls on a photoemissive surface and the number of photoelectrons that the surface emits?

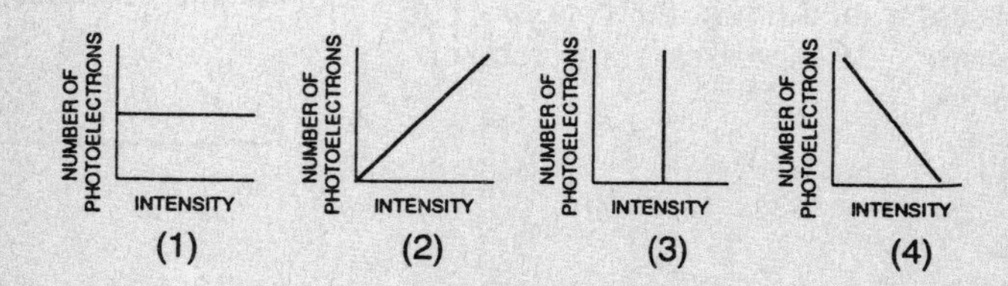

50. The work function of a certain photoemissive material is 2.0 electronvolts. If 5.0-electronvolt photons are incident on the material, the maximum kinetic energy of the ejected photoelectrons will be (1) 7.0 eV (2) 5.0 eV (3) 3.0 eV (4) 2.5 eV

51. Alpha particles fired at thin metal foil are scattered in hyperbolic paths due to the (1) attraction between the electrons and alpha particles (2) magnetic repulsion between the electrons and alpha particles (3) gravitational attraction between the nuclei and alpha particles (4) repulsive forces between the nuclei and alpha particles

52. The momentum of a photon with a wavelength of 5.9×10^{-7} meter is (1) 8.9×10^{26} kg•m/s (2) 1.6×10^{-19} kg•m/s (3) 1.1×10^{-27} kg•m/s (4) 3.9×10^{-40} kg•m/s

Note that questions 53 through 55 have only three choices.

53. As the resistance of a lamp operating at a constant voltage increases, the power dissipated by the lamp (1) decreases (2) increases (3) remains the same

54. Circuit *A* and circuit *B* are shown below.

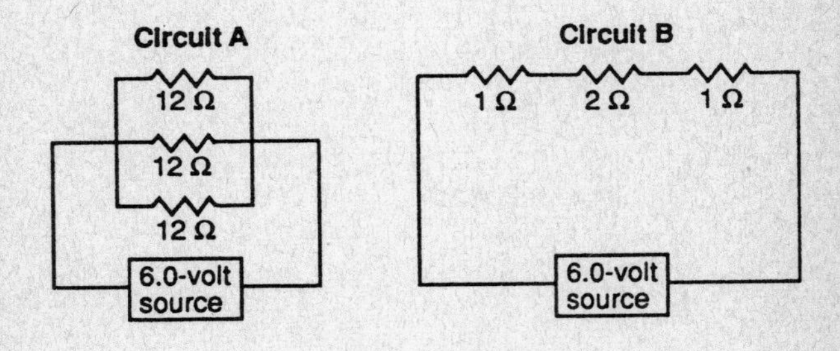

Compared to the total resistance of circuit *A*, the total resistance of circuit *B* is (1) less (2) greater (3) the same

55. The diagram at the right represents the path of periodic waves passing from medium *A* into medium *B*. As the waves enter medium *B*, their speed (1) decreases (2) increases (3) remains the same

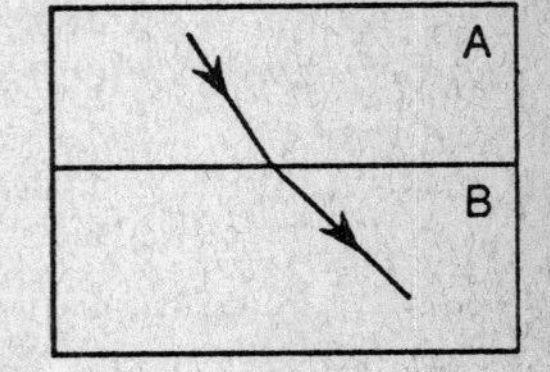

Part II

This part consists of six groups, each containing ten questions. Each group tests an optional area of the course. Choose two of these six groups. Be sure that you answer all ten questions in each group chosen. Record the answers to these questions in accordance with the directions on the front page of this booklet. [20]

Group 1—Motion in a Plane

If you choose this group, be sure to answer questions 56-65.

56. A ball is thrown horizontally at a speed of 20. meters per second from the top of a cliff. How long does the ball take to fall 19.6 meters to the ground? (1) 1.0 s (2) 2.0 s (3) 9.8 s (4) 4.0 s

57. A book is pushed with an initial horizontal velocity of 5.0 meters per second off the top of a desk. What is the initial vertical velocity of the book? (1) 0 m/s (2) 2.5 m/s (3) 5.0 m/s (4) 10. m/s

58. The diagram below shows a baseball being hit with a bat. Angle θ represents the angle between the horizontal and the ball's initial direction of motion.

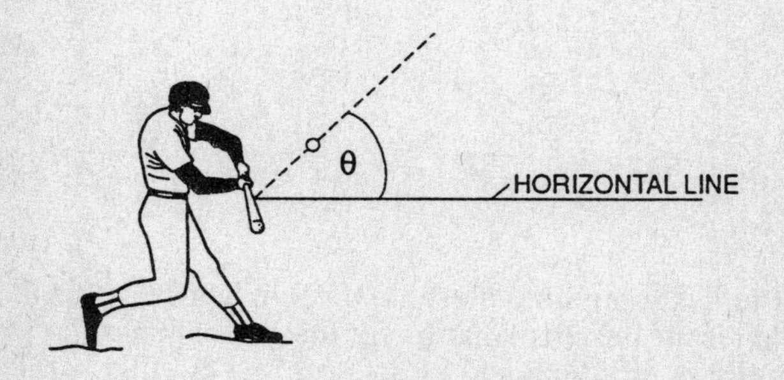

Which value of θ would result in the ball traveling the longest horizontal distance? [Neglect air resistance.] (1) 25° (2) 45° (3) 60° (4) 90°

59. The diagram below represents a bicycle and rider traveling to the right at a constant speed. A ball is dropped from the hand of the cyclist.

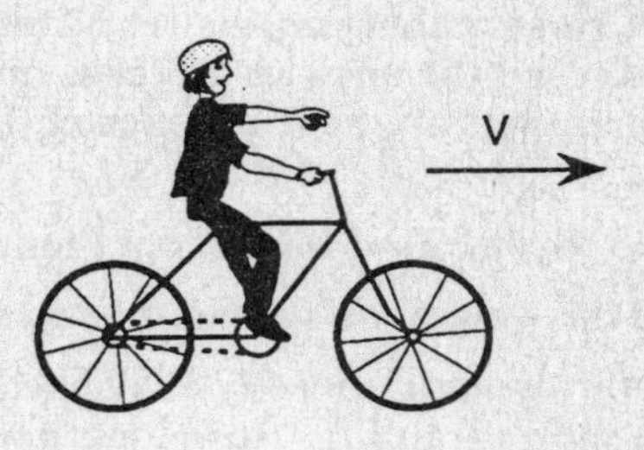

Which set of graphs bets represents the horizontal motion of the ball relative to the ground? [Neglect air resistance.]

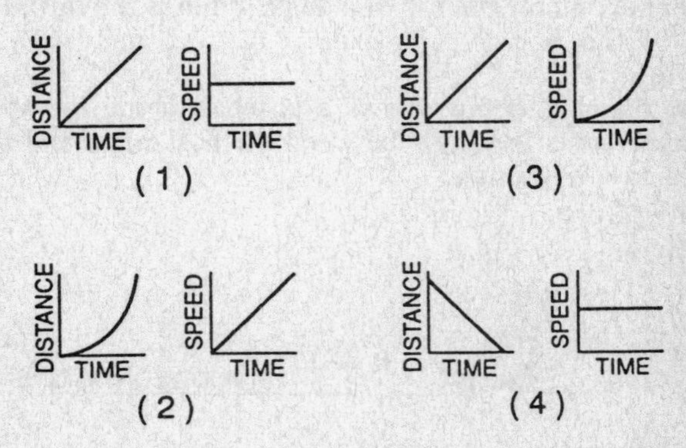

60. Pluto is sometimes closer to the Sun than Neptune is. Which statement is the best explanation for this phenomenon? (1) Neptune's orbit is elliptical and Pluto's orbit is circular. (2) Pluto's orbit is elliptical and Neptune's orbit is circular. (3) Pluto and Neptune have circular orbits that overlap. (4) Pluto and neptune have elliptical orbits that overlap.

Base your answers to questions 61 through 63 on the diagram below which shows a 2.0-kilogram model airplane attached to a wire. The airplane is flying clockwise in a horizontal circle of radius 20. meters at 30. meters per second.

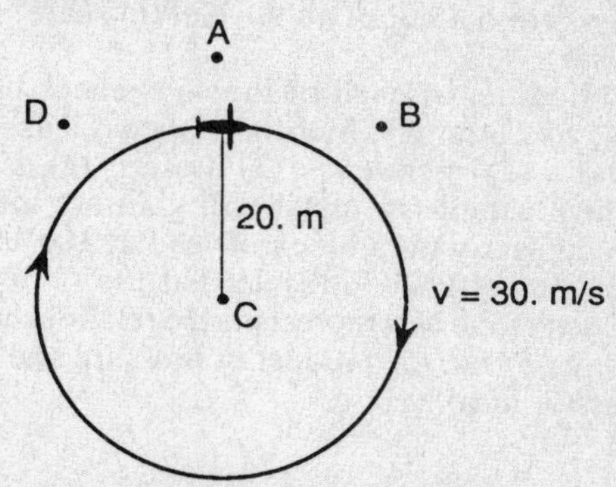

61. The centripetal force acting on the airplane at the position shown is directed toward point (1) *A* (2) *B* (3) *C* (4) *D*

62. What is the magnitude of the centripetal acceleration of the airplane? (1) 0 m/s² (2) 1.5 m/s² (3) 45 m/s² (4) 90. m/s²

63. If the wire breaks when the airplane is at the position shown, the airplane will move toward point (1) *A* (2) *B* (3) *C* (4) *D*

Note that questions 64 and 65 have only three choices.

64. A motorcycle travels around a flat circular track. If the speed of the motorcycle is increased, the force required to keep it in the same circular path (1) decreases (2) increases (3) remains the same

65. The diagram represents the path taken by planet *P* as it moves in an elliptical orbit around sun *S*. The time it takes to go from point *A* to point *B* is t_1, and from point *C* to point *D* is t_2.

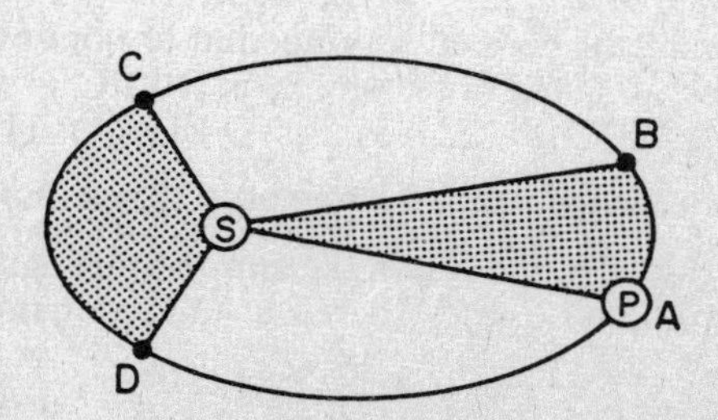

If the two shaded areas are equal then t_1 is (1) less than t_2 (2) greater than t_2 (3) the same as t_2

Group 2—Internal Energy

If you choose this group, be sure to answer questions 66-75.

66. What is the difference between the melting point and boiling point of ethyl alcohol on the Kelvin scale? (1) 38 (2) 196 (3) 352 (4) 469

67. A kilogram of each of the substances below is condensed from a gas to a liquid. Which substance releases the most energy? (1) alcohol (2) mercury (3) water (4) silver

68. Which sample of metal will gain net internal energy when placed in contact with a block of lead at 100°C? (1) platinum at 60°C (2) iron at 100°C (3) lead at 125°C (4) silver at 200°C

69. Which graph best represents the relationship between absolute temperature (*T*) and the product of pressure and volume (*P*•*V*) for a given mass of ideal gas?

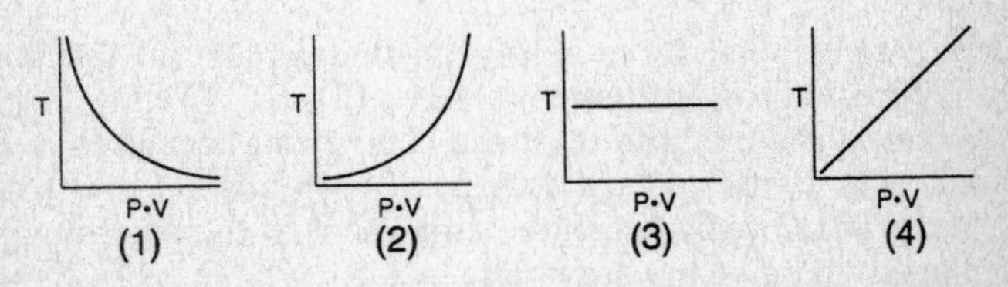

Base your answers to questions 70 through 72 on information below.

> Ten kilograms of water initially at 20°C is heated to its boiling point (100°C). Then 5.0 kilograms of the water is converted into steam at 100°C.

70. What was the approximate amount of heat energy needed to raise the temperature of the water to its boiling point? (1) 840 kJ (2) 3,400 kJ (3) 4,200 kJ (4) 6,300 kJ

71. The amount of heat energy needed to convert the 5.0 kilograms of water at 100°C into steam at 100°C is approximately (1) 1,700 kJ (2) 2,100 kJ (3) 5,500 kJ (4) 11,000 kJ

Note that question 72 has only three choices.

72. If salt is added to the water, the temperature at which the water boils will (1) decrease (2) increase (3) remain the same

73. The graph below shows temperature versus time for 1.0 kilogram of a substance at constant pressure as heat is added at a constant rate of 100 kilojoules per minute. The substance is a solid at 20°C.

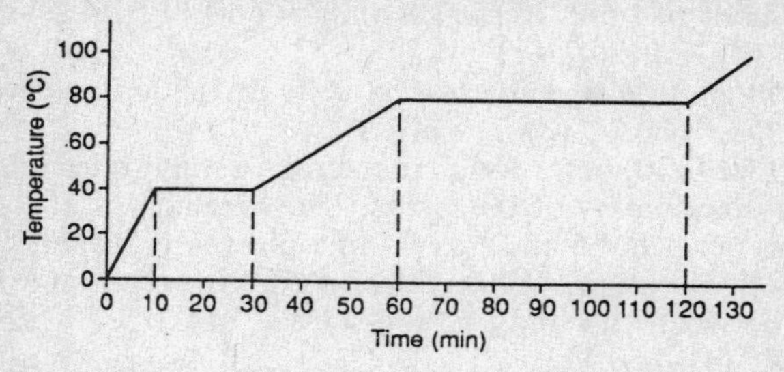

How much heat was added to change the substance from a liquid at its melting point to a vapor at its boiling point? (1) 3,000 kJ (2) 6,000 kJ (3) 9,000 kJ (4) 11,000 kJ

Note that questions 74 and 75 have only three choices.

74. As pressure is applied to a snowball, the melting point of the snow (1) decreases (2) increases (3) remains the same

75. Oxygen molecules are about 16 times more massive than hydrogen molecules. An oxygen gas sample is in a closed container and a hydrogen gas sample is in a second closed container of different size. Both samples are at room temperature. Compared to the average speed of the oxygen molecules, the average speed of the hydrogen molecules will be (1) less (2) greater (3) the same

Group 3—Electromagnetic Applications

If you choose this group, be sure to answer questions 76-85.

Base your answers to questions 76 through 78 on the information and data table below.

During a laboratory investigation of transformers, a group of students obtained the following data during four trials, using a different pair of coils in each trial.

	Primary Coil		Secondary Coil	
	V_p (volts)	I_p (amperes)	V_s (volts)	I_s (amperes)
Trial 1	3.0	12.0	16.0	2.0
Trial 2	6.0	3.0	8.0	2.2
Trial 3	9.0	4.3	54.0	0.7
Trial 4	12.0	2.5	5.0	9.0

76. What is the efficiency of the transformer in trial 1? (1) 75% (2) 89% (3) 100% (4) 113%

77. What is the ratio of the number of turns in the primary coil to the number of turns in the secondary coil in trial 3? (1) 1:6 (2) 1:9 (3) 6:1 (4) 9:1

78. In which trial was an error most likely made in recording the data? (1) 1 (2) 2 (3) 3 (4) 4

79. A wire 0.50 meter long cuts across a magnetic field with a magnetic flux density of 20. teslas. The wire moves at a speed of 4.0 meters per second and travels in a direction perpendicular to the magnetic flux lines. What is the maximum potential difference induced between the ends of the wire? (1) 2.5 V (2) 10. V (3) 40. V (4) 160 V

80. Compared to the resistance of the circuit being measured, the internal resistance of a voltmeter is designed to be very high so that the meter will draw (1) no current from the circuit (2) little current from the circuit (3) most of the current from the circuit (4) all the current from the circuit

81. A proton and an electron traveling with the same velocity enter a uniform electric field. Compared to the acceleration of the proton, the acceleration of the electron is (1) less, and in the same direction (2) less, but in the opposite direction (3) greater, and in the same direction (4) greater, but in the opposite direction

82. The diagram below shows an end view of a straight conducting wire, *W*, moving with constant speed in uniform magnetic field *B*.

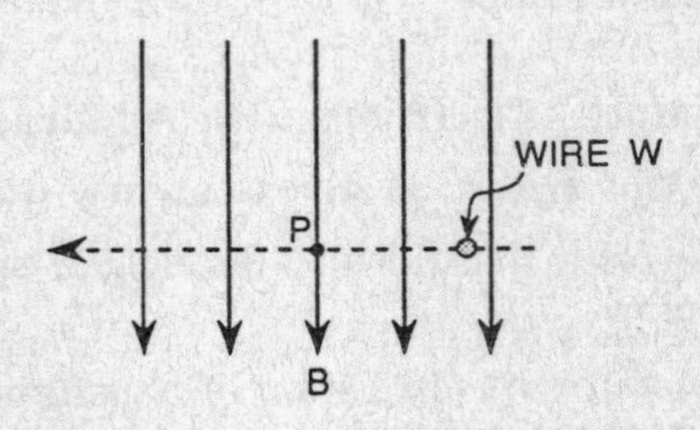

As the conductor moves through position *P*, the electron current induced in the wire is directed (1) toward the bottom of the page (2) toward the top of the page (3) into the page (4) out of the page

83. An electron moves at 3.0×10^7 meters per second perpendicularly to a magnetic field that has a flux density of 2.0 teslas. What is the magnitude of the force on the electron? (1) 9.6×10^{-19} N (2) 3.2×10^{-19} N (3) 9.6×10^{-12} N (4) 4.8×10^{-12} N

84. In each diagram below, an electron travels to the right between points *A* and *B*. In which diagram would the electron be deflected toward the bottom of the page?

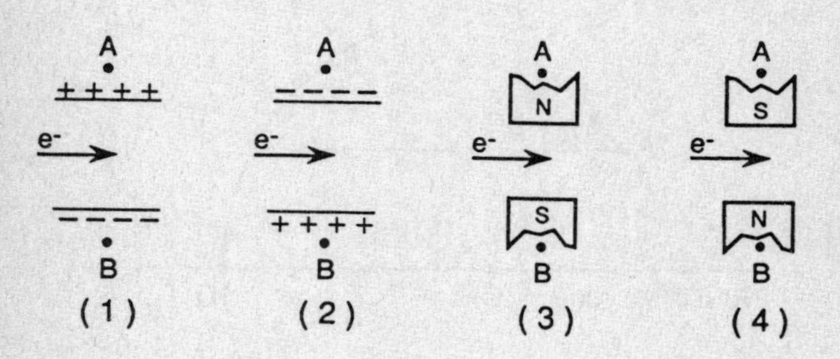

85. What is one characteristic of a light beam produced by a monochromatic laser? (1) It consists of coherent waves. (2) It can be dispersed into a complete continuous spectrum. (3) It cannot be reflected or refracted. (4) It does not exhibit any wave properties.

Group 4—Geometric Optics

If you choose this group, be sure to answer questions 86-95.

86. An object is placed in front of a plane mirroı as shown in the diagram at the right. Which diagram below best represents the image the is formed?

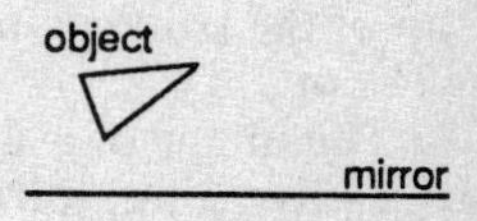

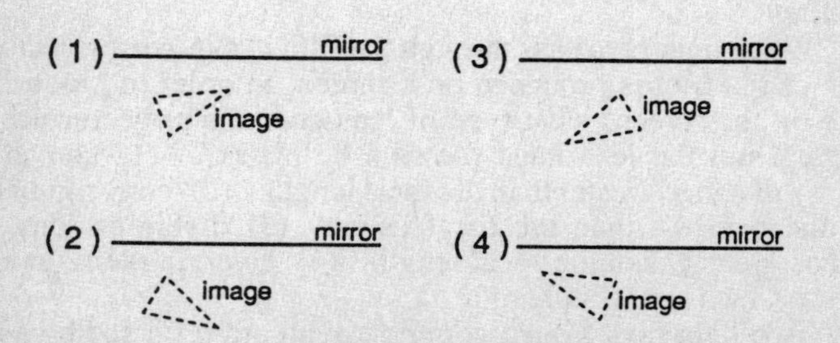

87. The diagram below shows light ray *R* parallel to the principal axis of a spherical concave (converging) mirror. Point *F* is the focal point of the mirror and *C* is the center of curvature.

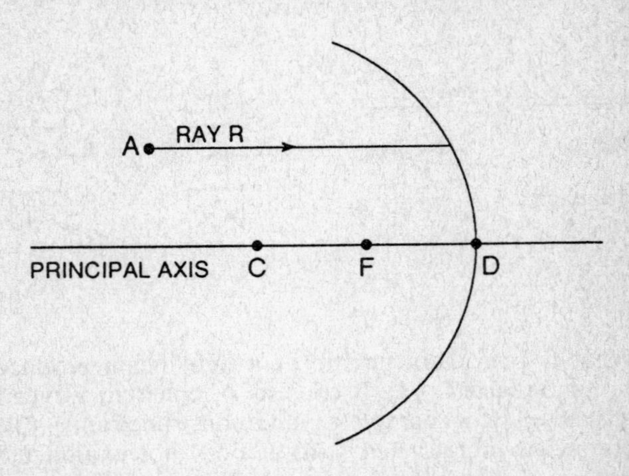

After reflecting, the light ray will pass through point (1) *A* (2) *F* (3) *C* (4) *D*

88. The tip of a person's nose is 12 centimeters from a concave (converging) spherical mirror that has a radius of curvature of 16 centimeters. What is the distance from the mirror to the image of the tip of the person's nose? (1) 8.0 cm (2) 12 cm (3) 16 cm (4) 24 cm

89. The image of a shoplifter in a department store is viewed in a convex (diverging) mirror. The image is (1) real and smaller than the shoplifter (2) real and larger than the shoplifter (3) virtual and smaller than the shoplifter (4) virtual and larger than the shoplifter

90. When light rays pass through the film in a movie projector, an image of the film is produced on a screen. In order to produce the image on the screen, what type of lens does the projector use and how far from the lens must the film be placed? (1) converging lens, at a distance greater than the focal length (2) converging lens, at a distance less than the focal length (3) diverging lens, at a distance greater than the focal length (4) diverging lens, at a distance less than the focal length

91. Two light rays from a common point are refracted by a lens. A real image is formed when these two refracted rays (1) converge to a single point (2) diverge and appear to come from a single point (3) travel in parallel paths (4) totally reflect inside the lens

92. The diagram below represents a convex (converging) lens with focal point F.

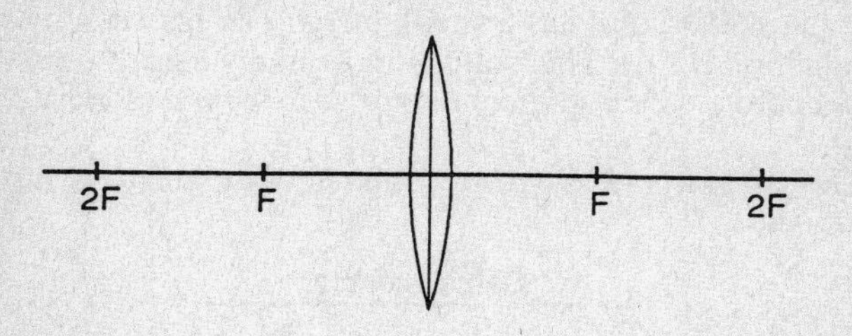

If an object is placed at $2F$, the image will be (1) virtual, erect, and smaller than the object (2) real, inverted, and the same size as the object (3) real, inverted, and larger than the object (4) virtual, erect, and the same size as the object

93. The diagram below shows the refraction of the blue and red components of a white light beam.

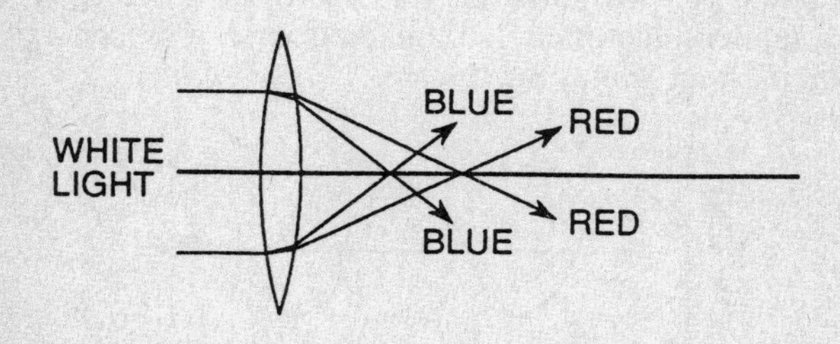

Which phenomenon does the diagram illustrate? (1) total internal reflection (2) critical angle reflection (3) spherical aberration (4) chromatic aberration

94. When a 2.0-meter-tall object is placed 4.0 meters in front of a lens, an image is formed on a screen located 0.050 meter behind the lens. What is the size of the image? (1) 0.10 m (2) 0.025 m (3) 2.5 m (4) 0.40 m

Note that question 95 has only three choices.

95. As the distance between a man and a plane mirror increases, the size of the image of the man produced by the mirror (1) decreases (2) increases (3) remains the same

Group 5—Solid State

If you choose this group, be sure to answer questions 96-105.

96. A particular solid has a small energy gap between its valence and conduction bands. This solid is most likely classified as (1) a good conductor (2) a semiconductor (3) a type of glass (4) an insulator

97. The diagram below shows an electron moving through a semiconductor.

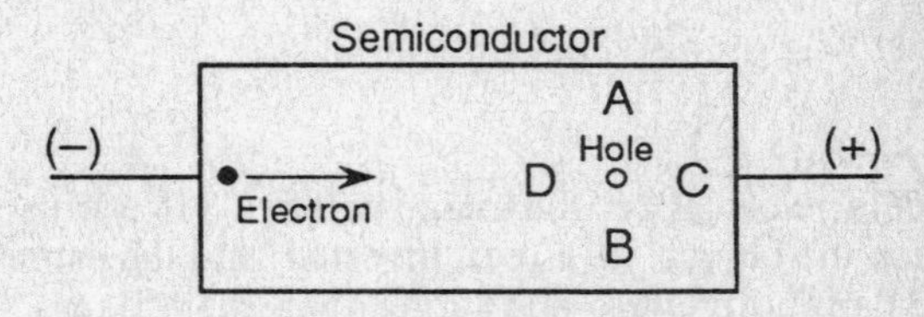

Toward which letter will the hole move? (1) *A* (2) *B* (3) *C* (4) *D*

98. Diagram *A* represents the wave form of an electron current entering a semiconductor device, and diagram *B* represents the wave form as the current leaves the device.

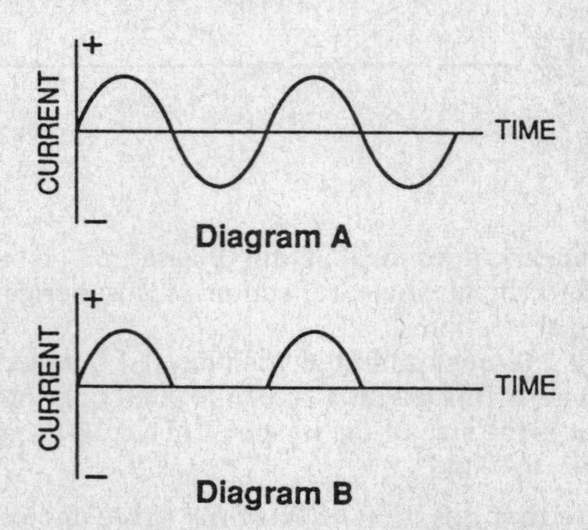

Diagram A

Diagram B

What is this device? (1) resistor (2) anode (3) cathode (4) diode

99. Compared to an insulator, a conductor of electric current has (1) more free electrons per unit volume (2) fewer free electrons per unit volume (3) more free atoms per unit volume (4) fewer free atoms per unit volume

Base your answers to questions 100 through 102 on the diagram below which represents a diode.

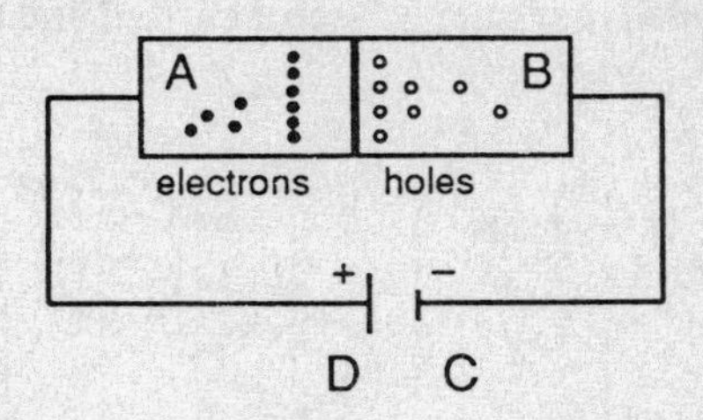

100. The *P-N* junction in the diagram is biased (1) reverse (2) forward (3) *B* to *C* (4) *A* to *D*

101. In the diagram, *B* represents the (1) *N*-type silicon (2) *P*-type silicon (3) cathode (4) diode

Note that question 102 has only three choices.

102. If the positive and negative wires of the circuit in the diagram were reversed, the current would (1) decrease (2) increase (3) remain the same

103. The graph at the right represents the alternating current signal input to a transistor amplifier. Which graph below best represents the amplified output signal from this transistor?

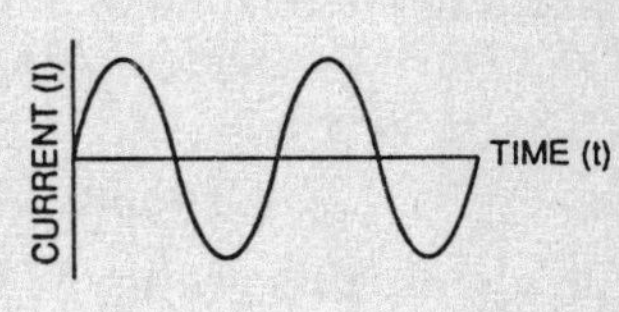

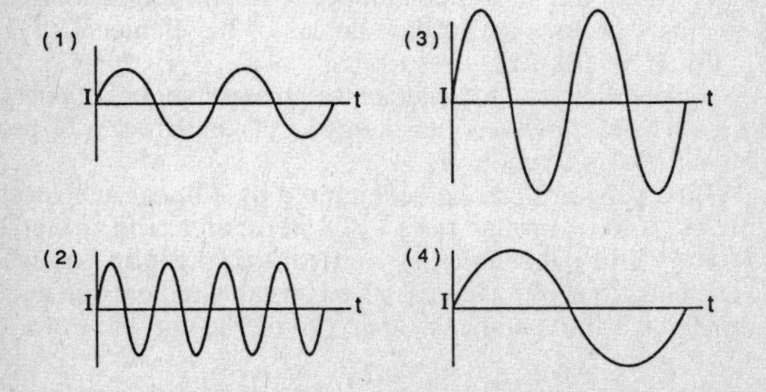

104. Which device contains a large number of transistors on a single block of silicon? (1) junction diode (2) conductor (3) integrated circuit (4) *N*-type semiconductor

Note that question 105 has only three choices.

105. The diagram below represents an operating *N-P-N* transistor circuit. Ammeter A_c reads the collector current and ammeter A_b reads the base current.

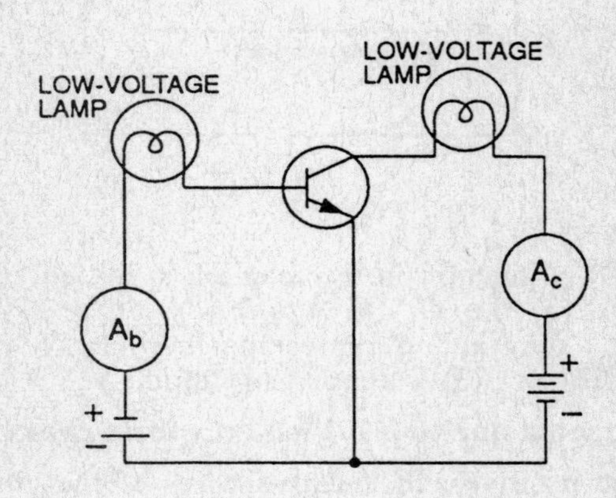

Compared to the reading of ammeter A_c, the reading of ammeter A_b is (1) less (2) greater (3) the same

Group 6—Nuclear Energy

If you choose this group, be sure to answer questions 106-115.

106. An element has an atomic number of 63 and a mass number of 155. How many protons are in the nucleus of the element? (1) 63 (2) 92 (3) 155 (4) 218

107. Which particle would generate the greatest amount of energy if its entire mass were converted into energy? (1) electron (2) proton (3) alpha particle (4) neutron

108. Which particles can be accelerated by a linear accelerator? (1) protons and gamma rays (2) neutrons and electrons (3) electrons and protons (4) neutrons and alpha particles

109. The equation below represents an unstable radioactive nucleus that is transmuted into another isotope (*X*) by the emission of a beta particle.

$$^{234}_{90}\mathrm{Th} \rightarrow X + {}^{0}_{-1}\mathrm{e}$$

Which new isotope is formed?

(1) $^{234}_{91}\mathrm{Pa}$
(2) $^{234}_{91}\mathrm{Th}$
(3) $^{235}_{90}\mathrm{Pa}$
(4) $^{235}_{90}\mathrm{Th}$

110. In 4.0 years, 40.0 kilograms of element *A* decays to 5.0 kilograms. The half-life of element *A* is (1) 1.3 years (2) 2.0 years (3) 0.7 year (4) 4.0 years

111. The subatomic particles that make up both protons and neutrons are known as (1) electrons (2) nuclides (3) positrons (4) quarks

112. Which equation is an example of positron emission?

(1) ${}^{226}_{88}Ra \rightarrow {}^{222}_{86}Rn + {}^{4}_{2}He$

(2) ${}^{210}_{82}Pb \rightarrow {}^{210}_{83}Bi + {}^{0}_{-1}e$

(3) ${}^{64}_{29}Cu \rightarrow {}^{64}_{28}Ni + {}^{0}_{+1}e$

(4) ${}^{14}_{7}N + {}^{4}_{2}He \rightarrow {}^{17}_{8}O + {}^{1}_{1}H$

113. Which process occurs during nuclear fission? (1) Light nuclei are forced together to form a heavier nucleus. (2) A heavy nucleus splits into lighter nuclei. (3) An atom is converted to a different isotope of the same element. (4) Transmutation is produced by the emission of alpha particles.

114. In order to increase the likelihood that a neutron emitted from a nucleus will be captured by another nucleus, the neutron should be (1) accelerated through a potential difference (2) heated to a higher temperature (3) slowed down to decrease its kinetic energy (4) absorbed by a control rod

115. The energy emitted by the Sun originates from the process of (1) fission (2) fusion (3) alpha decay (4) beta decay

Part III

You must answer all questions in this part. Record your answers in the spaces provided on the separate answer paper. Pen or pencil may be used. [15]

116. Base your answers to parts *a* through *c* on the information below.

A newspaper carrier on her delivery route travels 200. meters due north and then turns and walks 300. meters due east.

a *On your answer paper,* draw a vector diagram following the directions below.

(1) Using a ruler and protractor and starting at point *P*, construct the sequence of two displacement vectors for the

newspaper carrier's route. Use a scale of 1.0 centimeter = 50. meters. Label the vectors. [3]

(2) Construct and label the vector that represents the carrier's resultant displacement from point *P*. [1]

b What is the magnitude of the carrier's resultant displacement? [1]

c What is the angle (in degrees) between north and the carrier's resultant displacement? [1]

117. The diagram below shows a spring compressed by a force of 6.0 newtons from its rest position to its compressed position.

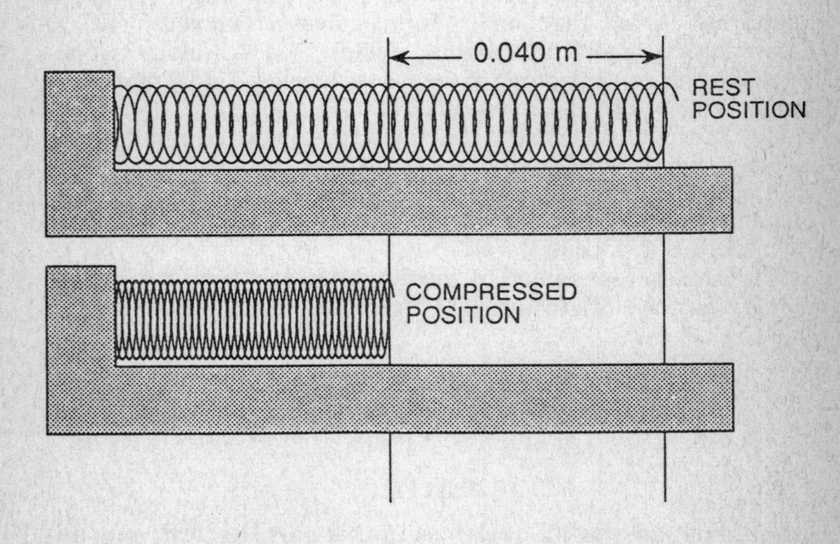

Calculate the spring constant for this spring. [Show all calculations, including equations and substitutions with units.] [2]

118. Base your answers to parts *a* through *c* on the diagram and information below.

Monochromatic light is incident on a two-slit apparatus. The distance between the slits is 1.0×10^{-3} meter, and the distance from the two-slit apparatus to a screen displaying the interference pattern is 4.0 meters. The distance between the central maximum and the first-order maximum is 2.4×10^{-3} meter.

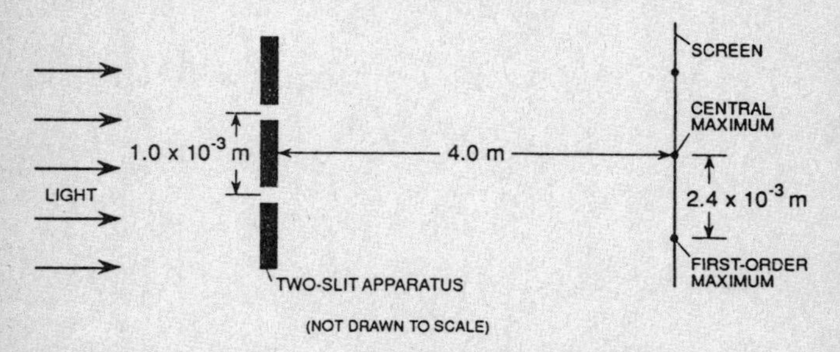

a What is the wavelength of the monochromatic light? [Show all calculations, including equations and substitutions with units.] [2]

b What is the color of the monochromatic light? [1]

c List *two* ways the variables could be changed that would cause the distance between the central maximum and the first-order maximum to increase. [2]

119. Infrared electromagnetic radiation incident on a material produces no photoelectrons. When a red light of equal intensity is shone on the same material, photoelectrons are emitted from the surface.

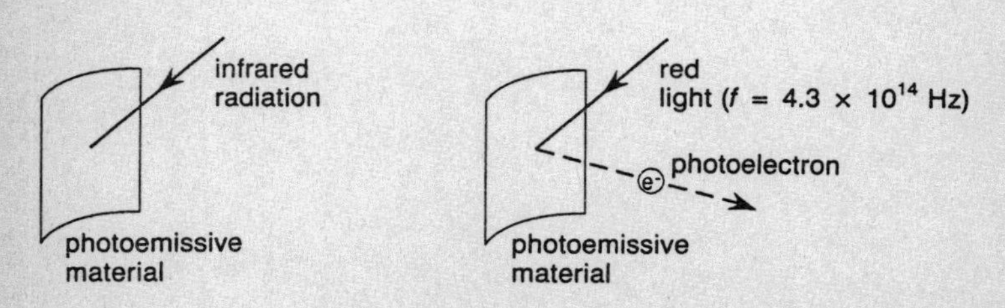

Using one or more complete sentences, explain why the visible red light causes photoelectric emission, but the infrared radiation does not. [2]